W9-DIW-988

DICTIONARY OF MILITARY TERMS

A GUIDE TO THE LANGUAGE OF WARFARE AND MILITARY INSTITUTIONS

DICTIONARY OF MILITARY TERMS

A GUIDE TO THE LANGUAGE

OF WARFARE

AND

MILITARY INSTITUTIONS

Compiled by

TREVOR N. DUPUY

CURT JOHNSON

GRACE P. HAYES

THE H.W. WILSON COMPANY

NEW YORK • 1986

Library of Congress Cataloging-in-Publication Data

Dupuy, Trevor Nevitt, 1916-
    Dictionary of military terms.

    1. Military art and science—Dictionaries.
2. Naval art and science—Dictionaries. I. Hayes,
Grace P. II. Johnson, Curt. III. Title.
U24.D87 1986   355'.003'   85-26592
ISBN 0-8242-0717-3

Printed in the United States of America

Introduction

No one hoping to acquire or maintain an understanding of world events can afford to be unfamiliar with the vocabulary of military affairs. Indeed, the military dimension enters any informed discussion of national and international policy issues, such as the defense of Western Europe or the Federal budget. And in the years since the Second World War the assimilation of military terms into the popular lexicon has continued apace, from the rhetoric of the Cold War, to Vietnam War-era slang and nuclear weapons' acronyms like MIRV and ICBM.

However, anyone interested in military affairs—whether scholar, professional, or student of military history—has certainly been frustrated by the absence of a comprehensive, reliable dictionary of military terms in English. The professional implications of this deficiency are acutely serious. Essential to communication among a community of professionals is a common understanding of the specialized terminology they use. The professions that have developed technical vocabularies—the legal and medical professions, for instance—can settle linguistic issues by consulting a variety of official dictionaries and encyclopedias. Not so the United States military, despite its history, traditions, and highly evolved body of speech. It is perhaps significant that there is no such problem for the Soviet armed forces; they have long had a standard and complete dictionary.

The charge that there is no standard reference work satisfactory both to the American military profession and a wider lay audience of students and historians is levied with the full realization that there has been for many years an official publication of the US Department of Defense—JCS Pub. 1: *Dictionary of Military and Associated Terms*. Yet this seemingly baseline publication has never been sufficiently comprehensive. For instance, the 1980 edition includes a definition for the obscure term "spasm war"—with the interesting caution that this term was not sanctioned for use by the Department of Defense—yet it does not define such basic military words as "attack," "combat," "defense," or "fortification," to name just a few significant omissions. Having taught military history, military science, and military affairs at Harvard, Ohio State, and Rangoon universities, I am keenly aware of the very real need of university faculties and students for an authoritative military lexicon. As a staff planner at SHAPE and in the Pentagon, I have experienced the equally urgent requirement for a comprehensive, universally accepted and respected military dictionary. And finally, as an author and theorist who has spoken with hundreds of readers about my books, I appreciate how a general audience needs to understand what a military professional means when such words as "attack" and "defense" are used.

My increasing awareness of the need for a sound military dictionary coincided with my participation in the establishment of an association of military scholars which has evolved into The Military Conflict Institute. A principal objective of the Institute has been to develop a theory of military combat, an endeavor requiring professional, intellectual rigor and precision in the formulation of concepts that are often quite abstruse. From the outset, one of the formidable problems these scholars faced was lack of precision in terminology. Consequently, I decided to try to produce—or to work with others to produce—the fundamental professional dictionary so badly needed. Fortuitously, in 1981, The H.W. Wilson Company approached me about compiling for them a dictionary of military terms.

This book provides concise definitions of some 2,500 terms related both to modern military affairs and to military history, from ancient times to the present. It covers all aspects of military and naval affairs, including strategy, tactics, weapons, fortifications, organizations, ranks, and administration. This is not a work of lexicography on historical principles; it is not, strictly speaking, an etymological dictionary. This *Dictionary of Military Terms* strictly emphasizes the military meanings and derivations of words and phrases, including those that enjoy popular usage. Thus, the military sense of both widely used and highly technical terms is accurately conveyed.

The team responsible for compiling the list of terms, and for providing them with definitions, includes military professionals of several services, military historians, and experienced operations research analysts. Our definitions have been reviewed by both specialists and generalists. In cases of doubt or disagreement, the authors selected what they considered to be the best definition, or we composed a new definition satisfactory to my coauthors and, ultimately, to me.

Thus no one can be held responsible for any omissions or errors in this work other than the undersigned.

Nevertheless I wish to mention the names of colleagues who have worked with me in this dictionary's preparation. First and foremost, of course, I am grateful to my coauthors—Grace P. Hayes and C. Curtiss Johnson—for their participation in this work with me. We received valuable assistance from Brian Bader, who carefully checked every word, term, and definition in at least one draft of this compilation. Also contributing were Gay M. Hammerman, John R. Brinkerhoff, the late John A.C. Andrews, and the late Paul Martell. Without in any way diminishing the responsibility for this dictionary, which is mine, Grace's, and Curt's, I must say that this work could not have been done without the help of these colleagues, and we are very thankful to all of them.

Finally, but not least, I would like to acknowledge and express thanks for the work of HERO's administrative and office staff in this effort. Particularly, I am grateful to Virginia Rufner, Mary Stolzenbach, Vera Phillips, and Margaret Marsh, all of whom contributed to the formidable task of typing the manuscript.

Trevor N. Dupuy
Fairfax, Virginia
1986

A

AA—Antiaircraft.

AAA—Antiaircraft artillery (q.v.).

abatis or **abattis**—A defensive obstacle formed of trees felled for the purpose, forming a kind of thicket, with butts embedded in the ground and trunks slanted outward so as to present the sharpened ends of their branches to the enemy. See also **plashing** and **zareba**.

ABC warfare—Atomic, biological, and chemical warfare.

Able—First letter of the old phonetic alphabet. See also **alphabet, phonetic**.

ABM—Antiballistic missile. See **missile, antiballistic**.

abort—v. To call off before completion, especially used of a bombing mission. n. The act of such cancellation.

about face—n. 1) A complete reversal of stationary position or attitude. 2) The command for such a reversal. v. 1) To execute such a reversal, usually in response to a command. 2) In the imperative, the command to do it. See also **face**.

abreast—Side by side.

abri—A shelter, such as a cave or a dugout.

absent without leave (AWOL)—Away from a place of duty without authorization. If absence is without intention of returning, under US military law it becomes desertion.

academies, military—Schools for the training of officers. The establishment of permanent armies and navies in the late 17th century in Europe made it necessary to estab-lish training programs and institutions for officers. In many countries (as in the USSR) military academies are schools for advanced training, comparable to war colleges (q.v.) in the United States. Academies in the United States provide the basic training and education for prospective military officers. The US Military Academy at West Point, New York, was founded in 1802; the US Naval Academy at Annapolis, Maryland, in 1845; the US Coast Guard Academy at New London, Connecticut, in 1876; and the US Air Force Academy at Colorado Springs, Colorado, in 1954.

accolade—Originating in medieval times, a ceremony (often performed by a monarch) conferring knighthood. In England the accolade is still given by a tap of the blade of a sword on the shoulder of a kneeling subject. Accolade has come to mean any recognition of special merit.

accouterment or **accoutrement**—Equipment (other than weapons and clothes) carried by a soldier, such as badges and emblems.

accuracy of fire—The measure of the deviation of fire from the point of aim, expressed usually in terms of the distance between the point of aim and the mean point of impact. It has two components: the inherent accuracy of the weapon and the proximity of center of impact (q.v.) to the point of aim.

ack-ack—Antiaircraft artillery (q.v.). The term originated with British signalmen in World War II to represent the initials AA, standing for antiaircraft.

acknowledgment—A response to a message informing its originator that the communication has been received and is understood.

acorn—A US Navy unit of administrative personnel, and the requisite material, needed to construct, operate, and maintain an advanced landplane and seaplane base. In the Pacific Theater during World War II, the personnel and matériel were assembled and then delivered as a unit to a captured island, where the base could be constructed and put into operation quickly.

acoustic—Activated by sound, as in acoustic mine or acoustic torpedo. Both of these, for example, are detonated by the sound of a ship's propeller.

acquire—To locate and define a target for artillery or other weapons. The term has also been applied to radars in two senses: for acquisition radars, acquire means to detect the presence and location of a target clearly enough so that it can be identified; for tracking radars, acquire means to position a radar beam so as to reveal a target and permit it to be attacked effectively.

act of hostility—An act that is unfriendly or belligerent.

act of war—An action by one nation intended to initiate a war, or considered by another nation to be sufficient cause for war.

acting rank—Temporary rank, generally assigned to an officer or non-commissioned officer who is performing duties normally performed by someone of higher rank.

action—1) A move on the part of a military force that imposes combat upon, or affects the condition of, an opponent. For example, when artillery "goes into action," it begins to place fire upon the enemy. 2) A minor combat encounter between two hostile forces (see also **engagement, battle, skirmish**). 3) Performance of activity for an administrative or operational purpose, as "remedial action is being taken." See also **conflict levels of intensity**.

action station—1) An assigned or prescribed position for a vessel in a naval formation or cruising disposition. 2) An assigned position to be taken by a person in case of air attack. 3) An assigned area in an approach, contact, or battle disposition.

activate—1) To establish, or put into existence by official order, a unit, post, camp, station, base, or shore activity which has been previously constituted and designated by

name or number or both so that it can be organized to function in its designated capacity. 2) To prepare for active service a naval ship or craft which has been in an inactive or reserve status.

active air defense—Direct defensive action taken to destroy or reduce the effectiveness of an air attack. It includes such measures as the use of aircraft, antiaircraft artillery, surface-to-air guided missiles, and electronic countermeasures. See also **defense**.

active defense—A posture of ground force defense in which limited offensive actions or counterattacks (by either front line units or mobile reserves) are used to deny a position to the attacker. See also **defense**.

active duty—Full-time duty in active military service. It may be permanent, temporary (especially for training purposes), or extended for the convenience of the government. The active services (Army, Navy, Air Force, Marine Corps, Coast Guard) include career people on permanent active duty, reservists on extended active duty, and reservists on temporary active duty for training.

active list—1) A list of all regular officers on active duty. 2) A list of air units assigned to a major air command for activation, and those that have been activated and are serving in active status.

active sectors—Areas (usually designated by boundaries) where combat is taking place.

actual range—In aerial bombardment, the horizontal distance a bomb travels from the moment of release until the moment of impact.

ADC—Aide de camp (q.v.).

adjust—To align a weapon, especially artillery, by firing so that its projectiles strike the target or its immediate vicinity. Such firing is often called "adjustment fire." See also **aimed fire**.

adjutant—An officer who serves as administrative assistant to the commanding officer of a military unit or formation below the level of division. The adjutant is principally concerned with correspondence, the issuing of orders, and most other non-logistical administration of the unit or formation.

adjutant-general—1) Adjutant of a unit with a general staff. 2) The officer in charge of a state's contingents in the US National Guard. 3) The chief administrative officer of the US Army, who serves as a principal assistant to the Army Chief of Staff. He is head of the Adjutant General's Corps, which publishes all orders, is responsible for all correspondence, personnel actions, awards, records, publications, and other non-logistical administrative functions.

administration—1) The management and execution of all military matters not included in tactics and strategy—primarily office management, logistics, and personnel management. 2) Internal management of units.

administrative chain of command—See **chain of command**.

administrative control—Exercise of authority in matters not directly related to combat, such as personnel management and services. For example, a unit temporarily under the operational control of another unit or formation often remains under the administrative control of its own higher commander.

administrative order—An order concerning aspects of military performance not directly related to combat operations.

administrative services—Activities essential for the functioning of an organization. These services include such activities as supply, communications, transportation, finance, personnel management, construction and maintenance, military justice, discipline, and public relations.

administrative troops—Troops in units designed to render supply, maintenance, transportation, evacuation, hospitalization, and other services required by air and ground combat units.

admiral—The highest rank in most navies, including the US Navy and Coast Guard. There are four degrees in the US Navy: fleet admiral, a rank given during or just after World War II to William D. Leahy, Ernest J. King, Chester W. Nimitz, and William F. Halsey; admiral (equivalent to general); vice admiral (equivalent to lieutenant general); and rear admiral (equivalent to major general); all known as flag officers. (The rank of commodore [q.v.], also a flag officer, is equiv-

alent to brigadier general.) The word admiral apparently originated before the 12th century with Arabs who combined *amir*, commander, the article *al*, and *bahr*, sea, in *amir-al-bahr* which was shortened by the Sicilians to *amiral*. Near the end of the 13th century the English commander of the Cinque Ports was given the Latin title *admirabilis*.

admiral of the blue, admiral of the red, admiral of the white, admiral of the yellow—By 1620 the English commander at sea was known as an admiral. The English fleet at this time comprised three squadrons. Ships of the squadron in the center flew red ensigns, and their commander was known as admiral of the red. The van squadron ships flew white ensigns, and their commander, a vice admiral, was known as admiral of the white. The squadron in the rear flew blue ensigns, and their commander, a rear admiral, was known as admiral of the blue. If a captain was promoted over others senior to him to become admiral of the blue, those passed over were made rear admirals on the retired list. The common expression was "yellowed," and they were called admirals of the yellow, a fictitious squadron.

Admiralty—Office of the British government responsible for the management of naval affairs and for commanding the Royal Navy. Prior to 1964, when the administrative offices of the British armed services were merged into the Ministry of Defence, the official title of the Admiralty was **Board of Admiralty**. The ten-man Board of Admiralty (four civilians and six admirals known as sea lords) performed its functions under the chairmanship of the First Lord of the Admiralty, a civilian member of Parliament and a cabinet minister. Since 1964 it has been known as the Admiralty Board of the Defence Council.

advance—v. 1) To move forward, used particularly in reference to an attacking force. 2) To raise in rank; to promote. n. The distance moved forward. See also **advance rate**.

advance force (amphibious)—A temporary organization within an amphibious task force which precedes the main body to the objective area. Its function is to help prepare the objective for the main assault by conducting such operations as reconnaissance, seizure of supporting positions, minesweeping, preliminary bombardment, underwater demolitions, and air support.

advance guard—A portion or detachment of a ground combat force sent ahead of the main body for any or all of several purposes: to gain information about the location and strength of the enemy's forces, to protect the main body against surprise, to remove obstacles and repair roads and bridges, to cover the deployment or conceal the movements of the main body by occupying the attention of the enemy. In France the term is *avant-garde*. In late medieval and early modern times the advance guard was generally about one-third of a field army; in battle formation it was deployed either in front of the main body, to initiate combat, or was arrayed on the right flank of the main body.

advance landing field—An airfield, usually having minimum facilities, in or near an objective area.

advance rate—The speed of the advance of a military force, usually expressed as miles or kilometers per 24-hour day.

advance to contact—Move forward in an effort to gain or reestablish contact with the enemy.

advanced base—A base located in or near a theater of operations, the primary mission of which is to support military operations.

AEF—Abbreviation for American Expeditionary Force, US troops under the command of General John J. Pershing who served in Europe during World War I, beginning in June 1917.

aerial barrage—See **barrage**.

aerial dart—A metal dart dropped from an aircraft, used in World War I against a target such as a zeppelin. Also called flèchette (*q.v.*).

aerial gunner—US Army enlisted specialty.

aerial mining—The act or process of laying mines (usually naval mines) from aircraft. See **mine**.

aerographer—US Navy warrant officer with meteorological duties.

aerographer's mate—US Navy enlisted specialty in meteorology.

aeromedical evacuation—The movement of patients to and between medical treatment facilities by air transportation.

aeronautical chart—See **chart**.

aeronautical rating—One of several ratings conferred on a member of the US Air Force (or Army, Navy, or Marine Corps air components) who has completed the requisite training and obtained the requisite competence in a given aspect of flying. In addition to pilots such persons include bombardiers, navigators, radio and radar personnel, engineers, and gunners with aeronautical specialties. Recipients of such ratings are usually authorized to wear a metallic emblem, in the form of wings, on the chest of the uniform tunic or shirt.

affirmative—A signal transmission response indicating agreement with, or confirmation of, the previous transmission. See also **acknowledgement** and **Roger**.

Afrika Korps (Deutsches Afrika Korps)—Designation of a renowned corps-sized force of German troops, commanded by General (later Field Marshal) Erwin Rommel, sent to North Africa in February 1941 to reinforce Italian forces which had been badly defeated by British Commonwealth forces during the winter of 1940–1941. Rommel assumed command of all Axis forces in North Africa, and the Afrika Korps became the mainstay of the German-Italian Panzer Army, which fought against the British Eighth Army in Libya and Egypt during 1941 and 1942. Rommel's defeat at El Alamein (23 October–4 November 1942) preceded a long retreat to Tunisia, where, in May 1943, the Axis forces, including the remnants of the Afrika Korps, surrendered to Allied forces.

AFV—Abbreviation for armored fighting vehicle. See also **IFV**.

agema—An elite corps of the Macedonian armies of Philip II and Alexander the Great. Composed of hypaspists (*q.v.*), this was the infantry bodyguard of the king.

agger—Originally a Roman term for a mound, or rampart, thrown up in front of a ditch to protect a camp or other position. The term also was used for a mound built in ancient siege operations to overlook the defenders' walls or to provide a direct approach to his ramparts. Now sometimes used for a military road, particularly one built up above the land it traverses. Typical historical examples were Alexander's agger at the siege of Gaza (332 B.C.) and the Roman agger at the siege of Masada (70 A.D.).

aggression—An unprovoked attack upon the territory or interests of a state by the armed forces of another state, termed the aggressor.

aggressor force—1) A force that initiates a combat action, particularly by invading the territory of another nation. 2) Euphemism often used by the US Army to designate the enemy in two-sided peacetime maneuvers in field exercises.

aid man—A medical corpsman who serves with a combat unit, usually of company size.

aide de camp (ADC)—An officer who serves as a personal assistant to a general or flag officer, performing mainly administrative duties. Usually referred to as an aide.

aiguille—A tool used to bore for buried land mines.

aiguillette—A loop of one or more strands, generally but not always of gold braid, worn on the shoulder of a uniform. In the United States they are worn by officers serving as aides to high-ranking officers or government officials. In France they are used to identify troops in units that have been awarded unit citations for exceptional or valorous conduct in battle. In England they are worn by non-commissioned officers of the Household Cavalry. In some countries they are worn by the gendarmerie. Aiguillettes were originally used to secure armor plates to the shoulder of the leather jerkin worn underneath.

aim—v. To point a weapon so that its effective component (missile, projectile, point, blade) will strike a selected target. n. The direction or line of fire or of sighting thus identified.

aimed fire—Fire that is placed on a target by direct fire weapons when the target is seen by the individual or crew firing the weapon, and by indirect fire weapons when the target is seen by an observer who can communicate with personnel manning the weapons to adjust the fire. Aimed fire can be directed at both point targets and area targets.

aiming point—A common reference point for the gunners of artillery pieces or mortars of a unit, toward which they aim the sights of their weapons. For indirect fire—and most artillery and mortar fire is indirect—all commands for direction or deflection are given in angular measurements (usually mils, sometimes degrees) with respect to the aiming point. When a field artillery unit (usually a battery) occupies a position, all weapons are immediately laid in a parallel sheaf (q.v.) in the general direction of fire, usually by the battery executive officer, who then selects the aiming point and orders the gunners to refer to it. If no prominent, readily identifiable fixed object is available for such reference, the cannoneers will set out aiming posts (q.v.) for reference as the aiming point for an individual gun or mortar.

aiming posts—Stakes or poles, two for each weapon, carried with each gun or mortar, and available for use as aiming points (q.v.), if needed. The stakes are painted in sharp, contrasting colors, usually red and white, so as to be readily identified by the gunner looking through his sight. When the command to refer to aiming posts is given, a cannoneer, carrying the stakes, runs out in the direction indicated by the gunner, who is looking through the sight of the gun. He aligns the aiming posts, which are placed about 100 feet and 50 feet respectively from the gun, usually to the rear or flank, by giving directions to the cannoneer, so that he can see only the closer of the two stakes through the sight. All firing commands for direction or deflection are then applied to the gun as angular measurements related to the aiming posts.

air attack—An attack on a surface target by combat aircraft.

air base—See **base**.

air burst—The explosion of a bomb or projectile above the surface of the earth, as distinguished from an explosion on impact with the surface or after penetration.

air-burst fuze—Either a time or proximity fuze (qq.v.). See **fuze**.

air cavalry—Units which make use of air

mobility (usually by helicopter) to perform missions, such as reconnaissance, screening, and surprise attacks, traditionally performed by horse cavalry before the early 20th century. See also **armored cavalry**.

air chief marshal—See **air marshal**.

air command—A major subdivision of the US Air Force, normally consisting of two or more air forces (q.v.).

air commodore—1) A commissioned rank in the air forces of Great Britain and other countries corresponding to that of brigadier general in the USAF. 2) A person holding this rank.

air corridors—Restricted air routes of travel specified for use by friendly aircraft; established to prevent friendly aircraft from being fired on by friendly forces.

air cover—1) Protection against attack, especially air attack, provided to surface or airborne forces by combat aircraft. 2) The aircraft providing such protection.

air crew—The personnel assigned to operate an aircraft.

air defense—Defense against attack from the air by hostile aircraft or missiles. Also the system itself. Air defense measures include antiaircraft artillery, surface-to-air missiles, and fighter aircraft. See also **active air defense**, **passive air defense**.

air defense area—A specifically defined and established territory (or air space over territory) that includes objectives of possible enemy air attack and for which air defense must be provided.

air defense artillery—Crew-served weapons and equipment that provide defense against attack from the air, including hostile aircraft and missiles. Usually applied to ballistic gunpowder weapons, as opposed to surface-to-air missiles (q.v.). See also **antiaircraft artillery**.

air defense direction center—An installation with the capability of performing air surveillance, interception control, and direction of air defense weapons. Some centers have an identification capability.

air delivery—Air transport of units, personnel, supplies, and equipment. The term includes both airdrops (q.v.) and air landings and covers both tactical and administrative movements.

air echelon—1) That part of an organization moved by air, either by military transport aircraft or by a unit's own aircraft, or both, as distinguished from the elements moved by sea or land. 2) That part of an air force organization composed of its air crew members. 3) The military force that operates in the air, as distinguished from ground or naval forces.

air force—1) The component of a country's armed forces or of one of its services concerned with military operations in the air, for example, US Air Force, Army Air Force, Naval Air Force. 2) A major organizational unit within a national air force, distinguished by a name or number, for example, Fifth Air Force.

air front—1) The direction from which aircraft attack, especially coming toward a nation's strategic targets. 2) A direction toward which an air attack is made.

air gun—A gun in which the projectile is discharged by compressed air, liquid carbon dioxide, or a combination of compressed air and an ignitible fuel. See also **dynamite gun**.

air interdiction—Interdiction fire on, or bombardment of, ground targets by weapons from one or more aircraft. See also **interdict** and **interdiction fire**.

air marshal—A commissioned rank in the British Royal Air Force and certain other national air forces (but not the USAF). There are usually four grades: air vice marshal equivalent to major general, air marshal equivalent to lieutenant general, air chief marshal equivalent to general, and marshal of the air force, equivalent to general of the air force.

air observation—Visual observation of an enemy's positions, forces, or installations from the air.

air observer—An individual whose primary mission is to observe an enemy's position or other target, or take photographs of them from an aircraft in order to adjust artillery fire or obtain military information.

air portable—Denotes equipment that can be carried in an aircraft with not more than such minor dismantling and reassembly as would be within the capabilities of user units. This term is usually qualified to show the extent of air portability in relation to cargo aircraft by weight, size, or specific aircraft designation.

air raid—An attack by aircraft against a surface target. Although air raids generally are thought of as bombing attacks, they include attacks with machine guns or cannon (strafing), rockets, torpedoes, or any other weapon by which aircraft can directly damage or destroy surface targets.

air room—A room or rooms at major US air commands where intelligence on the current situation is given to authorized personnel in the form of briefings, film showings, maps, drawings, charts, written summaries, etc. It corresponds to the Army's war room.

air sortie—See **sortie**.

air strike—An offensive attack by one or more aircraft on a specified target or targets.

air strip—An unimproved or partially improved surface that has been adapted for takeoff or landing of aircraft, usually having minimum facilities.

air superiority—Preponderance in air strength of one force over another such that the weaker force cannot interfere prohibitively with the other's conduct of air, ground, or sea operations. Although air superiority provides one side with general domination of the air space over surface combat operations, it does not mean total denial of air operations by the weaker side. Such denial, however, is implicit in air supremacy (q.v.).

air supply—The delivery of cargo by airdrop or air landing.

air support—All forms of support, operational or administrative, given by air forces to forces on land or sea. See also **close air support**.

air supremacy—That degree of air superiority wherein the opposing air force is incapable of effective interference with ground, sea, or air operations.

air-to-air missile. See **missile, air-to-air**.

air-to-surface missile. See **missile, air-to-surface**.

air transport—1) The movement of personnel or cargo by aircraft, as distinguished from surface or ground transport. 2) An aircraft used for such movement.

air transportable units—Ground units, other than airborne, which are trained and whose equipment is adapted for movement and delivery by transport aircraft.

air vice marshal—See **air marshal**.

air wing—A unit of the US Air Force and of other air forces, usually composed of one air group (which may be combat, training, transport, or service) with its support units, including at least supply, maintenance, and medical.

airborne—Transported by or carried in aircraft. This includes airborne divisions, airborne command posts, and airborne troops. Airborne units now form integral parts of the armed forces of all major nations.

airborne attack—See **attack**.

airborne landing area—See **landing area**.

airborne warning and control systems (AWACS)—Aircraft equipped with search radar, height-finding radar, and communications equipment for controlling weapons, generally other aircraft, especially fighters. AWACS are used for surveillance, early warning, and control.

aircraft—A vehicle capable of traveling through the air. Aircraft types include heavier-than-air (the airplane and helicopter) and lighter-than-air (balloon, blimp, dirigible). Military aircraft are further categorized by their functions (fighter aircraft, bomber aircraft, reconnaissance aircraft, etc.), by their means of locomotion (jet, turboprop, etc.), and by their configuration (for example, fixed wing, rotary wing, helicopter, biplane).

aircraft arresting barrier—An apparatus or system for slowing, and ultimately stopping, the progress of an aircraft landing on the flight deck of an aircraft carrier. The first such arresting gear was emplaced on the planked-over afterdeck of the US armored cruiser *Pennsylvania* in January 1911, when pioneer naval aviator Eugene B. Ely made the first landing on an "aircraft carrier." Most arresting apparatuses employ the same principles—a steel cable stretched across a carrier's deck catches a hook dropped from the tail section of the aircraft's fuselage.

aircraft arresting system—A series of components, whether barriers or other gear, used to engage an aircraft and absorb the forward momentum of a routine or emergency landing (or aborted takeoff).

aircraft carrier—A large naval vessel that serves as a mobile air base at sea. It has a control center at one side and a long, broad deck that serves for takeoff and landing of the aircraft it carries. The US Navy designation for carrier is CV. Designations for types of carriers are CVA, attack; CVN, nuclear; CVL, small; CVS, seaplane.

aircraftman—Any of the four lower enlisted grades or ranks in the Royal Air Force, including, aircraftman second class, aircraftman first class, leading aircraftman, and senior aircraftman.

airdrop—n.The act of unloading personnel or matériel from aircraft in flight. This is used as a means of supply in difficult terrain, where land lines of transport are difficult or nonexistent, as in Allied operations in Burma in World War II. An airdrop may be with or without parachute; in the latter case it is sometimes called a "free drop." *v.* To unload from an aircraft in flight.

airhead—1) A tactical base for an airborne or essentially air mobile operation. Usually seized as the first step in an airborne operation, an airhead constitutes an operational base of maneuver and a secure logistical base for supporting the operation. 2) A location used for supply and reinforcement and evacuation of wounded in an operations area.

airland battle—1) Combat waged by ground forces with air support. 2) A doctrinal concept of aggressive, mobile, air-supported ground combat introduced by the US Army in the early 1980s.

airlift—*v.*) To transport passengers and cargo by use of aircraft. n. 1) The carriage of personnel and/or cargo by air. 2) The total weight of personnel and/or cargo, that is, or can be, carried by air, or that is offered for carriage by air.

airman—Enlisted grade in the US Air Force, first, second, and third class; corresponds to private in the Army and seaman in the Navy.

airmobile operations—Operations in which combat forces move about the battlefield in air vehicles under the control of a ground force commander to engage in ground combat.

airship—A lighter-than-air craft, such as a blimp or dirigible.

airship rigger—US Navy enlisted specialty. He maintains lighter-than-air ships.

airspeed—The speed of an aircraft relative to the air mass through which it is flying. **Calibrated airspeed** is airspeed shown on an instrument (**indicated airspeed**), corrected for errors in the installation of the instrument. When corrected for compressibility errors it becomes **equivalent airspeed**. Equivalent airspeed corrected for error due to air density, i.e., altitude and temperature, is **true airspeed**.

aketon (hacqueton)—A type of early medieval body armor resembling a padded coat; it was made of buckram, stuffed with cotton, and stitched longitudinally. It was worn as padding under the hauberk (q.v.) and covered the body from the neck to the knees.

ala, alae—A Roman cavalry organization consisting of twelve turmae (q.v.). Its modern equivalent would be a cavalry squadron. The ala formed up in two or three lines.

ALCM—See **missile, air launched cruise**.

alert—A condition in which all members of a unit stand by in a state of readiness for active operations, either because of an anticipated attack, or the possibility of such an attack, or because of contemplation of other combat operations. The seriousness of the alert is frequently identified by numbered categories (as "alert status one") or code designations (as "red alert"); an alert may also be related to the forces affected, such as "air alert," or "ground alert."

alert force—A force maintained in a state of

readiness for immediate operational performance.

Alfa—First letter of NATO phonetic alphabet. See also **alphabet, phonetic**.

alignment—1) Relative positions of military forces in tactical operations; can be abreast (usually referred to as in line), echeloned, or one behind another (usually referred to as in column). 2) The bearing of two or more conspicuous objects (such as lights, beacons, etc.) as seen by an observer. 3) Representation of a road, railway, etc., on a map or a chart in relation to surrounding topographic details.

all hands—All members of a ship's company or of any naval unit, including the entire Navy.

alliance—1) A formal agreement by two or more nations to support and assist each other. 2) The collective members of such a union.

allotment of pay—A deduction from the paycheck of a member of the armed forces, paid directly to another person or organization.

allowance—1) A payment, the amount of which is usually specified by rank, to a member of the armed services in lieu of provision of quarters, food, or other items or as recognition of performance of special duty, such as sea duty, overseas duty, or submarine duty. 2) Specified quantities of various forms of supply, to include rations, to be provided on a regular basis to a unit or an individual.

allways fuze—An impact fuze designed to function regardless of the part of the fuze or projectile that makes contact with the target.

all-weather fighter—See **fighter aircraft**.

ally—A nation joined with another, in an alliance, by treaty or by common interests. In both World War I and World War II the nations fighting together against Germany and its cobelligerents were called the Allies.

almain armor—See **armor**.

alphabet, phonetic—*World War II*: Able, Baker, Charlie, Dog, Easy, Fox, George, How, Item, Jig, King, Love, Mike, Nan, Oboe, Peter, Queen, Roger, Sugar, Tare, Uncle, Victor, William, Xray, Yoke, Zebra. *NATO*: Alfa, Bravo, Charlie, Delta, Echo, Foxtrot, Gold, Hotel, India, Juliet, Kilo, Lima, Mike, November, Oscar, Papa, Quebec, Romeo, Sierra, Tango, Uniform, Victor, Whisky, Xray, Yankee, Zulu.

Amazon—A mythological female warrior from ancient Scythia.

ambuscade—An ambush.

ambuscader—One who takes part in an ambush.

ambush—n. A sudden, carefully prepared, surprise attack by a force whose position had been carefully concealed against an advancing hostile force. v. To make such an attack.

ammo—Ammunition (*q.v.*).

ammunition—Destructive materials intended to damage, destroy, or suppress hostile personnel and matériel, and the ancillary consumable equipment required to make them function. Projectiles fired from a gun are the principal form of modern ammunition. A round (*q.v.*) of ammunition includes not only the projectile but a primer, fuze, propellant charge, and a case or jacket that holds the propellant and (usually) connects it with the projectile. Projectiles are usually in the form of bullets or artillery shells. Ammunition also includes missiles, torpedoes, mines, and bombs. Nonexplosive ammunition (generally pregunpowder) includes arrows, bullets, shot, or stones for slings. (Nondestructive materials designed for specialized use in weapons [as blanks, or subcaliber training projectiles, and ancillary items] are considered to be ammunition.) Types of modern ammunition include the following: **fixed ammunition**—The most usual kind of ammunition for all types and calibers of guns, in which the projectile (or bullet) is firmly attached to the propellant and primer by means of a cartridge (usually called cartridge case for artillery weapons). **semifixed ammunition**—This type is similar to fixed ammunition, to the extent that the projectile is assembled with cartridge case, propellant, and primer, but the projectile can be lifted off to permit removal of part of the propellant (powder charge in separate bags) to permit a variety in charges with consequent differences in muzzle velocity and range. **separate-loading ammunition**—The projectile and propellant are packaged separately. The primer is sometimes assembled with the propellant, sometimes attached just before firing, or is put into a receptacle in the breech block. This type of ammunition (without cartridge case) is standard for larger calibers of medium and heavy artillery.

ammunition, armor-piercing (AP)— Ammunition designed to penetrate resistant materials, in particular armor on warships, tanks and other vehicles, or installations. A characteristic of certain bombs, bullets, shells, and other projectiles. Types of modern armor-piercing ammunition include: **armor-piercing discarding sabot (APDS)**—A thin, long, arrowlike projectile encased in a sabot (q.v.) which falls away after firing; the projectile obtains its penetration capability from high velocity. **high explosive antitank (HEAT)**—A projectile utilizing a shaped charge (q.v.) detonated on impact by a base fuze to produce a very hot jet of flame which sears its way through armor. **high explosive squash-head (HESH)**—A round which achieves penetration by a combined action of the collapsible head (to prevent ricochet) and a shaped charge. **hypervelocity armor-piercing (HVAP)**—A projectile with an extremely hard core of tungsten carbide. Fired at high velocities, it can penetrate many inches of armor.

ammunition belt—See **belt**.

ammunition box— See **box, ammunition**.

ammunition consumption—1) The process of expending ammunition by firing in combat. 2) The rate (usually daily) of such expenditure.

ammunition dump—A depot for storage of ammunition.

ammunition supply point—Storage location of ammunition close to the front lines, from which the combat units obtain replenishment for ammunition consumed in battle.

amphibious assault—See **amphibious operation**.

amphibious assault ship—A ship designed to transport troops, equipment, and supplies to the landing area. In World War II troops transferred from the assault ship to the shore on **amphibious assault craft** (landing craft). Since development of the troop-carrying helicopter, the troops are generally carried in

them. Some of the ships in use in the 1980s are converted aircraft or seaplane carriers.

amphibious attack—See **attack**.

amphibious command ship—A ship from which a commander exercises control of amphibious operations.

amphibious force—The entire force trained, organized, and equipped to perform amphibious operations, including personnel, vessels, and supporting forces. The naval elements form an organizational unit of a fleet; the ground elements are generally designated as the landing force (q.v.).

amphibious group—A command within an amphibious force under a headquarters designed to exercise operational command of assigned units in the execution of a division-size amphibious operation.

amphibious lift—The total capacity of assault shipping used in an amphibious operation, expressed in terms of personnel, vehicles, and measurement or weight tons of supplies.

amphibious operation—An operation involving the transfer of troops between ships and land. There are several types of amphibious operations: 1) an opposed landing on a hostile shore (**amphibious assault**) to obtain a beachhead from which to carry out further combat operations; 2) an unopposed landing of forces; 3) a withdrawal of land-based forces.

amphibious striking force—A seaborne force capable of projecting military power upon adjacent land areas.

amphibious task force—The task organization formed to conduct an amphibious operation. The amphibious task force always includes naval forces and a landing force, and, when appropriate, air forces.

amphibious vehicle—A vehicle capable of operating on both land and water.

anchor watch—A detail of men on a warship in port that stands deck watches of from two to four hours in turn from 2100 to turn-to in the morning, about 0530.

angel—1) A type of confusion reflector having the reflecting material suspended from parachutes or balloons to delay the descent. 2) One thousand feet of altitude.

annihilate—To take action that causes an enemy force to be totally incapable of further resistance. It is important to note that the military sense of this word is not synonymous with its nonmilitary sense of complete destruction.

antiaircraft artillery (AAA)—Guns used for attacking aerial targets from the ground or shipboard, and the radar, searchlights, and other equipment and devices related to them. Antiaircraft artillery weapons are classified as: light—20-57mm, medium—58-99mm, and heavy—100mm or larger.

antiballistic missile (ABM)—See **missile, antiballistic**.

antipersonnel mine—See **mine**.

antisubmarine warfare (ASW)—Operations conducted against submarines, including detection and attack.

antitank artillery—See **artillery**.

antitank grenade—See **grenade**.

antitank mine—See **mine**.

antitank weapons—Weapons, such as bazookas (q.v.), guns, and guided missiles, which may or may not be mounted on tanks or other armored vehicles, whose main purpose is combat against tanks and other armored vehicles.

ANZAC (Australian and New Zealand Army Corps)—Acronym for Australian and New Zealand Army Corps formed in World War I. Commonly used in both World Wars I and II for a member of the Australian or New Zealand armed forces.

AOP—Air observation post. See **air observation** and **observation post**.

AP—Armor-piercing. See **ammunition, armor-piercing**.

appreciation—British term for estimate of a situation (q.v.).

approaches—1) Regions through which an attacker draws close to any enemy position. 2) Trenches or other works erected to protect troops that are laying siege to a fortified position. The classical procedure was to dig a line of trenches parallel to the defenses, and outside artillery range of the defender, and then to dig one or more zigzag approach trenches or saps (q.v.) leading partway toward the enemy's position, at which point another trench line was dug parallel to the original line. The process was repeated until a line of trenches was dug close enough to the enemy fortification to allow an assault. Thus the word "parallels" is often used—incorrectly—as a synonym. **Counterapproaches** are works erected by defenders to interfere with the besieger's approach works. See also **parallel** and **siege works**.

apron—A paved or otherwise hard-surfaced area, generally located near airport or airfield hangars, terminals, or control areas, where aircraft stand or park for purposes of loading and unloading, refueling, arming as well as rearming, and servicing. Aprons are also called **hardstands** and **ramps**.

arbalest—A medieval missile launcher designed on the crossbow principle. See **crossbow, catapult, ballista**.

arc of fire—The segment of a horizontal circle through which fire may be directed from a weapon without shifting its position. This, with the weapon's range, sets the limits of the area it can cover by fire.

archer—One who shoots with a bow and arrow. The origins of the bow, and consequently the archer, are unknown. Evidence of bows, in the form of arrowheads, is found from Palaeolithic times and in many parts of the world. Archers appear in the paintings and sculpture of many ancient cultures, and mounted archers in those where horses were used. The Persians of the 1st millenium B.C. combined foot archers and horse archers in the opening moves of battle. By the beginning of the Christian era the horse archer dominated warfare in all of Asia except India. The Mongols were outstanding horse archers. In Roman times mounted soldiers began to carry lance, sword, and shield as well as a bow. In late medieval and early modern times the term archer acquired two new meanings. It was used to refer to personnel in certain bodyguard units who may or may not have been armed with bows and may or may not have been mounted (for example, the Scottish Archers of the French Royal Household), and was also used to refer to heavy, lance-armed cavalry, which, until the 1540s, were a component of gendarme units, whether lance (q.v.) or company. Gentlemen usually enlisted as archers in gendarme companies and, after serving a period of military apprenticeship, advanced to the rank of gendarme, or man-at-arms.

architecture, military—1) Class of structures uniquely designed for military purposes. The term is usually applied to castles or other permanent fortifications. 2) The design of such structures.

architecture, naval—The art of designing seagoing vessels to perform military missions.

area—n. A surface section of land or sea (or of a building or ship) that has generally defined boundaries. The term is also applied to intangible entities such as responsibility, study, or power—which can be delimited for a specified purpose. adj. Referring to a large space rather than a specific point. An **assembly area** is an area where the units of a military command come together in preparation for an operation; an **area of operations** is where military operations are conducted pursuant to an assigned mission; a **forward area** is the area nearest to and including the front line; a **combat area** is the site of combat; a **rear area** is behind the area of operations, where reserves of men, equipment, and supplies are held. A **bombing area** is an area within which bombing is to take place; **area bombing** is bombing intended to destroy a large area rather than specific targets or types of target. A **command area** is the area assigned to a command (army, division); an **area command** in US forces is a command related to a geographical area (North Pacific, Southeastern United States) in which all forces of all services are under a single commander. A **defense area** is an area to be defended; an **area defense** is a defense with units deployed to prevent attacks on a geographical area rather than a specific installation or target. A **target area** is an area designated for attack; an **area target** is a target consisting of an area rather than a single point.

area bombing—See **area**.

area command—See **area**.

area defense—See **defense** and **area**.

area fire—Fire that is directed at a general area, rather than at a precision target. Area fire can be either observed or unobserved fire (qq.v.).

area of burst—See **burst**.

area of operations—See **area**.

area target—See **area**.

argoulet—A French light cavalryman of the Renaissance period.

arm—n. 1) A weapon of war. Two categories are a) small arms—weapons carried and fired by one person; b) side arms—weapons, such as revolvers, knives, or swords, which are worn at the side or on a belt (included in small arms). 2) A combatant branch of an army, in a modern army usually including infantry, cavalry, armor, artillery, and air defense, and sometimes also including engineers. 3) An organized branch of a nation's armed forces, in modern times usually including army, navy, and air force. In addition, the United States has a marine corps, the Soviet Union has rocket and missile forces, and the Soviet Union and others have air defense forces. 4) Heraldic device of a family or a government. v. 1) To furnish with weapons of offense or defense. 2) To adjust the firing mechanism of a mechanical or explosive device—a bomb or a mine, for example—so that when triggered the device will explode or do whatever it is designed to do.

armada—Any large fleet of warships or aircraft. In 1588 Philip II of Spain sent a fleet, known to history as the Spanish Armada, against England. Although it was considered to be invincible, in fact it was weaker in firepower than the waiting and defending English vessels and suffered a disastrous defeat.

armament—1) The weapons and other equipment, including defensive equipment and supplies, of a nation's armed forces, or any portion thereof. 2) All of the offensive weapons of a ship. 3) The process of arming.

arme blanche—A blade weapon, usually a sword, saber, or bayonet, used in hand-to-hand combat.

armed conflict—A war, battle, engagement, or other meeting between forces of two or more nations or factions in which weapons are used to inflict physical damage.

armed forces—All the military forces of a nation or a group of nations.

armed merchantman—A commercial vessel equipped with guns or other weapons. In World War I ships were armed as protection against surface raiders and submarines. In World War II the practice was general, with emphasis on protection by antiaircraft guns against air attacks.

armed neutrality—The condition of maintaining a nation equipped and prepared for war while refraining from taking sides in a conflict. The term was apparently first used in 1780, during the American Revolution, when Empress Catherine the Great of Russia proclaimed a state of Armed Neutrality for the protection of shipping of neutral nations. The belligerents were called on to accept principles of neutrality embodied in a defensive treaty signed by Russia, Denmark, and Sweden. Spain and France accepted; Great Britain disregarded the principles.

armed reconnaissance—An air mission flown with the primary purpose of locating and attacking targets of opportunity, that is enemy matériel, personnel, and facilities, in assigned general areas or along assigned ground communications routes, and not for the purpose of attacking specific briefed targets.

armet—A type of medieval light helmet having a movable visor and protection for the neck.

armiger—An armor-bearer for a knight; generally a squire. One of the many duties of the squire was to assist the knight in donning his armor.

armistice—A suspension of hostilities agreed upon by warring forces; a truce. Terms under which armistices are negotiated vary from complete cessation of all action to temporary halt of battlefield action for a specific length of time. Best known is the Armistice of 11 November 1918, signed by representatives of the Allies and Germany, which ended fighting on the Western Front in World War I.

armor—1) Material or equipment used to protect men, animals, ships, vehicles, aircraft, and other equipment from hostile destructive agents (arrows, swords, bullets, shells, bombs, mines, etc.). Includes a) **body armor**—Some sort of body protection was used by warriors from the earliest times. Increased technological capability permitted the use of more and more durable materials, animal hides, leather, layered or quilted fabrics, bronze, iron, steel, and, in modern times, fiberglass and synthetic materials, of which kevlar is the most recent development. Among the major types are: **almain armor**—Light armor for the upper body, consisting of helmet, gorget, breastplate, backplate, and arm splints. This half-armor was common during the 16th century. **bronze armor**—The capability of working bronze permitted the use of this metal for helmets and for body armor. Bronze armor varied greatly from country to country and from age to age. The earliest known whole body armor dates from the 15th century B.C. It covered the wearer from shoulder to knee, and had articulated shoulder pieces. Greek armor generally consisted of a cuirass, which covered the chest, with a shirt of various materials, separate leg pieces, greaves, and helmets of various designs. **fabric armor**—Various fabrics, usually quilted, have been used for protection. Linen, padded with tow, was commonly used in medieval Europe. Heavy silk shirts were worn under armor by the Mongols in the 13th century to improve protection, and by some in Britain at the end of the 17th century. During World War I, silk was tested and found to protect against shell splinters and shrapnel traveling at speeds of up to 1,000 feet per second. The Chinese also made use of fabric armor, sometimes lacquered. Combinations of man-made fibers, including fiberglass and metallic fibers and metals, have been used in modern times. **lamellar** or **scale armor**—Used from antiquity, it consisted of pieces of metal or other material sewed in overlapping rows like shingles to a garment of cloth or leather. **leather armor**—Hides, and then tanned leather, provided a fairly inexpensive and readily available material that could be shaped to the human form. Leather helmets were cheaper than bronze and worn by the ordinary soldier in ancient and medieval armies. Garments of leather were common, and sometimes used as a base for attaching metal or other protective materials. **mail**—Also called chain mail. It consisted of interlocking metal rings forming a sort of fabric normally in the form of a shirt. Used by the Romans, it survived into medieval times, sometimes in combination with plate armor.

plate—Armor made in a solid piece, as opposed to mail or lamellar. Plate armor was carried to its ultimate form in the early 16th century, when knights were armored from head to foot in plate and fought on horses which also had their own armor protection. b) **ship's armor**—The armor on ships is usually specially treated case-hardened steel. It is thickest in the most vulnerable areas, particularly gun turrets, the deck, and below the waterline. c) **tank** and **vehicle armor**, including **self-propelled artillery**—Armor is applied to many kinds of vehicles whose purposes may take them into areas subject to enemy fire. It may be added as protection to an otherwise regular automobile, truck, construction vehicle, etc., or the design of the basic combat vehicle, such as tanks or self-propelled artillery (q.v.), may call for construction with armor protection. Most self-propelled artillery is protected by armor; most towed guns have armor shields. d) **armored fortifications**—Rifled siege guns, as demonstrated in the American Civil War and Franco-Prussian War, made brickwork forts obsolete; therefore, in the latter decades of the 19th century, modern permanent fortifications were constructed of concrete and earth. Cast steel armor, used in gun turrets and to protect underground magazines and living quarters, was first used in forts designed by General Séré de Rivière to protect the French frontier with Germany. Construction of this system was begun in 1873. 2) A general term for all combat vehicles, including tanks, that are armored. 3) The general designation given to armored force components of an army or command.

armor belt—See **belt**.

armored—Equipped with special physical protection, generally, but not always, made of metal.

armored artillery—See **artillery**.

armored car—A light armored vehicle, used primarily for reconnaissance and security missions.

armored cavalry—Motorized units that replaced horse cavalry after World War I. Also called mechanized cavalry, armored cavalry

units have been employed in reconnaissance and screening missions in much the way horse cavalry had primarily been used. See also **air cavalry**.

armored cruiser—A late 19th- and early 20th-century warship with both the armored deck of the protected cruiser (q.v.) —sometimes two decks—and belts of armor on both sides, as well as armored bulkheads fore and aft.

armored division—A combined arms ground force of division (q.v.) type and size, the principal combat components usually being tank units.

armored force—A military ground force in which most elements, or a preponderance of the major fighting elements, are armored vehicles (q.v.), or are units built around armored vehicles.

armored fortifications—See **armor**.

armored infantry—The infantry in an armored brigade or division. These units generally move into combat in armored personnel carriers (q.v.) or infantry fighting vehicles (q.v.).

armored personnel carrier (APC)—An armored vehicle, usually fully enclosed, used to transport infantry into combat, from which the troops usually debark to fight.

armored vehicles—Armor-protected vehicles, such as tanks, infantry fighting vehicles, many self-propelled artillery weapons, armored cars, personnel carriers, cargo carriers, recovery vehicles, and fire direction centers.

armorer—1) One who makes or repairs armor or weapons or has the care of them for his unit. 2) A person, especially a ground crew member, who repairs, loads, and handles aircraft armaments and bombs.

armor-piercing ammunition (AP)—See **ammunition**.

armor-piercing discarding sabot (APDS)—See **ammunition, armor-piercing (AP)**.

armory—1) A place for storing weapons and equipment. The term is often used in the United States for an Army Reserve or National Guard installation where the principal purpose is drilling and training the part-time military units, and where the storage of weapons and equipment is only incidental. 2) A factory for making arms. 3) An arsenal.

arms—See **arm**.

arms control—The concept of restricting the number, type, or performance of weapons systems, including command and control, logistics support, and related intelligence-gathering mechanisms, or the strength, organization, equipment, deployment, or employment of armed forces, through international agreement. The concept differs from disarmament in its focus on limiting, rather than eliminating, war-making capabilities.

Armstrong gun—A gun invented by Sir William Armstrong in the mid-19th century and commonly used by the Royal Navy in the 1860s. The Armstrong gun was a rifled breech-loader.

army—1) That element of a nation's defense establishment designed and organized especially for ground combat, including supporting arms. 2) A tactical and administrative element (often called a field army) within a national army. The size and composition of an army vary: in the US Army a field army (normally numbered First, Second, etc.) consists of a headquarters, two or more corps, and auxiliary forces. In the latter half of the 20th century the US Army has eliminated field armies from its command hierarchy, with corps commanded directly by army groups (q.v.). Types of armies include: **combined arms**—An army including all major arms and branches of the nation's army (particularly USSR); **field**—another term for a combined arms army (particularly US); **standing**—a permanent army of professional, paid soldiers; **consular**—a Roman army commanded by one of the two consuls of Republican Rome. It usually had 18,000–20,000 men and a combat front of about a mile and a half. If the two consular armies were joined the consuls would alternate in command, usually each 24 hours.

army corps—See **corps, army**.

army group—Two or more field armies under a designated commander. In the US Army in the late 20th century an army group

consists of two or more army corps. Primarily a tactical (or operational) command.

army of occupation—The units of an army stationed in the territory of a defeated enemy.

army reserve—See **reserve**.

army reserve corps—See **reserve corps**.

army service area—An administrative region in the rear area of a combat zone, usually behind the rear boundary of the combat corps of an army or army group, and in front of the combat zone rear boundary. Most of the administrative establishment and service troops in the combat zone are usually located in this area, and administered and commanded by an army service area commander.

arquebus—A 15th-century matchlock gun (q.v.), the first handgun to have a stock shaped to fit the shoulder. The arquebus was equipped with an S-shaped clamp (serpentine) that held the match. A trigger would move the serpentine so that the match would ignite the powder in a pan and fire the bullet. The weapon was also called harquebus or hackbut. It was first used extensively in the Spanish Army and became common in all European armies in the 16th century. See also **caliver**.

arquebusier—A foot soldier who carried and fired an arquebus. Arquebusiers wore a variety of dress and armors. In the late 15th and the 16th centuries, when arquebuses became a regular part of an army's weaponry, arquebusiers served with pikemen in mixed foot units of "pike and shot." Throughout this period, generally, the proportion of arquebusiers or "shot" to pikemen in infantry units rose gradually. In the mid- to late 16th century and early 17th century, all European armies replaced the arquebus with the matchlock musket (q.v.), and the arquebusier of the earlier period became the musketeer of more recent times.

arquebusier, mounted—An arquebusier mounted on a horse.

array—n. 1) A body of soldiers drawn up in battle order. 2) Any deployment of troops or weapons, as a target array. v. To draw up troops in battle order.

arrest—v. To seize and hold under authority of the law. n. A state of restriction upon the movement and freedom of a military person, who is then "under arrest." In this use of the word, it means a restriction less than confinement.

arrow—A straight thin shaft, usually made of light wood with a pointed head at one end and flight-stabilizing feathers at the other, and usually propelled from a bow.

arrowhead—The tip of an arrow, usually made of a hard material, like stone or metal, and shaped to facilitate penetration.

arsenal—1) An installation where weapons, ammunition, and other military matériel is made, repaired, or stored. 2) An armory, but without drill facilities. 3) A stock of weapons, from that of a single person to the total weapons holdings of a nation.

art of war—The conduct of combat operations by means of imagination, skill, and professionalism.

Articles for the Government of the Navy—Regulations for the disciplinary governance of the US Navy. For many years the articles were displayed in all ships and stations and were read to the assembled crew once a year. They were superseded on 5 May 1950, with the adoption of the Uniform Code of Military Justice, and issuance of combined Articles of War for all services.

Articles of War—A code of regulations governing the conduct of military forces. Those in existence today in the US and British military establishments derive from articles that were in use by the British before the American Revolution. In 1950 the Articles of War of the US Army and US Air Force and the Articles for the Government of the Navy were combined into an all-service code of Articles of War. The articles describe the offenses for which a member of the armed forces may be tried under the Uniform Code of Military Justice.

artificer—A specialist, dating at least from the 18th century, concerned primarily with the repair of weapons or equipment. The term is also used for an enlisted specialist rating in the military services. In the US Navy artificers repair a range of equipment, from engines to gun sights and optical instruments.

artillery—1) A general term, usually referring collectively to gunpowder weapons (cannon) too large to be hand-carried, although all large, pre-gunpowder devices for firing missiles may be considered artillery. 2) The branch or arm of an army that is equipped with such weapons.

Artillery may be categorized in many ways, although there are two principal types: **mobile artillery**, which includes cannon (q.v.) and comparable weapons that can be shifted from one place to another, and **fixed artillery**, which includes all such weapons permanently emplaced in fortifications or other installations. Types of artillery include: **antiaircraft** or **air defense**—Weapons which may be cannon or missiles designed for employment against hostile aircraft; **antitank**—Cannon or other crew-served weapons specifically for use against tanks and other armored vehicles; **armored**—Artillery designed to support armored forces. Usually, armored artillery weapons are self-propelled, and the crews and weapons are often protected by armor; **field**—1) Cannon and other mobile artillery weapons designed for use against most surface targets; this includes armored artillery; in some nations this includes antitank weapons. 2) The branch or arm of an army that is equipped with such weapons. Field artillery is usually categorized by weight and by means of mobility. Weight categories include light, medium, and heavy. **Light artillery** is intended primarily for direct support of infantry and armor, and includes cannon with calibers up to 105mm; the largest (and only) light artillery piece currently in the US Army is the 105mm howitzer. **Medium artillery** includes weapons in size from 105mm guns to 155mm howitzers. **Heavy artillery** includes guns of at least 155mm. Mobility categories include towed and self-propelled. **Self-propelled artillery** includes weapons that are permanently mounted on vehicles each with its own means of propulsion; **towed artillery** includes cannon incapable of independent movement, thus drawn by horses or by a prime mover (q.v.).

ASDIC—Anti-Submarine Detection Investigation Committee, commonly written Asdic. A British name developed before World War II for the undersea echo-sounding submarine detection device also known as sonar (q.v.).

as foragers (en fourrageurs)—A command of the late 19th and early 20th centuries for a military unit to deploy from close order formation to an irregular open order line.

assagai (assegaye)—A light, pointed, throwing or stabbing weapon like a javelin or short spear, usually pointed at both ends, used by the Zulus and other African tribes. A similar weapon had been used by the Albanian stradiots (q.v.) of the 16th century.

assault—n. 1) A planned, sharp attack upon a defended position, such as a gun emplacement or a fortified hill. 2) In an amphibious operation, the period from the time the first waves cross the line of departure to the seizure of the initial objectives. 3) In an airborne operation, the period in which the airborne force is delivered to the objective area and attacks designated objective positions. v. To make a violent attack on a local objective.

assault aircraft—Aircraft, including helicopters, which move assault troops and cargo into a combat objective area and provide for their resupply.

assault craft—Small craft and amphibious vehicles employed for landing troops and equipment in the assault waves of an amphibious operation.

assault gun—A cannon mounted on an armored, tracked vehicle, designed for close engagements and direct fire at fortifications. Traverse is usually limited and speed is usually slow. The German *Sturmgeschuetz* of World War II was an assault gun originally designed to provide antitank and close fire support at less cost in money and production time than a tank, because of its relative simplicity. It was often used as a tank by the Germans, particularly late in the war. It was more heavily armored than, and had a different combat role from, self-propelled artillery. See also **tank destroyer**.

assault landing—In amphibious operations, the debarkation of assault troops directly into combat.

assault shipping—Shipping assigned to an amphibious task force to transport assault troops, vehicles, equipment, and supplies to the objective area.

assault troops—Troops assigned to carry out an assault.

assault wave—A formation of forces, including assault troops, landing craft, amphibious vehicles, and assault aircraft, scheduled to arrive at a beach at the same time during an amphibious assault.

assembly—A bugle call calling troops to form ranks.

assembly area—1) An area in which a command is assembled preparatory to further action. 2) In a supply installation, the gross area used for collecting and combining components into complete units, kits, or assemblies. See also **area**.

assign—To integrate units or persons into an organization.

assignment—1) The status of being assigned. 2) The process of being placed in an assigned status. Unlike attachment (q.v.), an assignment is considered permanent and is terminated by a new assignment.

Assize of Arms—An important royal decree, given by King Henry II of England in 1181, which reestablished the old Saxon national militia, the *fyrd*, by directing all freeholders to provide themselves with arms and armor according to rank; it was also decreed that all should be ready to answer a summons to arms by the sheriff.

ASW—Antisubmarine warfare (q.v.).

Atlantic Wall—In World War II, Hitler's string of coastal defenses stretching from the northern tip of Denmark to Brittany and southward to the Franco-Spanish border. It was a line of permanent fortifications at intervals along the coast. In areas where attack was anticipated, it was often laced by strongpoints and field positions, as well as protected by minefields and underwater obstacles.

atom bomb— See **atomic weapon**.

atomic warfare—Warfare involving use of atomic weapons.

atomic weapon—A weapon whose explosive power is derived from atomic fission. However, the term is also loosely applied to all nuclear weapons, including those whose power is based on fusion. Atomic weapons include missiles, rockets, bombs, artillery shells, and landmines.

attach—To place people or units in or under the command of a larger organization on a temporary basis.

attaché, defense—See **attaché, military**.

attaché, military—A military officer attached to a diplomatic office, usually an embassy.

attachment—1) The status of being attached. 2) The process of being placed in an attached status. An attachment is temporary, and upon its conclusion the individual or unit returns to his or its regular assignment (q.v.).

attack—n. The initiation of combat by one force attempting to gain an objective held by another; a strike against a target to seize or destroy it. v. To launch such an action. Special types of attack include: **airborne**—An attack carried out by troops landing by means of parachutes, gliders, or helicopters. **amphibious**—An attack on a beach by forces delivered by landing craft. **deliberate**—An attack characterized by coordinated employment of firepower and maneuver to close with and destroy or capture the enemy. **flank**—An attack delivered against the side or flank of an enemy's forces rather than against their immediate front. **frontal**—An attack directly against the front of an enemy's force. **holding**—An attack designed to pin down a portion of an enemy's forces while a major effort is being made in another sector. **main** or **major**—A principal attack made on a portion of an enemy's forces, while lesser attacks are engaging other portions of those forces in other sectors. **piecemeal**—An attack begun by a limited force, with reinforcements joining the attack as they arrive on the battlefield or otherwise become available. **secondary**—Same as holding attack (see above).

attack aircraft carrier—A warship designed to support and operate aircraft, engage in attacks on targets afloat or ashore, and engage in sustained operations in support of other forces. Designated as CVA or CVAN. (CVAN is nuclear powered.)

attack cargo ship—A naval vessel designed or converted to transport combat-loaded cargo in an assault landing (q.v.). These ships have greater capability for carrying landing craft, more speed and armament, and larger

hatches and booms than comparable cargo ships. Designated as AKA.

attack carrier striking force—A naval force organized around one or more aircraft carriers. Other vessels accompany the carriers primarily to support them and screen against attacks by submarines or aircraft. Carrier-based aircraft are the main offensive weapons.

attack column—See **column**.

attack group—In the US Navy a subordinate task organization of an amphibious task force. It is composed of assault shipping and supporting naval units designated to transport, protect, land, and initially support a landing group.

attack transport—A naval vessel equipped with landing craft and designed to carry a battalion landing team with its equipment and supplies to an offshore point from which an amphibious landing can be made. Designated as APA.

attention—1) The posture assumed by a member of the armed services characterized by the body erect, shoulders square, eyes to the front, arms at the sides, and heels together. 2) The command to assume such a posture.

attitude—See **spin**.

attrition—Reduction of the effectiveness of a force caused by loss of personnel and matériel. Sometimes used synonymously with "loss rate." The **attrition rate** is usually expressed in percentages or in casualties or losses per thousand.

authentication—1) Evidence by proper signature or seal that a document is genuine or official. 2) A security measure designed to protect a communication system against fraudulent transmissions.

authenticator—A letter, numeral, or groups of letters or numerals, or both, attesting to the authenticity of a message or transmission.

authority—The power to direct action or to use resources to accomplish assigned missions.

automatic weapons—Firearms that fire continuously upon activation without the necessity for repeated triggering. They include machine guns, automatic pistols, automatic rifles, etc. See also **BAR**.

auxiliaries—1) In the ancient Roman army, soldiers with special capabilities who were not Roman citizens, such as archers from Crete and slingers from the Balearic Islands. These auxiliaries (*auxilia*) served in special units, originally in the frontier areas of the empire from which they came, but later abroad as needed. After 25 years of service an auxiliary was granted Roman citizenship. 2) Military personnel not regular members of a force or service.

auxiliary force—1) An aggregation of military personnel, weapon systems, vehicles and necessary support, or a combination thereof. 2) A force set up or established to assist in periods of need, as an auxiliary air force, auxiliary unit, etc.

aviation—Pertaining to air forces or to aerial activity.

aviation boatswain's mate—US Navy rating.

aviation electrician's mate—US Navy rating.

aviation machinist's mate—US Navy rating.

aviation metalsmith—US Navy rating.

aviation ordnanceman—US Navy rating.

aviation pilot—US Navy enlisted rating.

aviation radio technician—US Navy rating.

aviation radioman—US Navy rating.

AWACS—See **airborne warning and control systems**.

AWOL—Acronym for absent without leave (*q.v.*).

axe—A weapon with a broad, bladed head mounted on a handle. From earliest times axes were used as weapons in hand-to-hand

fighting, and there are many types known. In the 8th century Vikings used a **bearded axe**, so called for its odd shape, in profile somewhat resembling a truncated adze. The Viking **broad axe** of the 10th–11th centuries flared from a narrow base. The Franks in the 6th–8th centuries used a **throwing axe**, known as the **francisca**. The **lochaber axe** was used in Scotland from about the 16th century to the mid-18th century. The blade, attached to the end of a pole, was a semiellipse, a hook facing the other way was used by the foot soldier who carried it for pulling mounted men off horses, and for tearing defenses apart. The Danish **battle axe** was carried by most soldiers in England in the 11th century. It was later combined with a pistol, like the **boarding axe** which, with or without the pistol, was used by seamen or marines in sailing warships to chop masts and rigging off enemy vessels as well as to behead enemy sailors.

axis—A line or route assigned for purposes of control; often a road, or group of roads, or designated series of locations, identifiable both on a map and on the ground.

axis of advance—An axis or route to be followed in advancing toward the enemy.

axis of communications—An axis or route designated for coordination of front-to-rear communications or movement during a combat operation. See also **line of communications**.

azabs—Turkish irregular light infantry of the late medieval and early modern period. Azabs were armed with a variety of weapons, including arquebus or bow, or spear or sword and shield.

azimuth deviation—The lateral angle of deviation of a projectile or bomb from its intended direction, measured in degrees or mils.

AZON—A kind of bomb used in World War II, with movable control surfaces in the tail adjustable by radio signal so that the missile's flight could be controlled.

B

backplate—A piece of plate armor (q.v.) that covers the back.

back-step—v. To march backward, away from the direction in which facing. n. The command to so march.

backsword—A sword with a broad blade and only one cutting edge. The term is also used for any heavy sword with a basket hilt or shell hilt.

backup force—Any organized force prepared to back up or reinforce units already engaged in, or committed to, an assault.

badges—Devices worn on a uniform, sometimes to indicate rank or membership in a group or office. However, the term is applied primarily to pin-on devices awarded for attainment of proficiency or excellence in marksmanship or other military skills, or for experience in such performance.

bag, duffel—A bag, generally of canvas or denim, although the name is applicable to bags of any material, with a drawstring closure, used for carrying personal belongings.

baggage—The equipment and supplies, other than weapons, accompanying a military force.

baggage train—The supplies and equipment required by a military force and the vehicles that transport them. Transportation may be by animal, truck, railroad car, or whatever means is available. A baggage train usually follows the main body, and can sometimes stretch for miles behind the advancing combat force.

bail/bailey—Both bail and bailey were used for the outer wall of a castle and for the court

it enclosed. In the plural, the term "bails" is synonymous with palisade (q.v.).

Bailey bridge—A type of bridge designed by British engineer Sir Donald Coleman Bailey, and first used by British and American forces in World War II. Carried to the river bank in pieces, it is rapidly assembled by fastening prefabricated latticed panels of electrically welded high-tensile steel to form girders or larger panels. The resulting components can be laid side by side or superposed, the quantity and procedure depending upon the span and the load for which the bridge is needed.

Baker—Letter of old phonetic alphabet. See also **alphabet, phonetic**.

balance of power—An approach to the conduct of international relations that aims at preserving stability by preventing any one nation or alliance of nations from becoming strong enough to dominate or destroy other nations. Under this approach, if one nation threatens to dominate the international system, the threatened powers realign themselves, throwing their combined weight against the threatening power and restoring the balance.

balanced collective force—A force comprising elements from more than one nation and in such variety and strength that the total force will fulfill the mission for which it has been created.

baldric—A shoulder belt for attaching a sword or a bugle. It was usually worn across the shoulder and chest, but could simply be slung over one shoulder.

balisage—The marking of a route by a system of dim beacon lights enabling vehicles to be driven at near daytime speed, under blackout conditions.

ball—A solid missile, commonly but not always spherical, usually propelled by an explosive device. The term is also used as a collective noun for such missiles. Balls are generally described by the weapon that fires them, e.g., cannonball, musket ball, pistol ball.

ball cartridge or **ball ammunition**—A small-arms bullet and its cartridge (usually brass). The term is carried over from earlier centuries when musket and rifle bullets were spherical in shape. In modern times the term is used principally to distinguish live ammunition from blank ammunition (q.v.).

ball, minié—1) Invented in 1849 by French Captain Claude Etienne Minié, it was a lead projectile shaped like a blunt-nosed bullet (cylindro-conoidal) and originally used in the Minié muzzle-loading rifle musket. The original minié ball had three spiral grooves and a conical depression in the base. The grooves were usually filled with tallow, and the depression was an iron cup that was forced further into the ball by pressure of the gas resulting from the explosion of the powder, causing in turn the lead to expand into the grooves of the rifle. Smaller than the bore of the rifle musket, minié balls could be dropped down the barrel, greatly decreasing the time required for loading the weapon. Most smoothbore muskets could be converted by rifling to use a minié ball. 2) Mid-19th century US slang for any bullet fired from a rifled musket.

ball, red-hot—See hot shot.

ball turret—A turret in the shape of a ball, designed to project from the belly of an airplane and to house a machine gunner. The turret rotates as the gunner brings his guns to bear.

ballista—One of the many kinds of ancient military machines designed originally to hurl rocks and stones at an enemy. It consisted of a large wooden frame, with two projecting arms, to which were attached the ends of a twisted cord, consisting of skeins of human hair and animal tendons. It operated much like a giant crossbow, in that the cord provided the power to propel missiles toward the enemy. The energy of the cord was derived from torsion, rather than tension, as was the case for the catapult (q.v.) or the later crossbow. The cord was pulled back to the end of a trough, on which the missile was placed. Release of the cord, by a trigger mechanism, propelled the missile with great force along a path delineated by the trough. The ballista was an antipersonnel weapon and most commonly weighed about 60 pounds, although some varieties were much larger and were used as siege weapons. The ballista was designed by Vitruvius in about 26 B.C. and seems to have originated in a Greek-speaking area, inasmuch as the name means "thrower" in Greek. Later the ballista, like the catapult, was used to propel large darts or arrows as well as rocks.

ballistic missile—See **missile, ballistic**.

ballistic missile early warning system (BMEWS)—An electronic system for providing detection and early warning of attack by intercontinental ballistic missiles.

ballistics—The branch of applied physics that studies the motion of missiles or projectiles of all types. **Internal ballistics** is concerned with the action of missiles as they are expelled through and from the tube of the propelling weapon; **external ballistics** with their action in flight; and **terminal ballistics** with their action upon approach to a target.

ballistite—A smokeless propellant or powder containing nitrocellulose and nitroglycerine, used in some rocket, mortar, and small-arms ammunition.

ballock dagger or **ballock knife**—A 14th-century short dagger with two rounded lobes at the end of the handle, carried between the thighs on a low-slung belt. In was used until the 17th century in northern Europe. Also known as a dudgeon-dagger, and in the 19th century called a kidney-dagger.

balloon—A bag of nonporous material filled with hot air or a gas lighter than air, causing the bag to rise above the earth's surface. Since their invention by the Montgolfier brothers in late 18th-century France, balloons have been used for aerial observation, either floating free or, more usually, anchored, where they could raise the observer high enough to see beyond enemy lines. By World War I most military balloons (or nonrigid airships) were cylindrical in shape. Because of the vulnerability of balloons to modern weapons, their use for reconnaissance and observation was discontinued after World War I. **Barrage balloons** are anchored and float at very high altitudes near vulnerable targets as protection from air attack, mainly because their mooring lines pose a danger to lower-flying aircraft.

ban—1) A medieval lord's summons of vassals to arms. 2) Collectively, those so summoned.

banderole—Derived from the Italian for "small flag," the term specifically applies to a small flag or streamer attached to a knight's lance or flown from a ship's masthead.

bandmaster—A warrant officer in the US Navy and the British Royal Navy.

bandoleer/bandolier—A belt, usually of webbing, worn by soldiers over one shoulder and across the chest, with loops or small pockets in which cartridges are carried.

bandon—The basic military unit of the Byzantine Army. Banda (plural) varied in size and organization at different periods and according to whether they were infantry or cavalry. Infantry banda at the beginning of the 10th century had 256 men, and cavalry banda 200 to 400. Also known as **numerus.**

Bangalore torpedo—A device for cutting wire entanglements or for exploding mines. It consists of a long piece of TNT-filled iron pipe, with a detonating cap and a long fuze, which allows the igniter to leave the vicinity after lighting it. The name comes from the city of Bangalore, India, where a prototype was first used in the British siege of 1799.

banneret—Probably derived from the word for a small banner, the term "banneret" referred to a knight, ranking above a knight bachelor and below a baron, who led men into battle under his own standard.

banquette—A raised platform area along the inner wall of a trench or a parapet where soldiers stand and fire.

banzai attack—A charge by Japanese troops in World War II, so named for the word they shouted as they approached their opponents. Literally translated as, "A thousand years," "banzai" is roughly equivalent to "Hurray!"

BAR—Browning automatic rifle. Invented by John M. Browning (1855–1926), the BAR, used in World War I and (in a slightly improved version) in World War II, was a .30 caliber, air-cooled rifle, gas-operated, and magazine fed. With a firing rate of 200–350 rounds per minute and serving the same function as a light machine gun, it was fired by a BAR man, with an assistant BAR man, an infantryman, who dropped his rifle to pick up the BAR if its operator was incapacitated.

barbed wire—Strands of wire with barbs of various types. Such wire was used for delineating boundaries long before the first military use in 1874. Its full military significance was realized in World War I, and exploited since. Barbed wire, in rolls or strung on pickets, in front of a defended position slows the advance of an attacker and holds him in line of defensive fire.

barbette—A support that raises a gun high enough so that its fire may be directed over the surrounding protective armor or parapet, rather than through an embrasure, or opening in the protecting material. A barbette may be a mound of dirt, a platform, a carriage, or any other supporting device. A gun so mounted is said to be en *barbette.*

barbican—A fortification, such as a tower, guarding the approach to a castle or town, commonly placed at a gate or drawbridge.

barbut/barbute/barbuta—An almost cylindrical helmet with a rounded crown, similar to the classical Greek Corinthian helmet. It protected the cheeks and neck and sometimes had only a narrow T-shaped opening for the eyes and nose.

bardings—Armor, leather or other material used to protect or decorate a war horse (q.v.). Bardings of various types and styles were slung on the forequarters or hindquarters, or both, of the horse.

barge—The boat used by an admiral or the commanding officer of a large warship, for transportation to and from his ship or flagship.

Bar Lev Line—A line of fortified observations posts built by Israel along the east bank of the Suez Canal in 1970–1971 to facilitate the defense of the Canal line in the event of an Egyptian attack from the west bank. This line served its purpose by delaying the Egyptian attack in October 1973, providing time for Israel to mobilize.

barracks—A permanent building used for housing military personnel. The term is often used specifically to designate housing for enlisted personnel, in distinction from separate quarters in which officers and noncommissioned officers live.

barracks bag—A duffel bag (q.v.).

barrage—1) A temporary means of preventing enemy movement or action, usually consisting of a concentrated, continuous massing of fire by any kind of weapon. 2) Most commonly a screen or curtain of high-explosive fire placed in front of friendly troops by artillery or mortars. Variations of barrages in type and purpose are: **normal barrage**— Preplanned defensive artillery barrage to be placed directly in front of a friendly position in anticipation of a possible enemy attack. It is fired automatically upon call from the supported troops. **box**—A curtain of continuous artillery fire around three or four sides of a position to deny access by the enemy. **rolling**—A preplanned offensive barrage placed in front of attacking friendly troops. By periodic adjustments of range and deflection (direction) settings on the weapons, the curtain of fire moves forward into the enemy positions at the same rate as the attacking troops' forward movement. **aerial**—An approximation of an artillery barrage by aerial bombardment, with the curtain effect sustained by sequential flights of bombers over the target area. **balloon**—A screen formed by the mooring wires of barrage balloons (see **balloon**) to inhibit the approach of enemy aircraft to a defended area. **electronic**—Jamming measures put into effect over a wide frequency spectrum to interfere with such enemy electronic devices as radios and radars.

barrage balloon—See **balloon**.

barrel—The tubular part of a weapon through which its projectile travels when the propellant charge has been detonated. See also **bore**.

barricade—n. A structure, usually temporary and constructed of whatever material is at hand, set up to prevent passage of traffic along a means of approach to a defended area. v. To erect barriers to deny access to an area.

barrier or **barrier system**—A coordinated series of obstacles designed or employed to canalize, direct, restrict, delay, or stop the movement of an opposing force, and to impose additional losses in personnel, time, and equipment upon the opposing force.

barrier forces—Air, surface, or submarine units and their supporting systems that are positioned along routes that enemy forces are expected to travel. The mission of barrier forces is to detect and report enemy movement, block it, and destroy as much of the enemy force as possible.

barrier minefield—A minefield laid as part of a barrier system to block the approach of attacking enemy units, particularly toward the flanks of the defending force, and to channel the attack to selected battle areas.

bar shot—A projectile consisting of two cast iron balls or hemispheres joined by an iron bar, thus resembling a dumbbell. Because its motion was erratic, and it caused more damage than ordinary round shot, bar shot was used particularly in naval cannon for destroying the rigging of enemy ships.

bascinet—A type of helmet, developed by the early 14th century, having a pointed skull, and extending below the ears. An aventail of mail was generally fastened along the bottom to protect the neck. The commonest form of visor was shaped like a snout and hinged at the temples, although later helmets had the visor hinged at the front.

base—n. 1) A location or area from which combat operations are projected and supported. 2) An area or location consisting largely of logistic, maintenance, and support installments. 3) An installation at which aircraft or naval vessels are assigned, and to which they return following any trips or combat or noncombat operations in the air or at sea. v. To assign or station military individuals or units or matériel at a location or installation.

base altitude—An altitude maintained during an air mission, especially on the flight to the target or rendezvous.

base command—1) An area containing a

military base or a group of such bases organized under one commander. 2) The command organization of a base as opposed to the command of the units it supports.

base ejection shell—A type of shell that ejects its load from its base.

base exchange—See **post exchange**.

base fuze—See **fuze**.

base line—A surveyed line, established with more than usual care, to which surveys are referred for coordination and correlation, usually to assure accurate artillery fire and military mapping, particularly construction of air photo mosaic maps.

base, military—An installation consisting of facilities for support of military service activities, including living quarters, means of security, internal lines of communication, utilities, and other elements essential to maintaining and operating armed forces units. There can be bases for all military services, as army base, air force base, naval base, marine corps base.

base of operations—An area or facility from which a military force begins its offensive operations, to which it falls back in case of reverse, and in which supply and support facilities are organized.

base pay—The pay of an officer or enlisted person according to rank and longevity before additional pay or allowances are added for quarters, subsistence, flying status, hazardous duty, sea duty, fogy, or other special bases for extra reimbursement.

base plate—A large, flat piece of metal to which the barrel of a mortar is attached, to provide solid, even footing for the weapon.

base point—A selected registration point for an artillery unit (usually of battalion size), usually easily identifiable on the ground and, if possible, on a map or firing chart. The base point is selected to be as close as possible to the center of the unit's sector of fire. If the situation permits, one gun from each battery is registered on the base point; otherwise batteries that do not register must be tied to the registering battery by survey. When in position a unit may register on several check points, but there is only one base point; all

targets and concentrations fired on, and marked on the firing chart, are related to the base point. The line on the firing chart from the guns of a battery to the base point is the battery's base line (q.v.). When, for any reason, it is not possible to register on a base point on the ground, the unit registers upon a point in the sky by means of a high burst adjustment (q.v.). See also **registration precision fire**.

base section—An area within the communications zone of an area of operations organized to provide logistic support to forward areas.

bashi bazouks—Volunteer irregular troops, both infantry and cavalry, in the Turkish armies from the late 18th to the late 19th century. These soldiers, who provided their own weapons and horses and served under officers attached to but independent of the army, were brutal and undisciplined. They fought in Egypt against Napoleon, in the Crimean War, and in the Russo-Turkish War of 1877-78, after which the Ottoman government was forced to stop using them because of their excessive barbarity. The name comes from *bashi bozuq*, Turkish for "one whose head is turned."

basic load (ammunition)—The quantity of ammunition, measured in rounds or other units of measure, issued to an organization for its all its weapons. Basic load is related to the calculated use per weapon for a period of time. A unit going into or in combat is generally issued several basic loads. Sometimes called "a day of fire."

basic (military) training—The initial instruction, including drill, given to former civilians immediately after entering a military force; it is designed to provide essential fundamental knowledge of military service and practices. The trainee is then generally assigned to further training in the specialty in which he is expected to serve.

basilard—A short sword or dagger of the late 13th century, named for the city of Basle. It was apparently a forerunner of the Swiss shortsword (q.v.).

basilica—A huge bombard (q.v.) designed by the Hungarian engineer Urban and used by the Turks in the siege of Constantinople in 1453. Built of wrought iron bars and hoops, with a bore 36 inches in diameter, it fired stone balls weighing 1,600 pounds. It was designed to fire about seven shots a day and had a range of more than a mile. Two hundred men and 60 oxen were required to move the gigantic weapon. However, after firing only a few shots it blew up, and was never again used.

basilisk—A 16th-century artillery piece weighing about 12,000 pounds, firing a projectile of about 90 pounds, with a bore of ten inches and a length of ten feet. Its effective range was 750 yards, with a maximum range of 4,000 yards.

basket hilt—See **hilt**.

basket leave—A kind of contingency leave taken by an officer who is away from his authorized post or station on a weekend or holiday. A request for leave is placed in the commander's in-basket before the officer departs, and is destroyed in the event he returns before expiration of the holiday. It is approved and charged as leave if the officer is delayed in his return.

bastard sword—A weapon developed in the 14th century, with a long blade and a grip so lengthened that two hands could be used to wield it. By the 16th century, it had acquired the name "bastard sword" because it was neither a two-handed nor a one-handed sword. It is sometimes called a "hand and a half" sword.

bastille—A defensive tower or fortress. Because the Bastille in Paris was used as a prison and captured by the revolutionaries on July 14, 1789, at the start of the French Revolution, the name has come to mean a prison or jail.

bastion—A projection, often triangular, from the main walls of a fortification, presenting two faces from which defenders could fire on enemies attacking the main wall or adjacent bastions. It has come to mean any well-fortified or well-defended position. A bastion with only one face and one flank is a demi-bastion.

bateau/batteau—A flat-bottomed, shallow draft boat, tapering toward both ends, used on American rivers. Colonel Benedict Arnold used them in the march to Quebec,

which nearly ended in disaster because they were poorly constructed.

batman—A soldier who acts as a servant to an officer in the British Army. See also **orderly**.

baton—An emblematic stick or staff carried by one who holds a high office or rank. Traditionally the emblem of a field marshal.

battalion—1) A tactical unit of one branch of a ground army's combat arms, varying in size and composition, but generally in the range of 500 to 1,000 troops, and usually commanded by a lieutenant colonel. It normally consists of three to five of the basic combat units of its branch (company for infantry, battery for artillery, company or troop for armor). Three or four battalions are usually combined to form a regiment (q.v.) or brigade (q.v.). 2) A unit of similar size and composition in one of the supporting branches of an army, such as engineers, signals, ordnance, etc.

battalion company—See **flank company**.

battalion landing team—In an amphibious operation, an infantry battalion reinforced by combat and service elements; the basic unit for planning an assault landing (q.v.).

batter—The receding slope of the outer face of a wall, from bottom to top.

battering ram—A device in various forms which was driven with force against a wall or other obstacle in order to break through. The simplest rams were long, heavy beams or poles or tree trunks with metal tips and were carried by troops who ran with them against the wall. The usual battering ram was a large pole, often a full-sized tree trunk tipped with an iron head. Rams could either be mounted on wheels or suspended within huge wooden assault towers. The largest tower-enclosed battering rams could be as long as 200 feet and were operated by as many as 1,000 men.

battery—1) A set of guns, torpedo tubes, searchlights, or missile launchers, usually of the same size or caliber. 2) The guns of a warship. 3) The place where one or more pieces of artillery are installed (hence the Battery in lower Manhattan). 4) The basic firing unit of artillery, usually comprising four to eight guns and administrative and combat personnel to operate the weapons. In the US Army three batteries, usually of six guns each, constitute an artillery battalion. In the British Army the unit comparable to such a US battalion is a field regiment. 5) The piece of steel struck by the flint in a snaphance lock (q.v.).

battery, floating—A barge or special warship designed as a floating platform for heavy guns.

battery ship—A vessel designed or modified for the bombardment of shore fortifications from the sea. Also called floating batteries (q.v.), battery ships were often converted old men-of-war.

battle—A major combat encounter between the military forces of two belligerents, each having opposing aims or objectives (assigned or implicit) and each seeking to impose its will on the opponent by achieving its objective, while preventing the enemy from achieving his. A battle ends when one side has clearly achieved its objective, or when one (or sometimes both) has clearly failed to achieve its objective. Modern land battles between large forces usually are made up of several engagements and can last many days. Naval battles tend to be short and, in modern times, decisive.

battle axe—A medieval hand weapon with a heavy wooden handle two to three feet long and an axelike cutting and smashing blade, usually curved. See also **axe** and **halberd**.

battle bill—A list of duties and responsibilities assigned to a ship's crew during battle. See also **battle station**.

battle casualty—A member of a military organization who is killed, wounded, missing, or captured in combat action. See also **casualty**.

battle cruiser—A warship with the main armament of a battleship, but less armor, which enables it to reach the speed of a cruiser.

battle cry—A yell uttered by troops about to engage in close combat, particularly by the attackers. It may be a scream, a chant, a word (like the Japanese "Banzai!" in World War II), or any other vocal utterance, such as the "Rebel Yell" of the American Civil War.

battle drill—Practice specifically for combat including close duplication of the sights, sounds, and sensations of battle, in order to accustom troops to performing under actual combat conditions.

battle fatigue—See **fatigue, combat**.

battle formation—The arrangement or deployment (q.v.) of the elements of a military organization preparatory to entering combat.

battle group—A tactical and administrative infantry or airborne unit of the US Army in use during the late 1950s and early 1960s. Consisting of four rifle companies and one mortar company, or five rifle companies. The battle group was larger than a battalion and smaller than a regiment or brigade. See also **pentomic division**.

battle, line of—See **line of battle**.

battle map—A map showing ground features in sufficient detail for tactical use by all forces, usually at a scale between 1:25,000 and 1:100,000.

battlements—The top portion of a fortified wall behind which defenders stood to fight, consisting of high portions for protection (**merlons**) alternating regularly with low portions (**crenels**) through which arrows or other missiles could be fired. Battlements appear in representations and remains of fortifications from earliest times and in many societies throughout the world.

battleship—A major fighting warship, heavily armed with powerful guns and, since about 1860, heavily armored. Modern battleships were the largest naval vessels until the introduction of the fleet carrier in World War II, and were so called because they were included in the line of battle (q.v.). Battleships have traditionally combined heavy armor and armament with extensive cruising radius at fairly high speeds. Since World War II most navies have eliminated the battleship because of its high cost and vulnerability to air attack. The United States maintains a few of modified World War II vintage in its active fleet, with others "in mothballs" (q.v.) in its reserve fleet. See also **capital ship** and **ship of the line**.

battle sight—A predetermined sight setting on a weapon that will enable the firer to engage targets effectively at close battle ranges when conditions do not permit exact sight settings.

battle station—The place and duty to which each member of a ship's crew is assigned during a battle. Every man has such a station assigned to him in the battle bill (q.v.).

battlewagon—Slang term for a 20th-century battleship.

bayonet (plug, ring, socket, knife)—A daggerlike weapon designed to be attached to the muzzle end of a rifle for use as a thrusting weapon in close combat. The etymology of the term is in dispute, but the term was probably derived from the French *bayoner*, meaning "to put a spigot in a cask." A plug bayonet, inserted in the muzzle of the musket and widely used toward the end of the 17th century, permitted musketeers to perform as pikemen, who were no longer needed. However, it rendered the weapon inoperable as a firearm and tended to jam or to fall out. About 1680 the ring bayonet was invented (possibly by French Marshal Sébastien le Prestre de Vauban), which fitted around the muzzle, leaving the bore free for firing. This was soon improved by a socket in the handle that firmly locked the bayonet to a stud on the musket barrel. The bayonet is still used in hand-to-hand fighting.

bazooka—A portable rocket launcher. First used in World War II, the bazooka was designed primarily as an antitank weapon, but it proved effective against other targets as well. It originally consisted of a 2.36-inch tube (later enlarged to 3.5 inches) 4½ to 5 feet long, open at both ends, with a shoulder rest and sights attached. Electrically fired, it launched a rocket-propelled armor-piercing (hollow-cone) projectile.

beach—n. In an amphibious operation, that portion of the shoreline suitable for the landing of a tactical organization. v. To run a vessel ashore deliberately, either as part of a landing operation or for such other purposes as preventing a vessel in distress from sinking.

beach party—The naval component of a shore party (q.v.).

beach support area—The area to the rear of an established landing force, operated by shore party (q.v.) units. It contains facilities

for unloading troops and equipment and maintaining them ashore, and for evacuating the wounded, prisoners of war, and captured matériel.

beachhead—A section of a hostile shore which, when seized and held, may insure the continuous landing of troops and matériel and which provides enough maneuver space to conduct subsequent operations ashore. The primary, initial objective of an amphibious operation is to establish a beachhead with adequate space and depth for logistical and troop buildup, and suitable as a base for subsequent offensive operations.

beachmaster—The officer who commands a beach party (q.v.) in an amphibious operation.

beacon—In current military usage, an apparatus that emits signals, electronically or by light, for the determination of location or bearing. **Beacon fires** were used for the same purpose (as well as for the transmission of messages) from earliest times.

beam rider—A missile guided by radar, radio, or laser beam.

bear—v. 1) To direct or aim weapons at a target by means of a bearing (q.v.), as in "bring to bear." 2) To lie in a certain direction, as in "bears to the southeast."

bearded axe—See **axe**.

bearing—The horizontal angle at a given point measured clockwise from a specific reference datum to a second point, which can be a target or an indication of a navigational direction. Bearings may be measured in mils (q.v.) or degrees.

bearskin—A tall hat of black fur worn as part of the dress uniforms of certain units of the British Army and other European armies. Use by the British Guards was authorized in 1768 after the Grenadier Guards had started covering the front of their stiff hats with fur during the Seven Years' War.

beat to quarters—n. A drum beat relaying an order to all hands on a naval vessel to go to their battle stations. v. To execute such a beat.

beaten zone—The area on the ground that is

hit by the shots fired from a gun or gun battery at a certain combination of range (or elevation) and deflection (or direction) settings. It is the area on the surface of the earth which intercepts the cone of fire (q.v.) of a gun.

beaver/bevor—A movable section of armor attached to a helmet, or sometimes to a breastplate, to protect the lower part of the face and the throat. Its first known use was late in the 13th century. Later the word was used for the visor.

beefeaters—A corps, originally serving as the personal bodyguard of the sovereign of England and now performing ceremonial functions as part of the royal household. The corps has been in existence since 1485.

Beehive projective—A direct fire, defensive antipersonnel artillery round, it was an advance over earlier canister-type (q.v.) ammunition. The Beehive projectile of the US 105-mm howitzer is a canister filled with 8,500 steel flèchettes (q.v.) that is fired in a flat trajectory and detonated by a time fuze. At detonation the flèchettes fan out, producing a shotgunlike effect. The first combat use of the Beehive round was by US troops in late 1966 during the Vietnam War.

beleaguer—To surround an enemy fortress or other defensive position with troops. Synonymous with besiege (q.v.) and invest (q.v.).

belfry (belfroi, bevredum, belgragium)—A movable tower built of wood, used particularly in Roman and medieval times as a watchtower or for attacking and scaling fortification walls.

belligerence—A state of war or hostilities.

belt—A band of a flexible material generally worn around the waist, sometimes over the shoulder, which in military use often serves to carry a sword or other weapon. An **ammunition belt** is a string of cartridges, either linked with metal or plastic or carried in the loops of a fabric belt, to be fed into a machine gun. Ammunition belts can be classified according to the way the cartridge is removed therefrom. The cartridge may be pushed directly from the belt into the chamber or may be withdrawn to the rear and then, by reverse movement, thrust into the chamber or stripped sideways from the link and then fed. An **armor belt** is a strip of ar-

mor that is added to the hull of a warship at the warterline. See also **Sam Browne belt**.

beneficiarii—In the Roman Army, soldiers with privileges granted by their commanders, including exemption from certain labors, such as digging trenches and fetching wood and water.

berm—A ledge at the base of the exterior wall of a permanent or field fortification, between the vertical or sloping wall and the top of the moat or scarp (q.v.).

Berserker—A member of the bodyguard of any of several mythological or ancient Scandinavian heroes; also a type of legendary warrior who fought in a frenzy. The original Berserk was a mythological warrior who fought without armor and with great fury.

besiege—1) To lay siege to. 2) To cut off all access to or egress from a town or fortress for the purpose of capturing it and of forcing the defenders to surrender. See also **beleaguer** and **invest**.

bevor—See **beaver**.

bicorne—A hat with the brim turned up on two sides so that two points are formed on opposite sides. Such hats were worn by both army and naval officers, either sideways or fore and aft, in the 18th and 19th centuries and are still used as part of the formal dress uniform in some countries. See also **cocked hat**.

"Big Bertha"—A 420mm German howitzer of World War I. This type of gun battered the forts of Liege into submission in August 1914. The nickname was given the guns by the Germans, for Bertha Krupp von Bohlen und Halbach, owner of the German manufacturing plant. The term has sometimes been applied, erroneously, to the German long-range guns that bombarded Paris in 1918.

big wing—A tactic proposed by British Squadron Leader Douglas Bader in 1940 to resist attacks by the German Luftwaffe on targets in the British Isles during the Battle of Britain. The proposal was to meet and overwhelm German attacks with from two to five squadrons of aircraft, instead of the single squadron or less which was being sent to intercept and ascertain the objectives of the raiders in accordance with the doctrine of the

Royal Air Force commander, Air Chief Marshal Sir Hugh Dowding. A long and serious controversy developed between supporters of the two tactics. Modern theorists agree that Dowding was right and Bader was wrong under the circumstances of the Battle of Britain.

bill or **bill-hook**—A halberdlike (q.v.) weapon with a long shaft and a head incorporating a single cutting edge which, at the top, divides into a forward-curving hook and a spike; at the middle of the back of the head was another spike, projecting at a right angle. The bill was common in 16th-century armies, which used it for hacking, grabbing, and ripping in close combat. Soldiers armed with the bill were called **billmen**.

billet—n. 1) Housing for troops in military or nonmilitary buildings, most often used with respect to nonmilitary housing. 2) In naval parlance, an assignment or position. v. To assign troops to housing, particularly to nonmilitary lodgings.

binnacle list—A list, generally circulated daily in naval commands, of those personnel excused from duty by a medical officer.

biological warfare—The employment of microorganisms or toxic biological agents to cause disease or death in humans, animals, or plants.

bireme—A galley propelled by two banks of oars on each side. See also **galley**.

bivouac—n. A temporary assembly or encampment of troops in the field, with temporary shelter or, more often, none. v. To so encamp.

"black gang"—Navy slang for sailors whose duties were below decks—machinists, engineers, and the like—whereas the seaman's branch (q.v.) served above decks.

black powder—A mixture of potassium nitrate, carbon, and sulfur which burns rapidly and, if confined, explosively. It was the original gunpowder (q.v.), developed in China before it was in Europe. It produces a large amount of whitish smoke and was replaced as a propellant for military use in the late 19th-early 20th century by smokeless powder. Black powder is still used for primers, fuzes, and blank fire charges.

black propaganda—See **propaganda**.

blank ammunition—Ammunition containing powder but no projectile, used in firing salutes, in signaling, and for training purposes.

blast effect—Destruction of or damage to structures and people caused by the rush of air from the center of an explosion or the point of impact of an explosive projectile.

blimp—A lighter-than-air craft consisting of a heavy air bag, shaped like a stubby cigar, with a cabin and two or more motors attached below. Operating at low altitudes and slow speed, blimps were used by the US Navy in World War II for antisubmarine patrol and convoy escort within short distances from shore and for various photographic and search-and-rescue operations.

blind—A cover, whether to conceal a force in ambush or a covert operation.

blind bombing—A World War II aerial bombardment technique conducted against targets that could not be seen visually or by radar (because of bad weather or successful camouflage or other passive measures). The location of these targets was estimated or calculated by means of navigation.

blinker tube or **blinker light**—A light covered with shutters that are opened and closed to signal from one ship to another, using Morse or comparable code.

Blitz, The—German for "lightning." The term applied to the German air bombardment campaign against Britain in World War II, 1940–1941, after the German failure in the Battle of Britain. Large numbers of German bombers flew over British cities night after night, causing great destruction, killing 43,000 civilians and seriously wounding 51,000.

blitzkrieg—In German, "lightning war." Blitzkrieg tactics were developed in the 1930s by the Germans as a result of their experiences in World War I, and were first practiced by them in the attack on Poland in 1939. General Heinz Guderian was a major proponent of such tactics. Primarily conceived as the adaptation of armored firepower, mobility, and protection, supported by ground-attack aircraft, to the German tactics developed in World War I, blitzkrieg was characterized by speed, surprise, maneuver, and overwhelming firepower.

blockade—v. To isolate a city, region, or country by cutting off communications to or from a city, harbor, or the entire coastline and frontiers of a nation by the posting of military forces to intercept all traffic. n. Such cutting of communications. The most common blockades are naval, imposed by warships, like the blockade imposed by the Union Navy on Southern ports during the US Civil War, or the total blockade of German ports in World Wars I and II, but there are also air and ground blockades.

blockade runner—A person, vehicle, or ship that crosses or circumvents a blockade. Maritime blockade runners in the Civil War, operating mostly at night in coastal waters, took war materials from Europe, usually via the West Indies, into Confederate ports, most notably Wilmington, North Carolina.

blockbuster—Any bomb considered powerful enough to destroy a city block.

blockhouse—A defensive structure of heavy timbers, concrete, or other substantial material with small openings or loopholes for observation and for firing weapons. In North America in the colonial period, blockhouses were often two-story log structures with the second overhanging the first on all sides. Blockhouses of reinforced concrete, often partially covered by earth, usually form part of modern fortified defensive systems.

blocking position—A defended position situated so as to thwart enemy penetration between positions or areas, or to cover enemy lines of advance that could otherwise bypass forward defended positions.

blockship—A stationary ship, often sunk, so placed as to deny or limit passage to a harbor or other shore objective.

blowback—1) The escape to the rear of gases under pressure formed during the firing of a gun. 2) A type of weapon operation in which the force of expanding gases acting against the face of the bolt furnishes all the energy required to push it back and to initiate the complete firing cycle of the gun.

blue water—The open sea.

bluejacket—A naval enlisted man, so-called for the navy blue uniform jumper, coat, shirt, or jacket.

blunderbuss—A short, wide-bore musket, often flaring at the muzzle. Commonly used in the 17th century, the gun seems to have been known in Holland as early as 1353, for Dutch documents of that period mention a **donrebusse**. Several balls or other small projectiles could be fired simultaneously. Blunderbusses of various calibers, as large as light artillery, were used in both armies and navies of the period. It was the weapon of Miles Standish, among many others.

board—v. To enter a vehicle, particularly a naval vessel. Commanders of sailing or rowed warships in a battle attempted to get close enough to an enemy vessel to put a **boarding party** of soldiers, sailors, or marines on it. Armed with **boarding pikes**, **axes**, or **pistols** the men would attack, attempting to destroy the masts and rigging of the enemy ship. In modern times a boarding party may be sent to examine the papers or cargo of a ship that has been stopped for such inspection. n. A group of people, usually officers, under a chairman or president, which has a temporary or permanent mission or function to perform, for instance, a board of investigation.

Board of Admiralty—See **Admiralty**.

boarders—Sailors in a boarding party.

boarding axe—See **axe** and **board**.

boarding party—See **board**.

boarding pike—See **board**.

boar's head—A square tactical formation established by Marshal Thomas R. Bugeaud in the French conquest of Algeria, c. 1840.

boat—A small craft, propelled by oars, sail, or motor, small enough to be carried on a ship. Of the larger vessels, the term is applied only to submarines.

boatspace—The space and weight factor used by planners to determine the capacity of boats, landing craft, and amphibious vehicles to carry men, supplies, and equipment required for an amphibious operation. With respect to landing craft and amphibious vehicles, it is based on the requirements of one man with his individual equipment, together assumed to weigh 224 pounds and to occupy 13.5 cubic feet of space.

boatswain—In the US Navy and the Royal Navy, a warrant officer in charge of the instruction and performance of the deck crew of the ship in matters related to practical seamanship. The boatswain is also responsible for care of the rigging and gear, and his insignia is crossed anchors. It is a rating whose origins antedate the establishment of the Royal Navy. Pronounced "bō´ sŭn."

boatswain's call or **boatswain's pipe**—A small silver whistle carried by a boatswain as a badge of office and used by him to sound commands to the deck crew. The pipe was known as early as the Crusades in 1248 and was considered a badge of rank by the English in the 15th century, when it was worn by the Lord High Admiral.

boatswain's mate Naval enlisted ratings (second and first class) concerned with seamanship and related duties. A boatswain's mate third class is called a coxswain in the US Navy.

bodkin—A light arrow with a stiff, straight point, used particularly in the 16th century.

body—A number of troops or units physically assembled together, as a body of troops. The **main body** is the principal portion of a force.

body armor—See **armor**.

body count—The number of enemy dead counted in the battle area after conclusion of combat. A term widely used in the Vietnam War.

Bofors gun—A 40mm light antiaircraft gun developed by the Swedish firm Bofors in the early 1930s. The Bofors gun which could fire 80 to 120 rounds per minute over a slant range of four miles, was built under license or copied by nearly all the major belligerents of World War II. It was adopted as the standard light antiaircraft gun by the US, Great Britain, and Russia, and copied by Japan and Germany.

bogey—1) An unidentified flying aircraft. 2)

A supporting roller on an endless steel belt, as of a tank.

boilermaker—US Navy enlisted specialty rating, first, second, and third class, and chief.

bolo or **bolo knife**—A broad, single-edged knife, similar to the machete (q.v.) used in the Philippines as a weapon.

bolt—1) The missile shot by a crossbow (q.v.) or catapult (q.v.) generally having the form of short, stout, blunt-headed arrow. 2) A device on a breech-loading rifle used to prevent the escape of gas from the breech, to eject a spent cartridge from the rifle, and to insert a fresh cartridge from the magazine into the barrel preparatory to firing the next round.

bomb—n. Originally an explosive shell fired from a cannon, but currently a device placed or thrown by hand or dropped from an airplane or a vessel. The shape, the explosive agent, and the purpose of a bomb vary. A **depth bomb** is an antisubmarine weapon exploded under water. A **fragmentation bomb** is designed to explode in the air and scatter sharp pieces of metal over a broad area. An **incendiary bomb** contains chemicals that will start raging fires. A **high explosive (HE) bomb** is designed to cause damage by the blast and the effect of metal fragments scattered by the explosion. A **time bomb** contains a device that delays explosion until some time after setting. A **propaganda bomb** explodes in the air and scatters propaganda material. v. To attempt to destroy a target with a bomb, especially by dropping a bomb from an aircraft.

bomb bay—The compartment in the fuselage of a bomber where the bombs are carried for release through the bomb bay doors.

bomb ketch—A small craft used in the early-to-middle 19th century; armed with mortars, it lobbed shells into enemy fortifications. Usually ketch-rigged.

bomb line—An imaginary line following, if possible, well-defined geographical features, prescribed by the ground troop commander and coordinated with the air force commander. Forward of the bomb line air forces are free to attack targets, without reference to the ground forces; behind this line all attacks must be coordinated with the appropriate troop commander.

bomb release point—The point in space at which bombs must be released to reach the desired point of detonation.

bomb run—The flight course of a bombing airplane, from an initial point where the bomb run officially begins to the bomb release point. More generally, the flight course just before the release of bombs; the action of flying this course.

bomb ship—See **battery ship**.

bombard—n. An early form of cannon, with a short, wide barrel. Bombards varied in size, from a weapon that could be hand-carried to Henry VIII's massive bronze bombards that were in the Tower of London, each of which required 24 horses to be moved. v. To fire large numbers of missiles, particularly artillery shells or aerial bombs, at a fixed target.

bombardier—1) A person responsible for the delivery of bombs to a target. 2) In American usage, the member of the crew of an aircraft who operates the bombsight and releases the bombs. 3) In British usage, a noncommissioned artillery officer.

bombardment—The act of bombarding.

bomber—An aircraft whose primary mission is the delivery of bombs to a target from the air. Bombers are classified by range or by function. Types of bombers classified by range include: **medium range**—bombers that have a maximum combat radius greater than 925 kilometers and less than 1,850 when flying at best speed and altitude and when taking off at maximum gross weight; **intermediate range**—aircraft that have a maximum combat radius of between 1,850 kilometers and 4,600 when flying at best speed and altitude and when taking off at maximum gross weight; **long range**—bomber aircraft that have a maximum combat radius of at least 4,600 kilometers when flying at best speed and altitude and when taking off at maximum gross weight. **very long range**—bomber aircraft designed to have a maximum tactical operating radius of up to 8,050 kilometers, flying at best speed and altitude when taking off at maximum gross weight. Types of bombers classified by function include: **dive bomber**—aircraft de-

signed to drop steeply from a high altitude, aiming bombs by diving toward the target; bombs are released at low altitude, just before the plane pulls out of the dive; **fighter bomber**—See **fighter aircraft**; **torpedo bomber**—aircraft, particularly naval aircraft, which fly low enough to release torpedoes at enemy warships.

bombfall—1) The fall of one or more bombs, either simultaneously or successively. 2) The bombs while falling. 3) The points of impact of bombs. 4) The pattern or configuration formed by both the fall of bombs through the air and their impact and explosion.

bombing—The dropping or delivery of bombs, particularly from an aircraft on a target. See also **blind bombing**, **carpet bombing**, **dive bombing**, **skip bombing**, **horizontal (level) bombing**, **pattern bombing**, and **precision bombing**.

bombing area—See **area**.

bombproof—adj. Designed to resist and if possible survive the explosive force of bombs. n. A shelter secure against bombing; dugout (q.v.) shelters in the US Civil War were called "bombproofs."

"bombs away"—1) A phrase used by the bombardier to indicate to the pilot and other members of an airplane crew that the bombs have just been released. 2) The instant when the bombs are released.

bombshell—Archaic expression for shell (q.v.).

bombsight—A device that determines or enables a bombardier to determine the point in space at which a bomb must be released from an aircraft in order to hit a target. US bombers in World War II were equipped with the highly secret Norden bombsight, the most accurate one then available. Using preset data and manual operation by the bombardier, the Norden bombsight computes the correct dropping angle and, in connection with an automatic pilot or pilot direction indicator, determines the proper course of the aircraft required to maintain the necessary line of sight to the target.

bonnet rouge—A revolutionist. The name derives from the red caps worn in the French Revolution.

booby trap—n. An explosive device, especially an apparently ordinary harmless object, so concealed that it will be triggered by an unsuspecting victim. v. To place a device, especially an explosive device, in or near an apparently harmless object or place.

boom—A floating structure, frequently made of logs, sometimes having a metallic net suspended from the floats, which reduces or closes access to a harbor, river, or berth to provide protection for vessels within, particularly protection from submarines.

boondockers—Ankle-high boots.

booster—A high-explosive element of a shell or other explosive device that is activated by a fuze or primer and in turn activates the more powerful main explosive.

boot camp—Where military recruits (usually naval or marine) receive their initial training.

boots—1) Heavy footwear, usually of leather, that cover all or most of the lower leg, and in some types the thigh as well. Various types have been worn through the ages. 2) Gaiters, originally khaki but bleached white by frequent washing, worn by naval recruits in boot camp (q.v.), by naval personnel on shore patrol or other official duty ashore, and in ceremonial functions, such as parades. 3) Naval recruits.

booty—Material items of value seized or captured from an enemy or in enemy territory either legally (as in prizes captured on sea or on land) or illegally (as loot or plunder).

BOQ—Bachelor Officers' Quarters.

bore—1) The cylindrical inner chamber of a gun, in modern weapons extending from the breechblock to the muzzle. 2) The inner diameter of a gun barrel; the caliber (q.v.).

boresight—To check the alignment of the tube of a gun with its sight or aiming device.

boss—A projection from the surface of a shield, generally round, usually decorative. The boss in the center of some shields provides extra protection for the hand grip inside.

bounce—To attack an enemy airplane from the air, usually from a higher altitude, taking the enemy by surprise.

"bouncing betty"—A German antipersonnel mine of World War II vintage, so-called because when activated it jumped as high as seven feet into the air before exploding. It was filled with high explosive and about 300 steel or lead balls which, together with fragments of the steel casing, were hurled in all directions by the force of the explosion.

bound—The distance covered in one movement by a unit that is moving from cover to cover.

boundary—The line separating areas of responsibility. In modern armies boundaries are usually indicated on a map, and they divide the territory to be occupied, or for which responsibility is assigned, among units and formations for purposes of command and control. Boundaries frequently follow natural features such as streams or ridges. They normally change as forces advance or withdraw.

bounds, fire by—Sequential changes of artillery firing data to move the center of impact of the rounds from one point to another. Fire by bounds usually is done by changing range, but can also be done by changing direction of deflection.

bounds, movement by—Advancement in a series of moves, usually from cover to cover, often made by troops under artillery or small arms fire. Also called "leapfrog."

bounty, enlistment—A bonus paid for signing up for military service.

bounty jumper—A soldier who enlists to receive a bounty, then deserts and enlists elsewhere to get another bounty. The practice was prevalent among soldiers of the Continental Army and state militias in the American Revolution.

bow—A weapon consisting of a narrow, curved piece of wood or other material, with a cord stretched between its two ends, by which an arrow is propelled. The bow was a basic weapon of most armies from earliest times until after the introduction of gunpowder. Bows have been made in many sizes and shapes, and apparently were invented independently in many widely separated cultures. The simplest bow, the **self bow**, is made of a single stave of wood or other flexible material, including horn, sinew, metal and, in modern times, plastic, usually thickest in the center, and tapering to the two ends. A bow made of two or more strips of the same material bonded together is called a **built bow**. **Composite bows,** made of a combination of materials, were used as early as the third millenium B.C. The earliest ones combined wood, horn, and sometimes sinew, and were particularly popular in the Near East and Asia. Most Asian bows were **reflex bows**; that is, they were naturally curved in shape, but when strung were bent in the opposition direction. The deadly **Mongol bow**, principal weapon of the cavalry armies of Genghis Khan and his successors, was comparable in length and power to the English longbow (q.v.) but, unlike the longbow, was composite and reflex in nature. See also **crossbow**.

Bowie knife—A knife, usually about 15 inches long, with a long single-edged blade, a hilt, and, originally, a handle made of horn. The first Bowie knife is said to have been made in Arkansas in 1825, after a design by the frontiersman James Bowie. The knife was very popular among pioneers who used it for a variety of purposes, including hunting and hand-to-hand combat. It was adopted as a weapon by the Texas Rangers and, subsequently, by many Confederate troops in the Civil War. A knife like the Bowie knife was used by British commandos and US Rangers during World War II.

bowman—See **archer**.

bowyer—1) An archer. 2) One who makes bows.

box—Term sometimes applied to a self-contained, all-around defensive position of field fortifications, with strongpoints linked by trenches.

box, ammunition—A box to hold ammunition, particularly one from which linked ammunition can be fed into a machine gun.

box barrage—See **barrage**.

box, cartridge—1) A small pouch, usually made of stiff leather, worn by a soldier on cross belts over his shoulders, or attached to his waist belt, in which musket cartridges

were carried. 2) In the 19th century a webbed pouch, holding two or more clips of rifle ammunition, usually attached to a webbed waist belt.

box magazine—See **magazine**.

bracket—Successive rounds (or volleys, or salvos) of artillery fire aimed so that one falls beyond the target and the next falls short. When a target is thus bracketed its range can be established, and the observer can quickly bring fire for effect directly on the target. The term can be applied to deflection (q.v.) as well as to range (q.v.). See also **straddle**.

bracketing—The process of seeking to bracket a target; this is the basis of normal field artillery fire adjustment.

branch—An element of a military service having a specialized function. It can include the combat arms or services (for instance, artillery, infantry, submarines) and the support elements (quartermaster corps, signal corps).

brassard—1) A strip of cloth worn around the upper arm to indicate affiliation with a particular organization, or performance of some specialized function, or otherwise to distinguish the wearer. Brassards are worn, for example, by personnel who are temporarily or permanently performing the function of military or shore police. 2) Plate armor that protects the upper arm.

Bravo—Letter of NATO phonetic alphabet. See also **alphabet, phonetic**.

breach—n. A gap or rift in a structure such as a fortification wall, an obstacle like a minefield, or an enemy line. v. To create such a gap.

break off—To terminate a combat activity unilaterally.

breakout—An operation whereby a force that is surrounded or besieged escapes and regains freedom of action by disrupting the lines of encirclement.

breakthrough—An attack in which the attacking force successfully disrupts the lines of the hostile defender and obtains freedom of maneuver in the defender's rear area.

breastplate—A piece of plate armor that covers the breast.

breastwork—A protective wall, usually one which is hastily constructed of earth or other available materials, roughly high enough to reach the breast area of the average soldier.

breech—The rear part of a gun, specifically, the opening at the rear where the projectile and charge are inserted.

breechblock—The part of a gun, especially a cannon, which closes the breech. The breechblock usually contains the firing pin, and in many types of guns it is also used to set the round firmly in the chamber.

breech-loader—Any firearm in which the ammunition is loaded at the breech. Breech-loading weapons had been used before the 18th century but were generally unsatisfactory because imprecisely fitted metal parts left seams in the breech through which gas and flames could leak. The technology of the 19th century cured this deficiency. By permitting the user to load and reload at a more rapid rate and to lie prone while doing so, the breech-loading rifle of the mid-19th century made the muzzle-loader obsolete. Mass adoption of breech-loading guns came first in the Prussian Army in 1848, although the so-called needle gun (q.v.) was not unveiled until the Danish War of 1864. Breech-loading cannon, which could fire more rapidly than muzzle-loading, soon followed.

Bren gun—A light machine gun used as a standard weapon by the British Army and others in World War II. Developed in Brno, Czechoslovakia, the gun was manufactured in Enfield, England, and its name was taken from the cities Brno and Enfield. It was a gas-operated, air-cooled, .303-caliber weapon and could fire 120 rounds per minute, effective up to 1,000 yards. It could be fired from the shoulder and a bipod or from a tripod, but was usually mounted on a light vehicle, called a Bren Gun Carrier, the term often being applied to the gun and carrier together.

brevet—A semihonorary, semitemporary rank, originally in the British service, usually awarded for meritorious conduct. Brevet rank was applicable in some circumstances, not in others, creating confusion in terms of precedence and authority. The practice was adopted in the US Army in the 19th century.

The last American officer to hold brevet rank was General Tasker Bliss, who was promoted by President Wilson to four-star general by brevet when he was appointed to the Supreme Allied War Council in 1918.

bricole—1) A type of catapult. 2) A piece of armor protecting the breast of a horse.

bridge—1) A means of crossing a stream, in military context often a temporary bridge. Bridges may be constructed of various materials, depending on what is available and how urgently one needs to be constructed. Temporary **floating bridges** may be put together from a variety of materials and objects—including small boats placed side by side (Xenophon described the crossing of the Tigris River by the Ten Thousand on a bridge of boats in 401 B.C.), logs, rafts, or gasoline barrels—lashed together and generally covered with planks or other solid material on which a force can move. See also **Bailey bridge, drawbridge**, and **ponton bridge**. 2) The area of a ship, usually forward and high, from which the vessel and its operations and the operations of other vessels in a naval command are controlled.

bridgehead—A piece of ground on the enemy side of a river or defile (q.v.) that is seized and held by advance troops of an attacking force as a base for further operations.

briefing—1) The act of providing instructions or information about a task or a mission to one who is to perform it or who has an interest in it. 2) A report, usually but not always oral, on a subject about which the recipient wishes to be informed.

brig—1) The prison on a naval vessel or station. 2) A sailing vessel with two masts, both square-rigged, carrying a quadrilateral gaff sail aft of the mizzenmast and at least two headsails. It is to be distinguished from a **brigantine**, which carries fore and aft sails rather than square ones on the mainmast. Brigs were used as naval vessels (usually sloops of war) and as commercial ships as well.

brigade—A military unit smaller than a division (q.v.), in use since the 18th century, whose composition has varied from time to time and army to army. In World War I the infantry brigades of most armies were composed of two infantry regiments of three battalions (q.v.) each. Two brigades, each commanded by a brigadier general, constituted a division. A British division consisted of three brigades of four battalions each. In World War II a British brigade consisted of three battalions, and was roughly the equivalent of an American regiment. The US Army changed its divisional organization in the late 1930s to three infantry regiments, and eliminated the infantry brigade as a combat formation. In World War II there were no more than three or four two-regiment brigades in the US Army. Under the combat arms regimental system, which the US Army adopted in 1961, a new structure was called for: three battalions made up a brigade, and three brigades, in various configurations of armored and mechanized battalions, constituted a division. The entire body of midshipmen at the US Naval Academy is called the Brigade of Midshipmen. At the US Military Academy the Corps of Cadets consists of two brigades.

brigade combat team—See **combat team**, **regimental**.

brigade major—A military position in the British Army corresponding to brigade or regimental adjutant in the US Army; the principal administrative officer of a British brigade, usually a major in rank.

brigadier—An officer rank between colonel and general. In the British Army it is a temporary appointment of an officer who is permanently a colonel. In most armies of other nations that follow British organization it is a rank comparable to brigadier general.

brigadier general—The most junior general officer in the US Army, Air Force, and Marine Corps. His insignia of rank is a single star. See also **general**.

brigandine—A type of armored jacket of the 15th century made of fabric with metal scales riveted to the inside in overlapping rows. In many instances, the heads of the rivets were gilded and thus showed through on the outside of the garment.

brigands/brigans—French medieval infantry armed with spears and shields.

brigantine—See **brig**.

briquet—A short, curved, cutting sword carried by infantry soldiers in the 18th and 19th centuries. Inscribed briquets, called **briquets d'honneur**, in Napoleonic France were presented for valor.

broad axe—See **axe**.

broadcast controlled interception—See **intercept**.

broadside—1) All the guns on one side of a warship or, in vessels with turrets, all the guns that can be fired on one side. 2) The firing of all the guns at the same time.

broadside gun—1) On a modern naval vessel any gun of less than eight inches in caliber, (excluding antiaircraft guns and guns for saluting purposes only). 2) On pre-20th-century warships, one of the guns placed for firing on either side of the vessel.

broadsword—Any of various cutting swords with wide blades, particularly those which had a large basket guard.

broken-back war—In nuclear weapons theory, hostilities continuing after an unrestricted nuclear exchange between major powers. Originally British slang for a conventional phase of global war following the exhaustion of stocks of nuclear weapons.

bronze armor—See **armor**.

"Brown Bess"—British version of the 18th-century flintlock (q.v.) musket with an offset ring bayonet. The origin of the name is unclear. It seems to have come from a combination of the color of the wooden stock and a corruption of the "buss" in "blunderbuss."

Browning automatic rifle—See **BAR**.

BuAer—Bureau of Aeronautics. The division of the US Department of the Navy responsible for all administrative and executive aspects of the aeronautic equipment, including aircraft. In 1959 the functions of BuAer were transferred to the Bureau of Weapons.

buck and ball—A combat musket load consisting of a musket ball and three or more buckshot commonly used by American troops in the Revolutionary War.

buckler—n. 1) A small round shield that could be carried in the hand or worn on the arm, allowing the hand freedom. 2) A person or thing that defends or protects. *v.* To defend.

BuDocks—Bureau of Yards and Docks. The division of the US Department of the Navy responsible for the administrative and executive aspects of the physical plants of naval shore activities and certain service craft.

buff coat—A coat made of cowhide or other heavy leather which resisted penetration or cutting by a sword. It was worn in the 17th century, particularly by cavalrymen.

bugle—A brass wind instrument without keys or valves, blown by a bugler in both armies and navies to transmit orders or announcements in the form of **bugle calls.** Calls, composed of different combinations of notes, signal time to get up (reveille) and times to perform the routine duties of a military force. In the modern US services bugle calls are generally broadcast from recordings.

buildup—n. The process of increasing the strength of military units and their supplies and equipment. *v.* (**build up**) To increase the strength of units and their supplies and equipment.

built bow—See **bow**.

built-up area—An urban or semiurban region.

built-up gun—A gun constructed by shrinking an envelope of iron or steel over an iron or steel barrel. Probably invented by a French naval officer, M. Thiery, the built-up technique produced a strong compressive tension on the barrel, so that the expansive force and heat of the exploding powder charge encountered the resistance of this compression from the first instance of the explosion.

bullet—The projectile, spherical or cylindrical with a pointed end, fired from a small arm. The term is often used for the complete round of ammunition—bullet, cartridge case, powder, and primer.

bullet, explosive—A small arms projectile that contains a small charge of explosive which detonates on contact with its target. See also **dum-dum bullet**.

bulwark—A strong, defensive wall.

BuMed—Bureau of Medicine and Surgery. The division of the US Department of the Navy responsible for the health and medical care of all naval and marine corps personnel.

BuNav—Bureau of Navigation. The division of the US Department of the Navy responsible for the administration of naval personnel. It was replaced by the Bureau of Personnel.

bunker—A small defensive structure, covering a gun emplacement or serving another purpose. It may be a permanent structure of steel and concrete with openings through which guns can be fired, sometimes sunk entirely or partially in the ground, and may be one of a series like those of the Siegfried Line (q.v.) in World War II. Or it may be a dugout roofed with logs and covered with dirt and various forms of concealment.

BuOrd—Bureau of Ordnance. The division of the US Department of the Navy responsible for all administrative and executive aspects of naval weapons and ammunition. In 1959 the functions of BuOrd were transferred to the Bureau of Weapons.

BuPers—Bureau of Naval Personnel. That division of the US Department of the Navy responsible for the administration of all naval personnel. It was formerly known as the Bureau of Navigation.

burgonet—A helmet worn from the 16th century into the 19th, having a visor similar to that on a baseball cap and a covering for the nape of the neck. Most burgonets had a comb and ear pieces. They were used particularly for protection of sappers (q.v.) in a siege operation.

burgs (Frankish forts)—Fortified posts or towns originating in the Frankish system and under Charlemagne, built along the frontier of newly captured territory. Burgs were connected to each other and to what had been the old frontier by roads. Stocked with supplies, these forts became bases for the maneuver of the disciplined Frankish cavalry, either to maintain order in the conquered territory or as a base in further operations.

burst—The explosion of a shell, or a bomb, marked by smoke and dust at the point of impact or explosion. The **area of burst** is the area, usually measured in square yards or meters, directly affected by the shell's explosion. A **burst interval** is the distance between the bursts.

BuSandA—Bureau of Supplies and Accounts. That division of the US Department of the Navy responsible for all aspects of procurement of supplies other than ammunition and payments. The Chief of BuSandA is Paymaster General of the Navy.

busby—A tall fur hat worn with the dress uniform by certain units of the British and other armies, principally hussars, artillerymen, and engineers.

BuShips—Bureau of Ships. The division of the US Department of the Navy responsible for all aspects of procurement and maintenance of naval vessels and for purchasing ships and smaller craft for the US Army and Air Force.

bushwhacker—1) Specifically, one of a group of Confederate guerrillas of the US Civil War. 2) A guerrilla (q.v.).

butt—1) The end of a hand-held weapon such as a spear, lance, or rifle, opposite the point or muzzle. 2) A mound of earth or other material behind practice targets, to catch projectiles that miss the target. 3) In archery it is the mound on which the targets are set.

butterfly bomb—A small fragmentation or antipersonnel bomb equipped with two folding wings that rotate and arm the fuze as the bomb descends. It was developed in Germany and adopted by the United States.

butternut—A slang term used of Confederate soldiers in the US Civil War because of their homespun brown clothes.

BuWeps—Bureau of Weapons. The division of the US Department of the Navy responsible for administrative and executive aspects of the aeronautic equipment and naval weapons and ammunition. In 1959 BuWeps superseded and combined the functions of the Bureaus of Aeronautics and of Ordnance.

buzz bomb—See **V-1**

byrnie/birnie—A coat of mail. Apparently used sometimes for a type of breastplate.

C

cab rank—A formation of fighter and fighter-bomber aircraft, usually in line ahead, awaiting call for close air support (q.v.) missions, while at the same time patrolling in the vicinity of the anticipated area of action. Cab rank was organized by the Royal Air Force's Desert Air Force in the Mediterranean campaigns of World War II.

cabasset—A type of helmet, dating from the 16th century, having a narrow brim and a pointed, keeled crown, and chin straps that were sometimes covered by plates to protect the ears. It was commonly worn by infantry and ceremonial guards.

cadence—A rhythmical beat. In military units a regular measured timing for simultaneous activity, such as marching in step.

cadet—A student in military training to become an officer.

cadre—A disciplined group of trained officers and enlisted men, individuals whose capabilities include the principal military specialties that will be required in a military unit or formation, and thus provide a nucleus for the establishment of such a unit or formation.

caisson—1) A horse-drawn, usually two-wheeled vehicle, used to carry artillery ammunition. Caissons are still used by the US armed forces to carry coffins in military funerals. 2) A large box designed to contain ammunition.

caliber—1) The diameter of a projectile, usually a bullet or a shell, or of the bore of a gun barrel. Caliber is usually expressed either in millimeters or, for some small arms in the US, in hundreds or thousandths of an inch. Thus, the diameter of the bore of a .50 caliber machine gun is one-half inch, that of a .357 Magnum a bit over one-third of an inch. Nations on the metric system almost always express caliber in millimeters. 2) A measure for the length of a cannon bore expressed in multiples of the diameter of the bore. For example, a 5-inch 38 caliber naval gun has a bore length of 190 inches.

calibrated airspeed—See **airspeed**.

caliver—Another name for an arquebus (q.v.). The term is an English corruption of the word "caliber." Calivers were arquebuses of a standardized caliber (q.v.). In the mid-16th century the English and the French began to standardize the caliber and lock mechanisms of their arquebuses in order to simplify ammunition supply and training procedures.

call for fire—A request from ground units for support from artillery or aircraft. The call includes the data necessary to identify the target.

call sign—Any combination of characters or pronounceable words that identifies a communications facility, a command, an authority, an activity, or a unit; used primarily for establishing and maintaining communications.

caltrop—A small device, dating at least from Roman times, that when thrown on the ground presented one or more protruding points which slowed the advance of troops or horses. The Roman caltrop was a wooden ball with a number of spikes protruding from it. The medieval caltrop was made by twisting and hammering two strips of iron. Leonardo da Vinci designed machines to throw caltrops. Settlers in Virginia used them against the Indians. Sometimes called "crow's foot," or "crowfoot."

camisade—An unanticipated night attack. The name, dating from the 16th century at least, seems to derive from the Spanish "camisado," meaning "wearing a shirt," inasmuch as night attackers often wore white shirts over their armor for identification.

camouflage—*v.* To disguise troops, matériel, equipment, or installations by covering them or changing their appearance so as to conceal from the enemy their existence, form, location, or movements. *n.* 1) The art of disguising military objects. 2) Any device that disguises or contributes to a disguise.

camouflet—1) A landmine placed in or near an enemy's mining tunnel so that its explosion causes the tunnel to collapse. 2) The underground cavity resulting from a deep underground explosion.

camp—1) A group of tents, huts, or other shelters for temporarily housing troops. A camp is more permanent than a bivouac (q.v.), and the term is sometimes applied to any military post. 2) Military service, particularly used in World War I and for periods of active duty training.

camp follower—Any nonmilitary person who follows an army as it moves from place to place. Camp followers have traditionally included merchants, families of soldiers and officers, laundresses, and prostitutes.

campaign—A phase or a stage of a war, involving a series of related operations aimed at achieving a single specific strategic result or objective. A campaign may involve a single battle but more often comprises a number of battles conducted over a protracted period of time or considerable distance but within a single specific theater or delimited area. A campaign is usually identified with respect to a region (as the Peninsular Campaign of the American Civil War) or a period of time (as the Campaign of 1806).

campaign medals—Medals awarded for participation in, or service during, a battle, a campaign, or an entire war. Sometimes called service medals.

canalize—To force a hostile attacker to limit his advance or attack to a narrow sector; achieved by the use of terrain, or of obstacles, or of defensive deployments, or some combination of these.

canister—A cylindrical metal container packed with small projectiles (**canister shot** or **case shot**). When fired from a gun the container bursts, scattering the shot. Canister was used by artillery from the early 15th century to the middle of the 19th century. A modern version of canister is the beehive projectile (q.v.).

cannibalize—1) To remove parts from one piece of equipment for use in repair of another. 2) To take personnel from one unit in order to build up another.

cannon—1) A large gunpowder weapon, too heavy and bulky to be transported by hand, and usually having a caliber of 20mm or more. The term has traditionally been used to distinguish all larger and heavier weapons, usually artillery pieces, from small arms. In modern terminology, there are three types of cannon: gun, howitzer, and mortar (q.v.). In general parlance the term is used interchangeably with gun or artillery piece. Most modern cannon are rifled, but some high-velocity tank cannon are smoothbore. In some instances cannon can be automatic (as in aircraft cannon, and some tank, naval, and air defense cannon). The name is derived from the Latin *canna*, meaning reed or tube. 2) A special type of 16th century artillery piece, 15 to 28 calibers in length and weighing from one to six tons. Subcategories included: quartocannon, demicannon, bastard cannon, cannon serpentine, cannon, cannon royal (the largest of 16th-century cannon, also known as "double cannon" in England), and basilisk (q.v.).

cannon lock—The part of a cannon that mechanically causes the propelling charge to be detonated.

cannonade—*n.* Bombardment using cannon only. *v.* To bombard with cannon.

cannoneer—1) One who fires a cannon or is a member of a cannon crew. 2) An artilleryman or gunner.

canteen—1) A bottle or flask in which soldiers carry drinking water. 2) A recreational area for soldiers or sailors, including such services as a clubroom, a recreation hall, and a place to buy refreshments. 3) A post exchange (q.v.). 4) A British club for soldiers. 5) In the 18th and 19th centuries, a chest in which British officers stored food and liquor.

canton—To billet troops.

cantonment—1) A group of temporary structures used for housing troops. 2) The act of assigning troops to such structures. 3) A military post or camp.

cap—See **percussion cap**.

CAP—See **combat air patrol**.

capital ship—The largest class of ship in a navy of first rank. In each period of history since the development of multimasted ships, the biggest ships with the most guns have been designated capital ships. From the mid-17th to mid-19th centuries these were ships of the line (q.v.). From the mid-19th century on, ships mounting a battery of a caliber greater than eight inches were considered capital ships. Since World War II aircraft carriers have also been designated capital ships.

capitulate—To surrender on terms agreed between the combatants.

capitulation—1) The act of surrendering on agreed terms. 2) The agreement containing the surrender terms. 3) Any formal agreement, such as the several concluded by the Swiss Diet or Swiss cantons separately with France, and with other European states as well, from the 16th to the 19th centuries, specifying the terms of mercenary service by Swiss regiments or corps in foreign armies.

caplock—The mechanism used to ignite a percussion cap (q.v.).

capping the T—See **crossing the T**.

captain—1) In most navies the senior officer below flag rank (q.v.), ranking between rear admiral (or commodore) and commander; comparable rank in other services: colonel or group captain. 2) In most navies and in some air forces the commander of a vessel or aircraft, regardless of rank. 3) In most armies and many air forces the senior company-grade (q.v.) officer, ranking between major and first lieutenant; comparable rank in other services: lieutenant in most navies, or flight lieutenant in many air forces.

captain-general—A term used from the 15th to the 18th centuries for the supreme commander or commander-in-chief of an army, or of a nation's armed forces.

captain's mast—The lowest disciplinary authority in the US Navy. For lesser infractions an enlisted man is normally required to appear before the commanding officer, who has the authority to assign certain types of relatively light punishment, within limits prescribed by the Uniform Code of Military Justice. Captains may also hold mast to give awards and commendations. The term carries over from the days of the sailing navy when captains performed such functions by the mainmast.

car, armored—See **armored car**.

carabinier—1) Term for cavalryman of 16th-century France during the reign of King Henry IV. Carabiniers (or carbineers) were so-called because they were armed with the **carabin**, a short arquebus (q.v.). 2) Elite light infantryman (sometimes called **carabinier à pied**) in 18th- and 19th-century armies.

caracole—A cavalry tactic developed by the Germans in the 16th century for their heavy cavalry, newly armed with wheellock (q.v.) pistols. The cavalry advanced at a slow trot in dense columns, with two horses' width between files or pairs of files. As they approached a foe, each front-rank horseman emptied his three pistols, then swung away sharply to the rear in a 180-degree turn, falling in behind the last rank in his file. While the first ranks were reloading and joining the rear of their respective files, the succeeding ranks continued the process of deliberate advance, pistol fire, and peeling off. Usually the caracole tactic was used prior to a general advance. It was a very difficult operation to carry out smoothly.

carbine—A light, short-barreled shoulder weapon primarily used by cavalrymen from the end of the 16th to the end of the 19th centuries. The name originated with the carabin (q.v.), a heavy smoothbore pistol about three feet long. The carbine developed through several stages, first becoming a shoulder weapon. Cavalry carbines were often rifled after the 17th century; cavalry carbines were among the first automatic shoulder small-arms in the mid-19th century.

carcass—An iron shell with three to five vents around a fuze hole. Developed in the later 17th century, it became standard incendiary ammunition for smoothbore cannon. The carcass was filled with pitch and ignited before firing, and the flames came out through the vents.

careen—To lay a ship over on its side in order to repair or clean the bottom. Before the invention of the drydock even large sailing vessels were careened periodically to remove grass and barnacles from the hull, replace caulking, and paint.

carpenter—US Navy warrant officer.

carpenter's mate—US Navy enlisted specialty rating, first, second, and third class.

carpet—1) Any of a group of airborne electronic devices for radar jamming. 2) Intensive and comprehensive coverage of an area with high explosive detonations, as in carpet bombing (q.v.).

carpet bombing—The laying down of bombs by a large number of aircraft on a target area in a close pattern intended to cover the area as with a carpet.

carriage—Term used for a supply train or transport in the 16th and 17th centuries. See also **gun carriage**.

carrier—See **aircraft carrier**.

carrier air group—Two or more aircraft squadrons on a single carrier, organized under one command for administrative and tactical control of operations.

carrier task force—A naval force composed of one or more aircraft carriers and support combatant ships capable of conducting strike operations.

carroballista—A large ballista (q.v.) mounted on wheels. This form of the ballista was in common use in the Roman world after 200 A.D.

carroccio—A medieval wagon, usually drawn by an ox, which carried the standard of an Italian free city, a bell, and usually a crucifix. In battle it served as a rallying force for the city's forces.

carronade—A short, large-bore, relatively thin-walled iron cannon developed in 1776 in the Carron Iron Works in Scotland, and used into the 19th century. The carronade fired a very heavy ball and was designed for use at close range. Primarily a naval weapon, it was also used ashore, particularly to guard fortress gates, where it was fired with grapeshot or case shot (qq.v.).

cart—A light, two-wheeled vehicle drawn by men, horses, mules, or by jeeps or other vehicles. Carts have many uses and have been employed since early times for hauling military equipment and ammunition. In modern armies they are commonly used for a specific weapon, such as a machine gun or light mortar, or for ammunition, in areas not traversable by heavy vehicles.

cartel—In its narrower sense, an agreement entered into by belligerents for the exchange of prisoners of war. In its broader sense, it is any convention concluded between belligerents for the purpose of arranging or regulating kinds of nonhostile intercourse otherwise prohibited by the state of war. Both parties to a cartel are honor-bound to observe scrupulously its provisions; but it is voidable upon proof that one party intentionally violated an important particular.

cartouche—A cartridge case made of heavy paper used from the early 17th to the mid-19th centuries.

cartridge—A cylindrical case containing the charge, primer, and projectile for use in a firearm. Paper cartridges were invented in the late 17th century; modern cartridges are usually made of brass. The cartridge greatly reduced the time required to load a weapon. See also **ball cartridge**.

case shot—See **canister**.

casemate—1) A large chamber, usually covered with a vault, built in the thickness of a fortress wall. Such casemates usually had embrasures in the outer wall and housed a gun. They might also be used as barracks or quarters for officers. 2) An armored enclosure for a ship's gun.

caserne—A barracks.

cashier—To dismiss with ignominy from military service.

casque—A helmet, particularly one with a pointed crown, sometimes with a nosepiece.

castellum—A small blockhouse-type Roman fortress, often integrated into long walls defending exposed frontiers of the Empire.

castle—1) A large fortification, usually composed of a number of fortified buildings within a walled enclosure, most often the residence of a ruler or his noble vassal in the feudal societies of medieval Europe. 2) A small defensive structure built on the deck of a warship in medieval times. 3) A small tower in which several men could ride with some protection, set on the back of a war elephant.

castrametation—The building of a camp. The word derives from the Latin *castrum*, a fortified place. The Romans developed castrametation into an efficient system and an integral element of military doctrine. Whenever a Roman military force halted overnight, each man performed a specific task for which he had been trained, and thus a fortified camp was quickly and efficiently constructed. It then provided the commander with a secure base for offensive or defensive combat.

casual—A military person awaiting duty assignment or transportation from a station at which he is not assigned or attached.

casualty—A person lost to his organization for reasons other than an ordered transfer. A **battle casualty** is one who, as a result of combat action, is dead, wounded, or missing (whether captured or simply unaccounted for). A **nonbattle casualty** is one who is sick, injured (other than through combat action), or missing in a noncombat situation.

casualty rate—The rate at which casualties are incurred in combat. Although it may be expressed in absolute numbers over time, a casualty rate is more commonly expressed as a percentage of troop strength over time, usually per day.

casus belli—An action taken by one nation toward another, or an event related to two or more nations, that is interpreted as justifying a declaration of war.

cataphract—A heavily armored horse archer. First introduced in the Roman Army in the second century, the cataphract, who rode an armored horse, became the mainstay of the Byzantine Army. He combined firepower, discipline, mobility, and shock-action capability. His arms might include bow, quiver of arrows, long lance, broadsword, dagger, and an ax. Men and horses were superbly trained and capable of complex tactical evolutions on drill-ground and battlefield.

catapult—1) A machine developed before the 5th century B.C.—apparently in Syria—for hurling arrows to distances of up to 500 meters; constructed like a large crossbow, it used the force of tension to propel the missile. It was distinguished from the early ballista (q.v.) mainly by the nature of the power (the ballista used torsion) and the nature of the missile (the ballista originally hurled rocks), though by the time of the Roman Empire distinctions between ballista and catapult had become blurred. 2) Any ancient war machine of the catapult or ballista type. 3) A device for launching an airplane without a runway, as from the deck of a ship.

cat-o'-nine-tails—A whip with nine knotted cords at the end of a short handle. This device was used for flogging as punishment for infractions, particularly in navies during the era of sail. The practice was abandoned in the Royal Navy and the US Navy by the mid-19th century.

cavalier—1) From medieval times, a cavalryman, equivalent to the French chevalier. 2) A tall structure inside a fort or other defensive work, high enough to permit fire over the wall. Because of its visibility it was an easy target and was abandoned after the introduction of high explosive shells in the late 19th century.

cavalry—A military force that fights on horseback, or, in modern times, in tanks, armored vehicles, or helicopters. Cavalry has been used since at least 1100 B.C., forming an important element of most armies because of its contribution to speed, mobility, and maneuver. The extreme vulnerability of the horse to high-powered rifles and automatic weapons caused the use of cavalry to diminish by the late 19th century. Its place has been taken by armored units, although some nations retain cavalry units for ceremonial purposes, and at least one, the People's Republic of China, maintains saber-wielding cavalry units. Horse cavalry units were sometimes distinguished as **heavy** and **light**.

Modern calvary units are generally designated as **mechanized** or **armored cavalry** (q.v.). The principal cavalry functions are reconnaissance (q.v.), screening (q.v.), raiding, and mobile shock action.

CBR—Chemical-biological-radiological. Refers to warfare using gas, germ bombs, and nuclear weapons.

ceasefire—An agreement between belligerents to stop firing. A truce.

center of burst—See **center of impact**.

center of impact—1) In gunnery, the center of a dispersion pattern of projectiles fired from a gun, or a group of guns, without change in range or deflection. 2) In bombing, the center of the area of greatest concentration in a bombing pattern.

centurion—The officer in command of a Roman century (q.v.). The legion (q.v.), principal formation of the ancient Roman Army, was divided into 60 centuries, each of 80 to 100 men. A centurion was generally promoted from the ranks. The senior centurion in a maniple (q.v.), which contained two centuries, or a cohort (q.v.), which contained six centuries, commanded all of the centuries in the units, much like a battalion officer in modern armies. The senior centurion in each legion participated in councils of war.

century—A unit of the Roman army, consisting of approximately 80 to 100 men, of whom two-thirds were heavily armed javelineers (q.v.) and a third were skirmishers (q.v.).

CEP—See **Circular Error Probable**.

chaff—Narrow strips of aluminum or other reflective material that are fired from the ground or dropped from aircraft to confuse radio and radar receivers. It was first used in World War II. Also called "window."

chain mail—See **mail**.

chain of command—The succession of commanding officers from a superior to a subordinate through which command is exercised. An **administrative chain of command** is permanent, and applies in all administrative matters. An **operational chain of command** may be established solely for a specific operation or a series of continuing operations.

chain shot—A type of projectile consisting of two balls or hemispheres joined by a chain. Fired into the rigging of a naval vessel, chain shot was a particularly effective device in naval battles, for the balls would rip through sails and wrap themselves around spars or lines, cutting, tearing, and snapping them.

challenge—Any process of identification carried out by one unit or person to ascertain the friendly or hostile character or identity of another.

chamade—A drumbeat or trumpet call signalling a desire to parley.

chamfron—The armor that protected the head of a warhorse.

chandelier—A solidly built wooden frame that can be filled with fascines (q.v.) or other material. Chandeliers can be carried with troops and set up as field fortifications, particularly where it is difficult to dig, in rocky, swampy, or frozen ground.

chaplain—A clergyman, commissioned as an officer, who is responsible for the religious welfare of service-connected personnel, and who advises their commander on all matters pertaining to religious life, morals, and morale.

charge—n. 1) An accelerated, concentrated attack, such as a cavalry charge. 2) That with which a bomb, shell, or mine is filled, including explosive, thermite, bacteria, or other material. 3) A given quantity of explosive, either by itself or contained in a bomb, shell, or mine, or used as a propellant for a bullet or shell. 4) A cartridge or round of small arms ammunition. See also **shaped charge**. v. 1) To dash rapidly against an opponent or a target. 2) To fill an explosive device—bomb, mine, shell—with a charge. 3) To move a round of ammunition into the firing chamber of a weapon.

charger—A horse used as the mount of a military man, particularly a knight or an officer.

chariot—A type of ancient horse-drawn battle vehicle, usually with two wheels, sometimes with four, drawn by one to four horses. Chariots are pictured in reliefs and vase paintings from as early as the 3d millenium in Mesopotamia, and were used in most armies until the development of cavalry (q.v.).

The Gauls, the Britons, and the Picts used them as late as the 3d century A.D. They usually carried two people, the **charioteer**, who drove the horses, and the warrior, armed with bow, lance, or spear, or some combination of these. In the West chariots were used primarily for transportation, and the warriors dismounted near the field of battle to fight on foot. In Asian armies chariots became the principal shock-action weapons, sometimes with scythes attached to the wheels to increase their deadliness.

Charlie—Letter of old and NATO phonetic alphabets. See also **alphabet, phonetic**.

chart—A map used for navigation. **Nautical charts** show information about the depth of water and the nature of the sea bottom, aids to navigation, the location of channels, hazards, and other matters required for safe ship handling. **Aeronautical charts** give equivalent information for the use of the aerial navigator. See also **firing chart**.

chase—The portion of a cannon from the trunnions (q.v.) to the muzzle.

chase gun—A cannon placed in either the bow or the stern of a vessel, used in pursuit or defense against pursuit.

Chassepot—A French breech-loading rifle invented by Antoine Chassepot and used by the French Army after 1866. This bolt-action needle rifle, whose caliber was 11mm, differed from the earlier Prussian needle gun (q.v.) in that the firing pin ignited the primer at the base of the cartridge. After modification to take metallic cartridges in 1874, the weapon was renamed the Gras rifle.

chasseur—A French soldier, lightly armed and equipped for reconnaissance and screening purposes. The term in now applied to light armor and light infantry units of the French Army. **Chasseur à pied** is a light infantry soldier, first so designated in 1743. Infantry troops trained for mountain warfare are called **chasseurs alpins**. **Chasseurs à cheval**, light-cavalry troops, were introduced in 1779.

Chatellerault gun—A French machine gun developed in the Chatellerault Arsenal in the 1920s. It had a very high rate of fire, 500 rounds per minute.

chausses—Medieval leg and foot armor made of mail.

checkerboard formation—A formation in which elements of a military force are positioned in a checkerboard or **quincunx** pattern. Such a formation was used in the Roman legion and has been employed in other tactical formations ever since.

checkpoint—1) A predetermined and clearly identifiable point on the earth's surface used as a means of controlling movement, or as a registration target for fire adjustment, or as a reference for location. 2) Geographical location on land or over water above which the position of an aircraft in flight may be determined by observation or by electronic means. 3) A place where military police check vehicular or pedestrian traffic in order to enforce circulation control measures and other laws, orders, and regulations.

chemical agent—A solid, liquid, or gas which through its chemical or incendiary properties produces lethal or damaging effects on man, animals, plants, or material, or produces a screening or signalling smoke.

chemical, biological, and radiological operations—A combining term used only when referring in the collective sense to the three areas of: chemical operations, biological operations, and radiological operations. These are operations involving nonconventional weapons that include poison gas, biological or bacteriological agents, and the radiation effects of nuclear weapons, and that cause damage through effects other than violent impact or explosive forces.

Chemical Corps—In the US Army, that branch of service charged with furnishing the means for and technical guidance in the employment of and defense against chemical and biological weapons. The functions performed by the Chemical Corps in support of combat operations are: a) operation of smoke generating equipment and special chemical weapons, and b) decontamination of areas and matériel and assistance to other units in carrying out such tasks.

chemical warfare—The employment in warfare of chemical agents other than explosives to kill or disrupt hostile forces, to contaminate areas, or destroy matériel.

cheval-de-frise—A military obstacle consisting of wooden beams, or frames, or barrels from which spears or pointed stakes protrude at intervals. Chevaux-de-frise were first used in the Dutch War for Independence in the 16th and 17th centuries, primarily as a defense against cavalry. The name derives from "horses of Frieseland." Benjamin Franklin devised a submarine version that was placed in the Delaware and Hudson rivers during the Revolution as an obstacle to British warships.

chevron—A mark or insignia in the shape of a caret or a V, worn on a uniform sleeve to indicate rank and class of service.

chief gunner—See **gunner**.

chief officer—Second in command of a vessel.

chief of staff—1) The senior or principal member or head of a staff, usually a general staff (see also **chief of the general staff**). 2) The principal assistant in a staff capacity to a person in a command capacity. 3). The head or controlling member of a staff, for purposes of the coordination of its work; a position, which in itself is without inherent power of command by reason of assignment, except that which is invested in such a position by delegation to exercise command in another's name.

chief of the general staff—The senior officer of a general staff, from the level of a division up to the level of a major service of a nation (as in the United States for Army and Air Force) or for a nation's armed forces (as in the Soviet Union).

chief of section—1) The senior noncommissioned officer in a small military sub-unit, usually part of a platoon. 2) In the US Army Field Artillery, the commander of a gun crew, usually a sergeant.

chief petty officer—In the US Navy, the highest enlisted rating below warrant officer. A sailor of that rating may be a chief yeoman, chief boatswain's mate, or chief in any of the other naval specialties.

chin turret—A turret installed immediately beneath and slightly behind the nose of a bomber.

Chindit—Member of "special forces" commanded by British Major General Orde Wingate in Burma in World War II long-range penetration (q.v.) operations in 1943 and 1944. So-called because their emblem was a mythical Burmese bird called a chinthe, and also because their operations took place near the Chin Hills. The entire force was often given the collective name of "Chindits." Objective military assessment suggests the Chindits' accomplishments were not commensurate with their strong effort or their own heavy casualties.

chivalry—The feudal institution of knighthood.

chlorine gas—That form of poison gas first used in modern warfare. Introduced by the Germans against the French in World War I, during the Second Battle of Ypres, April 22, 1915. See also **gas, poison**; **chemical warfare**; **phosgene**.

CHOP or **Chop Line**—From Change of Operational Control, the dividing north-south line in the North and South Atlantic oceans during World War II in which convoy and ship movements changed from American to British control or vice versa.

cipher—Any cryptopgraphic system in which arbitrary symbols or groups of symbols represent units of plain text, usually single letters, or in which units of plain text are rearranged, or both, in accordance with predetermined rules.

Circular Error Probable (CEP)—An indicator of the accuracy of a missile or projectile, used as a factor in determining probable damage to a target. CEP equals the radius of a circle within which half of the projectiles fired from a single gun or launcher at given settings for range and deflection are expected to fall. Also known as **probable error**.

circumvallation—A wall, trench, or other barrier erected during siege operations to protect the besieging forces from the interference of relief forces; thus the barrier faced outward, away from the place or area besieged, in contradistinction to a barrier of contravallation (q.v.).

citadel—A central fortress, often on a hill overlooking a city or larger fortification. The term was formerly used to designate the heavily armored central portion of a warship, including the engines, guns, and other vital portions.

citation—1) A specific mention of a person or unit in orders or other documents for commendable or meritorious action. 2) A written narrative statement of a meritorious action or deed for which a decoration is awarded.

civic action (military)—The use of military forces on nonmilitary projects useful to local populations.

civil affairs—Those phases of a commander's activities that embrace the relationship between the military forces and the civilian population, particularly civil authorities, in a friendly or occupied country or area.

Civil Air Patrol—In the United States a federally chartered, nonprofit corporation that was designated by Congress in 1948 as a volunteer auxiliary of the US Air Force. Its mission is to provide service during local and national emergencies, motivate youth to high ideals of leadership, and to generate interest in and support of the Air Force through aerospace education and training.

civil defense—All those activities and measures designed or undertaken to: a) minimize the effects upon the civilian population of an enemy attack; b) deal with the immediate emergency conditions created by any such attack; and c) facilitate emergency repairs to, or the emergency restoration of, vital utilities and facilities destroyed or damaged by any such attack.

civil-military relations—Relations between military forces and the civilian people and agencies in areas where troops are stationed.

civil war—A war between indigenous factions in a single country.

clandestine operation— Intelligence-gathering activities sponsored or conducted so as to assure secrecy or concealment. (It differs from covert operations [q.v.] in that emphasis is placed on concealment of the operation rather than on concealment of the identity of the sponsor.)

classified information—Official information that has been determined to require, in the interests of national defense, protection against disclosure to those with no need to see or know about it.

claymore— 1) A Scottish Highland two-handed, double-edged broadsword. 2) The name of a US Army antipersonnel mine, which can also be used as short-range, remote-control weapon fired by an operator-observer; particularly useful for final protective fire and against an enemy advancing through a defile (q.v.).

clear—n. 1) To approve or authorize or to obtain approval or authorization for a proposed action or activity. 2) Specifically to give one or more aircraft a clearance. 3) To give a person a security clearance (q.v.). *Adj.* Of military messages, not encoded or encrypted.

clip—A device holding several rounds of ammunition to be fired either single-shot or automatically in an automatic pistol or a magazine rifle.

clipeus—A flat oval Roman shield.

close air support—Air attacks against hostile surface targets that are in close proximity to friendly ground forces. Such air attacks require detailed integration of each mission with the fire and movement of the ground forces. Usually close air support is flown to meet specific requests from ground forces during the course of a battle.

close controlled interception—See **intercept**.

close-in security—The physical security (defensive) measures taken inside a restricted area or national defense area in direct support or protection of the operational resources in the area.

close order—Deploying or positioning of soldiers in a unit with a minimum of space between them. The amount of space is usually regularized and dependent upon the requirements or characteristics of the weapons they carry.

close order drill—Practice of units in formation in administrative or ceremonial movements. See also **drill**.

close station— 1) To cease ongoing activities or operations, particularly for an artillery unit, and to withdraw from a signal net. 2)

The order to cease activities and withdraw from the net. The usual complete command is, "Close station, march order (q.v.)."

close stick—A stick of bombs released more or less simultaneously.

close support—Supporting action by one force for another at a distance so slight that the supporting action must be integrated or coordinated directly with the action of the supported force. Close air support (q.v.) is that supporting action by combat aircraft (usually fighter bombers) by which fire from the air (by bombs, strafing, napalm, etc.) is directed at hostile forces immediately in front of supported ground forces units.

close supporting fire—Fire placed on enemy troops, weapons, or positions to support a unit that is in close proximity or even engaged in combat with the enemy. Such supporting fire must be coordinated with the fire and movements of the supported unit to avoid shelling friendly troops.

close train—One after the other, as in "The bombs were released in close train."

club the musket—To use a musket or a rifle as a club, usually wielding it by the barrel.

cluster—1) Fireworks signal in which a group of stars burn at the same time. 2) Groups of bombs released together. A cluster usually consists of fragmentation or incendiary bombs. 3) Component of a pattern-laid minefield. 4) Two or more engines coupled together to function as one power unit. 5) Use of two or more parachutes for dropping a single load.

cluster bomb—An aerial bomb designed for use against personnel, armor, airfields, and other ground targets. Composed of many small explosive bombs packed into one container, a cluster bomb is designed to open in flight and release its payload, thereby increasing the dispersion and lethality of the weapon's submunitions.

coast artillery—1) Fixed or mobile artillery weapons emplaced close to harbors, cities, or strategic waterways, in positions from which they can fire on approaching enemy ships. 2) The personnel or units manning such weapons.

coast guard—A maritime force whose functions in peacetime include such duties as the protection of life and property at sea and the enforcement of laws governing customs, immigration, and navigation. In time of war the US Coast Guard functions as part of the US Navy.

coastal frontier—A geographical division of coastal area, established for organization and command purposes to insure the effective coordination of military forces engaged in military operations within the area.

coastwatcher—A person inside enemy-held territory who observes coastwise shipping and enemy movements and reports by radio. The United States used coastwatchers on Pacific islands in World War II, landing them clandestinely or providing equipment for islanders to use.

coat of mail—A defensive garment of metal scales or chain mail.

cock—1) To draw back the hammer, bolt, or plunger of a firearm to ready it for firing. 2) To set a bomb release mechanism so as to make ready to drop the bomb load.

cockade—A rosette or knot of ribbon worn as a badge, especially on the hat.

cocked hat—A hat with a wide brim, turned up in two or three flaps. When two flaps are formed such a hat may be called a bicorne (q.v.). With three it may be a tricorne.

code word—A word which conveys a meaning other than its conventional one, prearranged by the correspondents. Its aim is usually to increase security; it can be used also for convenience because of brevity.

codes—Any system of communications in which arbitrary groups of symbols represent units of plain text of varying length. Codes may be used for brevity or for security. See also **cipher**.

coehorn—A popular small mortar of the 17th and 18th centuries, invented by Menno, Baron van Coehoorn, in 1673, which threw shells weighing up to 24 pounds.

cohesion—The quality or characteristic of a unit whereby its members work well together and demonstrate loyalty to each other and to their unit under all circumstances. Cohesion is achieved by training together, living together, and strong leadership. Sometimes called **cohesiveness**.

cohort—1) A unit of the Roman Army. From the 2d century B.C. a cohort consisted of approximately 420 men, organized in six centuries (q.v.). There were ten cohorts in a legion (q.v.). In the 1st century A.D. the size of the first cohort was increased to about 800 men, organized in five centuries. Cohorts were also formed in special military organizations, such as the Praetorian Guard, which was not a legion but was composed of cohorts. 2) The annual manpower increment of a nation reaching a given age, usually that for military service. 3) A unit equivalent to a battalion in the French National Guard during the First Empire.

coif, mail—A skullcap, made of fabric, mail, or leather, worn by a medieval soldier under a hood of mail.

Cold War—A state of extreme tension involving rivalry between two or more nations, with each employing measures short of overt military action to gain ascendancy over the other. The term was coined in the period after World War II to describe the situation—military preparedness, mutual suspicion of motives, worldwide political rivalry—between the United States and the Soviet Union.

collateral damage—Damage to nonmilitary structures and facilities resulting from a strike on a nearby military target.

collecting point—A point designated for the assembly of casualties, prisoners of war, stragglers, disabled matériel, salvage, etc., for further movement to collecting stations or rear installations.

colonel—The highest military field-grade rank, that between lieutenant colonel and brigadier or brigadier general. In 1505 King Ferdinand of Spain created 20 units called "colunelas," or columns. This was the first clear-cut tactical formation in Western Europe since the time of the Roman legion. This unit, prototype of the modern battalion, regiment, or brigade, was commanded by a *cabo de colunela* (chief of column) or colonel. The French soon copied the colunela (q.v.) concept, and also adopted the military rank,

which persists to this day in the French and English languages.

colonel-commandant—A British term referring to the commander (usually honorary) of a regiment (q.v.), which, in British usage, is an administrative formation as opposed to a combat unit. Sometimes called **colonel-in-chief**.

colonel-general—This grade is normally identified with the commander of a field army or an army group, and is comparable to a four-star general.

colonel-in-chief—See **colonel-commandant**.

colonial troops—Troops of colonial powers, raised in, or raised for employment in, a colony. For instance, colonial units were raised in many of the colonies belonging to Great Britain. Indigenous colonial units were usually commanded by officers from the colonial power.

color bearer—See **standard bearer**.

color guard—A guard assigned to protect the colors (q.v.) in a ceremony or on the march.

colors—1) A flag or banner, particularly a national flag or that of a military unit. The term was first used in Elizabethan times. 2) Military service. In this sense it is used in phrases like "called to the colors."

color-sergeant/corporal—A noncommissioned officer selected to carry the national flag and (in the US) unit flag of a unit, or as a guard (usually a corporal, called a color-guard) for the colors.

Colt revolver—A handgun invented by Samuel Colt about the middle of the 19th century. He designed a cylinder that could hold several cartridges and could be rotated by releasing the squeezed trigger, so that a new cartridge was aligned with the barrel to replace one that had been fired. The Colt revolver, a .45 caliber weapon, was first used by the US Army in the Mexican War.

columbiad—A 19th-century American muzzle-loading cannon, usually used for coast defense or siegecraft, with a heavy, long, iron barrel, of 8-, 10-, or 12-inch caliber. The range of these guns, which replaced similar bronze ones, was more than 5,000 yards.

column—1) A military formation in which a number of like units, whether troops, squads, battalions, ships, vehicles, or aircraft, are deployed one behind the other. In ground combat the **attack column** came into general use during the French Revolution. Perfected by Napoleon, it was an adaptation of the linear system, the deployment of a number of linear units (usually battalions) in depth to provide physical and psychological weight to an attack (see **linear tactics**). The individual units could still operate in a linear formation if desired. The great tactical value of the column lay in its flexibility and versatility. It permitted the commander to move large numbers of men over the battlefield far more rapidly and with better control than had been possible before. 2) Term used for independent Boer forces, also called commandos.

column cover—Protective cover of a ground column furnished by aircraft in radio contact with the column, thereby providing reconnaissance and protection (by attack of air or ground targets which threaten the column).

colunela—A unit created in 1505 in the army of Ferdinand of Spain. Consisting of from approximately 1,000 to 1,250 men (mixed pikemen, halberdiers, arquebusiers, and sword-and-buckler troops [q.v.] organized in five companies. For all practical purposes the colunela was the genesis of the modern battalion, regiment, or brigade.

combat air crew—The crew assigned to a combat aircraft.

combat air patrol (CAP)—An aircraft patrol flown over an objective area, over a protected force, over the critical area of a combat zone, or over an air defense area, for the purpose of intercepting and destroying hostile aircraft before they reach their target.

combat area—1) The area in which combat units, including close support units, are operating. 2) A restricted area (air, land, or sea) established to prevent or minimize mutual interference between friendly forces engaged in combat operations. See also **area**.

combat boot—A high shoe with a leather cuff, often worn outside the trousers.

combat box—A rectangular formation used with certain airplanes during, or when expecting, combat. The term was especially applied to a heavy bomber formation during World War II.

combat command—A major tactical unit of combined arms within an armored division, corresponding to an infantry regiment or brigade. A combat command consists of a headquarters and headquarters company and a variable number of units or elements attached.

combat control team—A team of specially trained air force personnel who airdrop (q.v.) into forward areas to advise the ground force commander on all aspects of landing area requirements and to provide local air control.

combat effectiveness—1) A term used to describe the abilities and fighting quality of a unit. 2) The quality of being effective in combat.

combat fatigue—See **fatigue, combat**.

combat forces—Those forces whose primary missions are to participate in combat operations.

combat information center—Area in a ship or aircraft manned and equipped to collect, display, evaluate, and disseminate tactical information for the use of an embarked flag officer, commanding officer, and certain control agencies.

combat intelligence—That knowledge of the enemy, weather, and geographic features required by a commander for the planning and conduct of tactical operations.

combat loading—The arrangement of personnel and the stowage of equipment and supplies for an amphibious assault, or follow-up to an amphibious landing, in a manner designed to conform to the anticipated tactical operation of the organization embarked. Each individual item of cargo is stowed so that it can be unloaded in the sequence of anticipated need.

combat, military—A violent, planned form of fighting, in which at least one party is an organized force, recognized by governmental or de facto authority. One or both opposing parties hold at least one of the following objectives: to seize control of territory, to prevent the opponent's seizure and control of territory, or to protect one's own territory.

The presence of weaponry creates an atmosphere of lethality, danger, and fear in which one party's achievement of objectives may require its opponent to choose among: continued resistance (and thereby risk destruction), retreat and loss of territories and facilities, or surrender.

combat model—See **model, combat; scenario**.

combat patrol—A detachment of ground, sea, or air forces sent out by a larger unit for the purpose of gathering information or carrying out a destructive, harassing, mopping up, or security mission.

combat power—The total means of destructive or disruptive force that a military unit or formation can apply against an opponent at a given time.

combat readiness—The capability of a military unit or ship to perform its assigned missions. The status of personnel, equipment, supplies, maintenance, facilities, and training is considered in determining this capability.

combat reserve—A fighting force held in reserve for immediate use if needed during a combat operation.

combat service support—The assistance provided to operating forces primarily in the fields of administrative services, chaplain services, civil affairs, finance, legal services, health services, military police, supply, management, maintenance, transportation, construction, acquisitions, engineering functions, food service, graves registration, laundry, dry cleaning, bath, property disposal, and other logistic services.

combat support—Operational assistance (including direct combat involvement) furnished combat elements by other designated units.

combat support troops—Those units or organizations whose primary mission is to furnish operational assistance for the combat elements.

combat team, regimental—A reinforced infantry regiment, normally composed of one three-battalion regiment of infantry strengthened by a battalion of field artillery, a company of engineers, and a battery of antiaircraft artillery. This composition may be changed to meet the demands of the tactical situation. In some situations, a combat team may need other attachments, such as an ambulance platoon, a clearing platoon, a signal company, additional tank units, additional field artillery, etc. A **brigade combat team** consists of a three-battalion infantry brigade similarly strengthened by attachments (q.v.).

combat theory—See **theory of combat**.

combat troops—Those units or organizations whose primary mission is destruction of enemy forces and/or installations.

combat vehicle—A vehicle, with or without armor, designed for a specific fighting function.

combat zone—1) That area required by combat forces for the conduct of operations. 2) The territory forward of an army's rear area boundary.

combatant—n. One who engages in combat. adj. 1) Adept at fighting. 2) Serving or ready to serve in combat. 3) Used in fighting. 4) Involving combat.

combined—Designating the presence or involvement of two or more forces, branches, agencies, or types of arms (for example, infantry, armor, and air). In World War II the term "combined" was used to designate committees, forces, or operations involving forces of two nations (principally the United States and Great Britain, and particularly including the Combined Chiefs of Staff [q.v.] and all the staff committees). See also **joint**.

combined arms army—See **army**.

combined arms force—A major field force that includes elements of various service branches, usually infantry, armor, and artillery, and frequently air support as well.

combined arms operation—A military operation performed by a combined arms force.

Combined Bomber Offensive—The round-the-clock bomber offensive conducted by the US Eighth Air Force and the British Bomber Command against targets in Germany in World War II. Bomber Command attacked the economic system and waged psychological warfare against civilian morale by large-scale nighttime saturation bombing; the Eighth Air Force concentrated on daylight precision bombing against industrial targets and the Luftwaffe itself. Later, both forces concentrated on vital industries to destroy Germany's ability to make war. It was codenamed Operation "Pointblank."

Combined Chiefs of Staff (CCS)—The World War II high command organization composed of the military service chiefs of the United States and Great Britain, whose function was to plan and direct global strategy, under the two national leaders.

combined operations—1) Operations involving two or more forces, branches, or agencies participating in the same war-related endeavor. 2) In British usage, amphibious operations.

command—n. 1) The authority vested in an individual of the armed forces for the direction, coordination, control, and administration of military forces. 2) An order given by a person with authority. 3) An organization or area under the direction and control of a duly constituted authority. 4) A US Air Force organization of three or more wings. v. 1) To exercise authority over a military unit or larger force. 2) To order. 3) To dominate an area, a hostile position, or a hostile force by a field of weapon fire or by observation from a superior position.

command and control—The exercise of authority and direction by a properly designated commander over assigned forces in the accomplishment of the mission. Although redundant, the expression "command and, control" has the sanction of common usage. Command and control functions are performed through an arrangement of personnel, equipment, communications, facilities, and procedures which are employed by a commander in planning, directing, coordinating, and controlling forces and operations in the accomplishment of the mission.

command area—See **area**.

command net—A communication net that connects headquarters with subordinate units for the issuing of commands and other messages.

command of the sea—The ability to protect the sea routes taken by a nation's shipping, and to deny to the enemy the use of those routes.

command pilot—1) The highest aeronautical rating in the US Air Force and Naval Air Force, based on years of service, flying time, and performance. 2) A pilot qualified by time and experience to hold this rating.

command post (CP)—The headquarters installation and personnel of a military unit or formation under combat conditions. A commander will sometimes maintain a **forward command post** as close as possible to the front line, as well as the **main command post** farther to the rear.

command post exercise (CPX)—A training exercise involving a commander, his staff, and the communications between headquarters and command posts.

command ship—A specially equipped ship carrying the admiral in command of fleet actions, or an amphibious operation, and his staff. Command ships, first used in the Pacific in World War II, carried electronic devices for communications, intelligence gathering, and the functions of command and control. In amphibious operations, such a floating command post enables close personal cooperation between the naval commander, responsible for putting the ground forces ashore, and the ground commander, who assumes control once the troops have gained the beach.

command vehicles—Vehicles designed primarily for command and communication; they may be tracked or wheeled and may or may not also be armored vehicles.

commandant—1) A commanding officer of a military organization or installation other than an operational combat force. The senior officers of the US Marine Corps and the US Coast Guard have the title of commandant. 2) In the French Army a rank equivalent to major.

commandeer—To take for military use without formality (and usually without customary procedures) as a house or a vehicle.

commander—One who exercises command.

commander in chief—The commander of a

major military force, as Commander in Chief, US Army, Europe, or Commander in Chief, US Fleet. In some nations the title is used for the supreme commander of the nation's armed forces.

commander's concept—See **concept of operations**.

commanding ground—Ground that overlooks neighboring terrain.

commanding officer—An officer who commands a unit or installation.

commanding position—A position, usually on higher ground than the enemy, in a ground action, or to windward in an action between sailing vessels, from which one force has a visual or tactical advantage over the other.

commando—1) Any soldier belonging to a specially selected group of combatants, particularly in the British Armed Services, trained for hand-to-hand fighting, surprise raids, and other shock tactics; term originally used for such troops in World War II and apparently derived from Boer War experience; see 3) below. 2) Often applied to similar troops of any nation. For example, the American equivalent of the generic "commando" is the Ranger. 3) A Portuguese word adopted for a unit of the Boer military forces, composed originally of all males of an electoral district between 16 and 60 years of age, later applied to any Boer independent command. Such units fought effectively in the Boer War. See also **ranger**.

commerce raider—A warship, usually fast and well armed, or a submarine, that operates against the enemy's merchant ships.

commissariat—An installation or department supplying food and other supplies, transport, etc.

commissary—1) In the United States, a market maintained on a military post where authorized service personnel and their families may purchase food and other provisions. 2) In the British and other armies (and the US Army earlier), an officer responsible for providing food and other supplies. In the 17th and 18th centuries commissaries oversaw the force's expenses and the conduct of officers and men.

commission—n. 1) A written certificate giving a person rank and authority as an officer in the armed forces. 2) The rank and authority given by such an order. v. 1) To confer such rank. 2) To put in or make ready for use or service, as to commission an aircraft or a ship.

commissioned officer—See **officer**.

commissioned warrant officer—See **warrant officer**.

commit—1) To assign, place, or use a military unit in such a way that it inevitably becomes engaged in combat. 2) To put oneself in a position where withdrawal or change in a course of action becomes impossible.

commodore—1) The lowest naval flag rank, between captain and rear admiral. It was abolished in the US Navy in 1899, reactivated for limited use during World War II, and reactivated again in 1982. It was modified to commodore-admiral in 1985. 2) A designation used unofficially in the British Royal Navy for a captain with temporary command of a fleet division or a squadron.

common item—1) Any item of matériel which is required for use by or in more than one service or activity. 2) Sometimes loosely used to denote any consumable item except repair parts or other technical items.

common servicing—That function performed by one military service in support of another military service for which reimbursement is not required.

common use—Services, materials, or facilities provided by a Department of Defense agency or a military service for two or more Department of Defense agencies.

communication trench—See **trench**.

communications—1) The system for transmitting messages from one point to another. 2) The route connecting one military force with another, in particular a force in a combat area with its rear headquarters and source of supply. See also **line of communications**.

communications center—A facility responsible for the reception, transmission, and delivery of messages. Its normal elements are a message center section, a cryptographic section, and a sending and receiving section, which uses electronic communications devices.

communications intelligence—A specialized form of intelligence (q.v.) whereby technical information is derived from hostile or foreign communications.

communications, lateral—1) The availability of means—by wire, radio, or messenger—for the headquarters or commanders of two adjacent, friendly forces to communicate with each other. 2) The employment of such means.

communications net—1) The various means, mechanical and other, of transmitting information, orders, and reports within an organization. 2) The system formed by the command posts at different levels of an organization and all lines of signal communication linking them.

communications zone—Rear part of theater of operations (behind but contiguous to the combat zone) that contains the line of communications, establishments for supply and evacuation, and other agencies required for the support and maintenance of combat forces. See also **rear area**.

compagnie d'ordonnance—A medieval French mounted organization, commanded by a captain, consisting of 600 men, organized in *lances* of six men each, each lance consisting of a knight or man-at-arms, a squire, two archers, a valet, and a page. Because the men were well paid and provided with liberal rations there was no need to pillage. Establishment of the compagnies d'ordonnance was the death knell of feudalism as a military and social system, and the beginning of standing armies and professionalism in modern European warfare.

Companions—An elite cavalry unit of the Macedonian army, so-called because Philip, and later Alexander, led them personally in battle.

company—A military organization—particularly in the infantry—consisting of a headquarters and two or more platoons (q.v.). The next higher unit is usually a battalion (q.v.). An infantry company usually consists of 100 to 200 men. **Ship's company** includes all the officers and men assigned to a naval vessel (but not, for instance, members of naval air squadrons on an aircraft carrier, who serve on the ship but have their own organization). In medieval armies, before organization was established, the men who accompanied a lord or knight were known as his company, Soldiers discharged from service, notably in the 14th century, sometimes formed **free companies** primarily for mercenary service, sometimes for plunder. See also **compagnie d'ordonnance**.

company grade—In the US Army and the Marine Corps, the designation for an officer of rank appropriate for service in or command of a company, usually second lieutenant, first lieutenant, and captain. See also **field grade**.

complement—All the crew required to man a ship or a naval station.

complementary angle of site—See **site, complementary angle of**.

composite bow—See **bow**.

compromise—To expose or reveal classified information or clandestine activities to unauthorized persons, whether done intentionally or inadvertently. A code known to be compromised is immediately discarded for another, unless it is kept alive to feed false information to the enemy.

computer-assisted model—See **model, computer-assisted**.

computer model—See **model, computer**.

concealment—1) Protection from observation. 2) The use of cover, blackout measures, darkness, camouflage, or smoke, or any combination of these, to hide personnel, installations, movements, etc., from enemy observation. 3) The removal of organizational insignia or other identifying marks and the protection of records to keep from the enemy the organizational identity of the troops opposing them. 4) An object, phenomenon, activity, or measure that conceals. The term is often and incorrectly used for cover (q.v.).

concentrate—To bring all elements of a combat force together, or in close vicinity to each other, generally in preparation for action.

concentrated fire—See **fire, concentrated**.

concentration—1) A principle of warfare embodying the doctrine that effective offensive action requires the massing of superior combat power or striking force at a decisive point. 2) The action of massing power or striking force at or against a single point or area. 3) An artillery fire mission against a specified target. See also **mass** and **Schwerpunkt**.

concentration area—1) An area, usually in the area of operations, where troops are assembled before beginning active operations. 2) A limited area on which a volume of gunfire is placed within a limited time. See also **concentration**.

concentric advance—An advance upon enemy forces from several directions, so as to bring the separately advancing forces into supporting distance of each other at approximately the same time as they engage the enemy. Usually one or more of the advancing forces is performing an enveloping maneuver.

concentric castles—Castles built with one or more outer walls surrounding an inner keep (q.v.) or citadel.

concept of operations—A verbal or written statement, in broad outline, of a commander's assumptions or intent in regard to an operation or series of operations. The concept of operations frequently is embodied in campaign plans and operation plans, in the latter case particularly when the plan covers a series of connected operations to be carried out simultaneously or in succession. The concept is designed to give an overall picture of the operation. It is intended to provide additional clarity of purpose, and is frequently referred to as the **commander's concept**.

concertina wire—A coil of barbed wire that is compressed for carrying but which expands when the constraining band is removed.

concussion fuze—A fuze that detonates a bomb in response to the concussion produced by the explosion of another bomb nearby.

condottiere—The leader of a mercenary band, or free company (q.v.), in 14th- to 16th-century Italy. Composed of as many as several thousand men, most of them heavily armored cavalry, such bands roamed over Italy, fighting for pay in the wars of the Italian states for whichever side would hire them. When not so employed, they often plundered and fought freely on their own. The term is frequently used for the soldiers as well as their leader.

conduct of fire—The procedure of giving commands to the crew of an artillery piece, battery, or larger unit for the purpose of bringing effective fire on targets.

cone of fire—1) The three-dimensional area on and just above the surface of the ground through and into which fall the bullets or other projectiles of a fixed weapon with a given range and deflection setting. The cone effect is created by the range and deflection dispersion of the projectiles. Usually applied to the lethal area covered by a fixed machine gun. 2) The three-dimensional area covered by the fire of two or more fixed weapons converging on a single point.

confidential—A security classification denoting that unauthorized disclosure of the contents of a document or the characteristics of an object would be prejudicial to the defense interests of a nation.

confinement—Physical restraint under guard in a jail, prison, guardhouse, stockade or other facility for incarceration of prisoners who are either awaiting trial or serving a sentence imposed by a court.

conflict—A war, or a discrete portion of a war.

conflict levels of intensity—The categories or levels of conflict intensity are major war, significant local war, small war, minor hostilities, contingency (both major and minor), engagement, action, high-intensity conflict, low-intensity conflict, and insurgency. A **major war** is a high-intensity conflict between two or more great powers, with the possible involvement of their key allies and with significant national interests of the antagonistic powers at stake. A **significant local war** is a high-intensity conflict in which a great power seeks to protect some aspect of its vital national interests; or in which conflict arises among important regional states, with vital national interests at risk. In a **small**

war armed hostilities involve: 1) a great power in a relatively low-intensity conflict in which there is no significant risk to vital national interests; 2) an important regional state in a relatively low-intensity conflict in which there is no significant risk to vital national interests. 3) civil war or prolonged insurgency in a regional state; or 4) hostilities between minor regional states. **Minor hostilities** are a form of low-intensity conflict other than a major war, a significant local war, or a small war. Minor hostilities can include: 1) limited military expeditions or excursions by major powers or important regional states; 2) Low-intensity insurgencies; or 3) any other low-intensity armed hostilities limited in time and space. **Contingency** is a term used in military planning to designate the possible employment of armed forces in hostilities that might break out at unexpected times or places. From the standpoint of the planner or planning agency, contingencies can be major or minor. A **major contingency** is a military operation involving high-intensity, and probably sustained, hostilities against the forces of one or more great powers or major regional states. Thus it is usually a major war or a significant local war. A **minor contingency** is a military operation involving armed hostilities (other than major wars) limited in time (no more than six months) and space (a maximum radius of 1000 kilometers). A minor contingency consists of one or more engagements or actions and can include: 1) significant local wars (limited in time and space as specified above), or discrete, limited campaigns or operations within such wars; 2) small wars (limited as specified above), or discrete, limited campaigns or operations within such wars; or 3)any minor hostilities (limited as specified above). (This definition of military planning for a minor contingency does not include peacekeeping or shows of force, inasmuch as these do not involve hostilities.) An **engagement** is a combat encounter between two forces, whose size is neither larger than a division (q.v.) nor smaller than a company (q.v.), in which each has an assigned or perceived mission which begins when the attacking force initiates action to accomplish its mission. An engagement ends when: 1) the attacker has accomplished its mission; or 2)ceases to attempt to accomplish its mission; or 3) one or both forces receive significant reinforcements, thus initiating a new engagement. An engagement is often part of a battle (q.v.) between elements of the division-sized or larger forces. An engagement may last only a few hours, and is rarely longer than five days. An **action** is a minor combat encounter between two small hostile forces, neither larger than a battalion (q.v.) nor smaller than a squad (q.v.). An action can last from a few minutes to a few hours, but never lasts more than one day; it ends when one force or the other withdraws, or both forces terminate significant combat activity. An action is often part of an engagement or battle between elements of larger forces. **High-intensity conflict** is sustained armed hostilities between organized military forces, or significant components of those forces, involving frequent combat encounters in which both sides employ a wide variety of weapons and weapons systems in combined arms operations. **Low-intensity conflict** involves sporadic or limited armed hostilities between military forces, which may or may not be conventionally organized, in which there may be prolonged lulls between combat encounters, or in which the weapons or weapons systems of one or both sides are limited in numbers, or in types, or in sophistication. An **insurgency** involves intrastate hostilities, between forces supporting the government and opposition forces which can be either organized or irregular (guerrilla), or a combination of both. The intensity of an insurgency can range from that of a significant local war (as in the case of a civil war in a major power) to that of a minor hostility. A military manifestation of internal political unrest or turbulence within a state, it can be brief in duration, but more often, because of the conditions that cause political instability, tends to be prolonged over months or years.

Congreve rocket—A type of rocket developed by British colonel William Congreve at the beginning of the 19th century. The Congreve rocket, given thrust by black powder, resembled a skyrocket, with a stick 12 to 16 feet long, and weighed from 25 to 60 pounds. It was launched from a thin copper tube or trough, had a range of almost three miles, and carried a warhead of incendiary material or steel balls for antipersonnel use. Rocket corps soon appeared in European armies. Such corps were used by the British in the War of 1812, and it was the "red glare" of Congreve rockets to which Francis Scott Key referred in "The Star Spangled Banner."

conical bullet—See **ball, minié**.

conoidal bullet—See **ball, minié**.

conquer—1) To defeat, particularly a large force or an entire nation, in battle. 2) To gain control by force of arms of territory belonging to another nation.

conquest—1) Conquering. 2) The gaining of territory through warfare.

conquistadores—Spanish for "conquerors." Used in English for the Spanish soldiers who conquered Mexico and western South America in the 16th century.

conscientious objector—One who, for reasons of conscience, including religious and moral principles, refuses to enter military service, or (if in service) refuses to enter combat.

conscript—n. One who is drafted for military service; a draftee. v. To cause compulsory enrollment of a person in the military service.

conscription—Involuntary drafting of certain sectors of a population into a nation's armed forces; the draft. Called in the United States **selective service** and administered at various times in the 20th century under a **Selective Service Act**. In the French Revolutionary Wars the term was first used for the enlistment of men for the national armies of France. Conscription was used by the major participants in the Franco-Prussian War of 1870 and in the two world wars.

consolidate—To organize or strengthen a force or a position, particularly a position newly gained in combat.

constable—In Western European monarchies during the Middle Ages, a high ranking officer. In the absence of the king, the constable generally served as military commander.

constitute—To provide the legal authority for the existence of a new unit of the armed services. The new unit is designated and listed but it has no specific existence until it is activated.

construction battalion—Units organized by the US Navy in World War II to construct bases on islands recaptured in the advance toward Japan. Known as "Seabees" after the initials, C.B., the units contained engineers, construction men, technicians, and other specialists, many of whom were over draft age.

consul—The title of the two highest magistrates of the Roman Republic, in whom were vested the executive and military authority of the state. Two consuls were elected each year. They shared responsibility for combat operations, usually exercising command on alternate days. In Imperial Rome the office became an honorary one.

consular army—See **army**.

contain—To engage an enemy in combat of sufficient intensity to prevent withdrawal of any part of that force for use elsewhere.

contamination—The deposit and/or absorption of radioactive material, biological or chemical agents on and near structures, areas, personnel, or objectives.

Continental Army—The army created in June 1775 by the Continental Congress to represent all the American colonies in the rebellion against Great Britain. In contrast to the colonial militias, this new organization was to be a standing army. George Washington was placed in command.

Continental Navy—The navy established by the Continental Congress in November 1775.

contingency—See **conflict levels of intensity**.

contingent—A quota or force, as men or bombers.

contingent effects—The effects, both desirable and undesirable, which are in addition to the primary effects associated with a nuclear detonation.

contingent zone of fire—A zone within the area of fire of an artillery unit other than the normal target zone, upon which the unit may have to fire under certain contingencies.

continuous fire—See **fire, continuous**.

contravallation—Defensive works erected around a fortification or walled city, to isolate the besieged installation and to protect the besiegers against sallies by the besieged. See also **circumvallation**.

control—Authority exercised by a commander over the activities of subordinate organizations.

control point—1) A position along a route of march where information and instructions for the regulation of supply and traffic are issued. 2) A position marked by a buoy, boat, aircraft, electronic device, conspicuous terrain feature, or other identifying object which is given a name or number and used as an aid in navigation of ships, boats, aircraft, or vehicles.

controlled effects nuclear weapons— Nuclear weapons designed to achieve variation in the intensity of specific effects other than normal blast effect.

convalescent—A military individual recovering from wounds or illness and not yet restored to full duty status.

convention—1) An agreement between two or more nations. 2) An agreement between the commanders of two opposing forces in relation to limitations on military operations.

conventional—1) Nonnuclear. This use of the term has become general, referring to all aspects of warfare that do not involve the use of nuclear weapons (although chemical and biological weapons may also be considered unconventional). 2) Noninsurgency related. Used to distinguish operations or forces of a normal combat nature, as opposed to operations or forces of an insurgent or in an insurgency (q.v.).

converged sheaf of fire—See **sheaf of fire**.

convergent lines of operations—Lines originating at different points and converging on a common objective area so that forces following these lines arrive within supporting distance of each other in the battle area.

convoy—1) One or more merchant vessels or naval auxiliaries escorted by warships and sometimes by aircraft for protection against surface and subsurface attack. 2) A group of military vehicles traveling together for the purpose of control and orderly movement, with or without escort.

convoy escort—1) The warships or combat aircraft traveling in or with a convoy to protect noncombat ships. 2) Armed vehicles accompanying a convoy of vehicles.

convoy loading—The loading of troop units with their equipment and supplies in ships of the same movement group, but not necessarily in the same ship.

coordinates—1) Lines drawn on a map to form a grid by which to locate areas or positions. Coordinates may be related to degrees or minutes of latitude and longitude. They are usually evenly spaced, and cross at right angles. 2) The grid values for a specific point on a map.

coordinating point—Designated point at which, in all types of combat, adjacent units or formations must make contact for purposes of control and coordination.

coordination—1) The execution of plans and combat actions of several elements of a command so that their interrelation brings the strongest possible advantage to the command as a whole. 2) Planning for such execution.

coordination of fire—See **fire, coordination of**.

cordon—A line of soldiers, ships, strongpoints, or other elements placed around an area to deny access to it.

cordon defense—The placement of forces along a defensive line with little or no depth and with insufficient forces available for reserve. See also **linear defense** and **defense**.

corned gunpower—A granulated mixture of sulfur, saltpeter, and charcoal invented in 1420. When the three elements were held in a steady relationship with an equal distribution of airspaces, corned gunpowder exploded uniformly and nearly instantaneously.

cornet—1) A British cavalry unit commanded by a captain. 2) The rank of the officer in a cavalry troop who carries the standard, also called a cornet.

corporal—1) In nonnaval military forces a noncommissioned rating above private first class and below sergeant. The corporal is the lowest noncommissioned officer. 2) In the British Royal Navy a petty officer who is assistant to the master-at-arms.

corps—1) A specialized branch of one of the armed forces of a nation, like the US Marine Corps, the Air Corps, Nurse Corps, or Supply Corps. The students at the US Military Academy are known as the Corps of Cadets. 2) Often used for army corps (q.v.).

corps, army—A tactical unit larger than a division (q.v.) and smaller than a field army (q.v.), usually consisting of two or more divisions together with auxiliary arms and services.

corps, elite—See **elite corps**.

corps reserve—See **reserve**.

corps troops—Troops assigned or attached to an army corps, but not a part of one of the divisions that make up the corps.

corpsman—An enlisted man (in the navy, usually with the rating of pharmacist's mate) who is trained to give first aid and some nursing or medical treatment.

correlation of forces—A Soviet concept of calculating the relative combat power of two opposing forces as a basis for military decision making. It is comparable to the American concept of estimate of the situation (q.v.), but is more comprehensive and entails a quantitative assessment of all considerations that could affect a combat outcome.

corridor—1) An opening in enemy-controlled territory through which forces may pass in reasonable safety. 2) An **air corridor** is a flight passage for aircraft, especially a passage either leading through a system of antiaircraft defenses or established by international agreement. 3) A **terrain corridor** is an area like a valley or an isthmus, which permits relatively unhampered movement between terrain features that inhibit movement.

Corsair—A World War II fighter airplane built by Chance Vought. It was flown by the US Navy and had the official designation F4U.

corselet—Armor that protects the upper torso.

corvette—1) Originally a sailing ship smaller than a frigate (q.v.), usually carrying guns on only one deck. 2) In modern navies a ship smaller than a destroyer (q.v.), lightly armed and fast. It is used primarily for antisubmarine missions.

corvus—A combined grappling device and gangway used on Roman war galleys. It was a narrow bridge some 18 feet long, mounted on a pivot near the bow, with a hook or a pointed iron stud on its outboard end to hold the structure to the deck of an enemy vessel. When not in use it was held in a near-vertical position by ropes and pulleys, and could be swung outward in any direction. When an enemy ship drew near, the corvus was let down with a crash on its deck, where the gangway was held fast by the iron stud. Immediately a boarding party stormed across the corvus onto the foe's deck.

cost effectiveness—A comparative evaluation derived from analyses of alternatives (actions, methods, approaches, equipment, weapon systems, support systems, force combinations, etc.) in terms of the interrelated factors of cost and effectiveness in accomplishing a specific mission.

council of war—A meeting of the senior officers of a military force in combat, called by the commander, usually in an emergency, to seek a majority opinion on tactical or strategic questions. Such a council was common in 18th-century warfare, but has been little used in modern times.

countdown—The step-by-step process leading to initiation of an action or an operation. One modern use is in missile testing, launching, and firing. A countdown is performed in accordance with a predesignated time schedule.

counter—To make a military move intended to offset a hostile action.

counter air—1) Weapons, tactics, or operations designed to be used against aircraft or hostile air operations. 2) Air attacks, operations, or missions directed against an enemy's air force in an effort to achieve control of the air.

counterapproaches—See **approaches**.

counterattack—Attack by a part or all of a defending force against an enemy attacking force, for such specific purposes as regaining ground lost or cutting off or destroying enemy advance units, and with the general objective of denying to the enemy the attainment of his attack objectives. In sustained defensive operations it is undertaken to restore the battle position and is directed at limited objectives.

counter-battery fire—Fire intended to destroy or neutralize enemy artillery weapons. Sometimes called **counterbombardment**.

counterbombardment—See **counter battery fire**.

counterespionage—A category of counterintelligence (q.v.), the objective of which is the detection and neutralization of hostile espionage (q.v.).

counterfire—Fire intended to destroy or neutralize enemy weapons. Includes counter-battery fire, counterbombardment, and countermortar fire.

counterforce—Employment of strategic nuclear weapons (usually in a first strike) against hostile strategic nuclear weapons to prevent or preempt retaliation.

counterinsurgency—Those military, paramilitary, political, economic, psychological, and civic actions taken by a government to defeat subversive insurgency (q.v.).

counterintelligence—The aspect of intelligence activity that attempts to destroy the effectiveness of hostile intelligence activities, to protect information against espionage and to prevent subversion against individuals and sabotage against installations or material. The character of counterintelligence activity can be both offensive and defensive. In its defensive mode it addresses the questions, "What does hostile intelligence activity seek to accomplish?" and "How are hostile espionage operations conducted?" In its offensive mode it undertakes operations aimed at thwarting the endeavors of hostile espionage agents.

countermand—To issue an order canceling or reversing a previous order.

countermarch—*n.* A march back toward the point of origin. *v.* To march back in the direction from which one has come.

countermeasure—A device or technique that operates to impair the operational effectiveness of enemy activity. Countermeasures may include such diverse operations as tacti-

cal maneuvers, electronic jamming procedures, and defensive field fortifications.

countermine—*n.* 1) A mine or tunnel dug to intercept and destroy a mine being dug by an opposing force. 2) An explosive device designed to cause an enemy's mines to detonate. *v.* 1) To dig a countermine. 2) To place a countermining device.

counteroffensive—A large-scale offensive undertaken by a defending force to seize the initiative from the attacking force.

counterpreparation fire—Intensive prearranged fire delivered when it is discovered or assumed than an enemy attack is imminent. It is designed to break up enemy formations, disorganize the enemy's system of command, communications, and observation, decrease the effectiveness of his artillery preparation, and impair his offensive spirit.

counterscarp—The side of a ditch, moat, or trench that faces the enemy, or the expected direction of attack.

countersign—The response to a secret sign or challenge given by a sentry. A password.

coup de main—A sudden, vigorous, and successful attack to seize a hostile position or defended locality.

coup d'état—Vigorous action, usually violent and usually military or involving the use of military force, to overthrow a legal government and replace it with a new governing authority.

coup d'oeil—The capability of a military commander to make a rapid and accurate estimate of a combat situation. See also **estimate of the situation**.

courier—A messenger responsible for the secure physical transmission and delivery of documents, messages, or materials.

course of action—1) Any sequence of acts that an individual or a unit may follow. 2) A possible plan open to an individual or commander that would accomplish or is related to the accomplishment of his mission. 3) The scheme adopted to accomplish a job or mission. 4) A line of conduct in an engagement.

court martial—1) A body of military personnel designated by the proper authority as a court to conduct the military trial of persons subject to military law. 2) Trial by such a court. There are three types of court martial, distinguished by the authority empowered to convene them and the punishments they can impose: **general court martial**, **special court martial**, and **summary court martial**. The **general court martial** is the highest type of court martial in the US armed services. It consists of not fewer than five members not including the law officer, and has the power to try any offense punishable by the Uniform Code of Military Justice. A **special court martial** consists of a law officer and at least three members and has jurisdiction over the trial for any crime or offense that is not capital but is punishable under the Uniform Code of Military Justice. Composed of one officer, a **summary court martial** may impose punishment of confinement for up to one month, hard labor without confinement for up to 45 days, restriction for up to two months, and forfeiture of up to two-thirds of one month's pay.

cover—n. Protection or shelter, whether natural or created by military action, from hostile fire. v. 1) To provide protection for a military force through action by land, air, or naval forces, such as artillery fire or close air support. 2) To maintain a continuous radio receiver watch.

covered way—A protected line of communication, such as a communications trench, or a concrete-walled trench or tunnel in a permanent or armored fortification system.

covering fire—1) Fire to suppress hostile weapons, used to protect friendly troops when they are within range of enemy small arms. 2) In amphibious usage, fire delivered prior to the landing to cover preparatory operations such as underwater demolitions or minesweeping.

covering force—1) A force operating apart from the main force for the purpose of intercepting, engaging, delaying, disorganizing, or deceiving the enemy before he can attack the main force. 2) Any body or detachment of troops that provides security for a larger force by observation, reconnaissance, attack, or defense, or by any combination of these methods.

covert operations—Operations so planned and executed as to conceal the identity of, or permit plausible denial by, the sponsor. They differ from clandestine operations in that emphasis is placed on concealment of identity of sponsor rather than on concealment of the operation.

coxswain—A naval enlisted rating corresponding to boatswain's mate third class.

CP—Command post (q.v.).

CPX—See **command post exercise**.

crater—The pit, depression, or cavity formed in the surface of the earth by an explosion. It may range from saucer-shaped to conical, depending largely on the depth of burst. In the case of a deep underground blast, when no rupture of the surface may occur, the cavity is termed a camouflet (q.v.).

C-ration—A US Army meal consisting of several prepared items, usually canned, issued when troops are out of range of field kitchens.

crease—See **kris**.

crenel—See **battlements**.

crest, military—1) A line on the forward slope of a hill or ridge from which maximum observation of the terrain in front can be obtained without exposure of a standing observer by being outlined against the geographic crest. 2) An emblem, such as a coat of arms, on the hat or helmet of a uniformed or armored soldier.

crew—1) All officers and men assigned to a naval vessel or an aircraft. Sometimes the term refers only to enlisted men. 2) A detail or team manning or serving a piece of equipment like an artillery piece, or performing specific duties.

crew-served—Manned or operated by two or more soldiers, as an artillery piece.

critique—A critical discussion of a mission just performed, emphasizing mistakes and successes.

crocodile—A World War II Allied tank with a flame gun in place of the forward machine gun. The tank towed a two-wheeled armored trailer that contained the flame fuel. Approximately 100 one-second shots of flame were possible before replenishment. Maximum range was 100 yards.

cross loading—A system of loading troops so that different units may be disembarked or dropped at two or more landing or drop zones, thereby assuring unit integrity upon delivery.

cross servicing—Any function performed by one military service in support of another military service for which reimbursement is required from the service receiving support.

crossbow—A weapon consisting of a bow permanently fixed transversely on a stock, with various devices for drawing back and releasing the cord or string, and a groove which holds and guides an arrow or bolt. The crossbow is known to have been used in China before the Christian era. The oldest surviving examples in the West are from France, dating to the first or second century A.D. Crossbows were widely used in the Middle Ages, from the 10th century on. Their use was prohibited against Christians (but not against infidels) by Pope Innocent II in the 12th century, but it continued in use well after the introduction of gunpowder, and is still used for hunting. The medieval crossbow had a range of up to 300 yards. The typical arrow was the **quarrel**, a short, heavy arrow with a quadrangular pointed tip. In metallic form (usually iron), the quarrel was called a **bolt**. The stabilizing fletchings were generally wood, leather, or brass. Fired from a crossbow, it could pierce even fairly heavy armor. See also **bow** and **longbow**.

crossbowman—A soldier armed with a crossbow.

crossfire—Fire from two or more weapons (which may be firing against each other) that crosses in transit to the target or targets.

crossing the T—A naval maneuver from the days of sailing ships in which a fleet sailing in line ahead (single file) crossed at right angles in front of another fleet. The fleet executing the maneuver could bring the full broadside fire of all its ships to bear on the enemy, whereas he could fire only the bow guns of his lead ships. It was a difficult maneuver to accomplish because the fleet under attack could turn away out of range before

the attacking fleet crossed the T. The maneuver figured in several modern naval battles between steam-driven battleships, including Tsushima (1905), Jutland (1916) and Leyte Gulf (1944).

crowfoot—See **caltrop**.

cruise missile—See **missile, cruise**.

cruiser—1) A fast warship, larger and heavier and more heavily armed than a destroyer, but smaller, lighter, and with less firepower and armor than a battleship. 2) A privateer of the 18th century. Modern cruisers are generally classed as light or heavy. A **light cruiser** in the US Navy is a warship with six-inch naval guns and/or surface-to-air face missiles as the main battery. It is designed to operate with strike, antisubmarine, or amphibious forces against air and surface threats. Full-load displacement is approximately 18,000 tons. Designated by the US Navy as CL. A **heavy cruiser** is a warship designed to operate with strike, antisubmarine warfare, or amphibious forces against air and surface threats. Main battery consists of eight-inch guns, and full load displacement is approximately 21,000 tons. Designated by the US Navy as CA.

cruiser, battle—See **battle cruiser**.

crusade—1) Any of several military expeditions launched from the 11th through the 14th centuries with the objective of recovering the Holy Land from the Moslems. Crusades were undertaken by European Christians with the sanction of the several popes. 2) A European Christian military expedition, during the same general period, directed at pagans, infidels, or heretics outside of the Holy Land. Examples were the operations of the Teutonic Knights against the Slavs in Prussia and Lithuania (13th and 14th centuries); Albigensian Crusade (1203–1226); Crusade of Nicopolis (1396).

crusaders—Warriors who were taking part in a crusade.

cryptanalysis—The deciphering of encoded or encrypted messages, particularly those of an enemy or another nation, and translating them into plain language.

cryptologic technician—A warrant officer specialist in the US Navy.

cub—1) In World War II, an advanced base unit consisting of all the personnel and material necessary for establishing a medium-sized advanced fuel and supply base. 2) Military slang in World War II for light artillery observation aircraft.

cuirass—Armor for protecting the upper torso. Normally the term refers to a piece that covers both front and back, but it is sometimes used for the breastplate alone.

cuirassier—A European mounted soldier equipped with a cuirass. Most cuirassiers were armed with a heavy saber; cuirassiers armed with lances were usually designated as heavy lancers.

cuisse—A thigh-protecting piece of plate armor, developed in the 14th century.

culverin—1) An early musket. 2) A long-barreled, thick-walled artillery piece of the 16th and 17th centuries, designed to fire accurately at long range. About 11 feet long and weighing about 4,800 pounds, with a bore 5½ inches in diameter, the culverin fired an 18-pound projectile a maximum of about 6,700 yards. Its point blank range was 1,700 yards.

cupola—A fixed fortress gun emplacement with overhead armored cover, like a naval turret, usually with 360-degree traverse. The Belgian engineer, Henri Alexis Brialmont, devised a **disappearing cupola** so that the entire gun with its protective casing could be withdrawn. The cupola traverses with the gun, which may be run out through an opening and withdrawn when not in use.

curtain—The portion of a rampart (q.v.), or parapet (q.v.), or wall between two bastions or gates.

curtain of fire—See **barrage**.

cutlass—A short saber or sword with a slightly curved single-edged, broad blade, especially used by sailors, who found it easy to carry while boarding an enemy vessel and effective in fighting at close quarters.

cutter—1) A small, single-masted sailing vessel. 2) A ship's boat, next in size to the launch, used for general ship duty. 3) A small, lightly armed vessel used by the US Coast Guard for patrolling and other duties.

cyclic rate—Rate of fire of an automatic weapon, usually in rounds per minute.

D

dagger—A thrusting hand weapon, with a short, usually double-edged, blade. The length varies greatly, and there is little distinction between a long dagger and a short sword. Modern equivalents of the military dagger are the bayonet and the trench-knife (*qq.v.*).

Dahlgren gun—A cast-iron cannon invented by US naval officer John A. Dahlgren in the mid-19th century, and used by the US Navy and the US Army in coast defense installations. A smoothbore (*q.v.*) gun, it derived its nickname, "Soda Water Bottle," from its shape.

"daisy cutter"—A 15,000-pound bomb used in the Vietnam War. Set to explode about seven feet above the ground, it could clear jungle areas the size of a baseball park.

damage assessment—1) Determination of the effect of attacks on targets. 2) A determination of the effect of a compromise of classified information on the national security.

damage control—In naval usage, measures necessary aboard ships to deal with damage caused by accident or hostile action. The objective is to preserve and re-establish watertight integrity, stability, maneuverability, and offensive power; to control list and trim; to effect rapid repairs of matériel; to limit the spread of, and provide adequate protection from, fire; to limit the spread of, remove the contamination by, and provide adequate protection from, toxic and radiological agents; and to provide for care of wounded personnel.

Damascus steel—Steel of great tensile strength and flexibility that was originally produced in Damascus and highly prized for sword blades. It was first imported to Europe from the Middle East around the 8th century

A.D. It is sometimes called damask steel. Similar sword blade steel was also produced in Toledo, Spain, following the Arab conquest of Iberia.

dart—A small missile with a sharp point and often with tail fins. Darts may be thrown by hand, or fired with a blowgun, crossbow, or other device.

date of rank—The date from which an officer or noncommissioned officer (NCO) holds a given rank. This is the basis for determining seniority among officers or NCOs holding the same rank.

date-time group—An item in military messages; the date and time are expressed in six digits followed by the time zone identification letter, used to identify the time at which a message was prepared for transmission through military communication channels. An example is 011210Z; the first pair of digits denotes the day, the second pair the hour, and the third pair the minutes; the letter is the international time zone suffix.

day of supply—The quantity of a given supply commodity adopted as a standard of measurement of expected consumption of that commodity by a unit; used for planning purposes in estimating the average daily expenditure under stated conditions. It may also be expressed in terms of a factor, for example, rounds of ammunition per weapon per day. For ammunition it is sometimes called a "day of fire."

day room—A room in barracks set aside for the use of soldiers or airmen during their leisure.

dazzle—Temporary loss of vision, or a temporary reduction in visual acuity.

D-day—In operation planning, the day on which an operation or an action is to begin. The best known D-day is 6 June 1944, the beginning of the Allied invasion of northern France in World War II.

dead angle—An angle from which an air gunner is unable to direct fire. This may be due to equipment failure or to a blind spot.

dead space—1) An area within the maximum range of a weapon, radar, or observer that cannot be covered by fire or observed from a particular position, usually because of the intervention of a barrier to fire or observation (as a hill mass), or the technical characteristics (usually ballistic) of a weapon. 2) An area or zone that is within range of a radio transmitter, but in which a signal is not received. 3) The volume of space above and around a gun or guided missile system into which it cannot fire because of mechanical or electronic limitations.

deadline—v. To remove a vehicle or piece of equipment from operation or use for one of the following reasons: a) that it is inoperative due to damage or malfunctioning, b) that it is unsafe, or c) that it would be damaged by further use. (Note: The term may be used in the sense, the "vehicle is deadlined.") n. The date or hour at which an action or undertaking must be completed.

debouch—To pass into an open area after marching through a narrow passage or defile.

decamp—To break camp and move on.

deception—Measures designed to mislead the enemy by manipulation, distortion, or falsification of evidence in order to induce him to react in a manner prejudicial to his interests.

decimation—In ancient Roman armies a punishment sometimes inflicted for mutiny or other serious misdeeds. It consisted of selecting every tenth man by lot and executing him. The term is frequently misused to describe heavy losses.

deck court—A disciplinary body in the US Navy for the trial of offenses by enlisted men considered too serious for decision at a captain's mast (q.v.), but not serious enough for a summary court martial (q.v.). The deck court was abolished when the Uniform Code of Military Justice went into effect in 1951.

decoration, military—Tangible recognition of extraordinary achievement. Decorations are awarded for extraordinary feats in battle or in noncombat activities and for outstanding accomplishment in line of duty. (Decorations should not be confused with service medals, which testify to participation in certain wars or battles.) In the United States most decorations include a medal on a ribbon of specific design, and a separate ribbon of the same design on a pin. On the ordinary uniform the ribbon alone is worn. If the wearer is entitled to wear more than one, the highest ranking one is placed at the right of the top row, with others following in sequence of rank of decoration, followed by service medals or ribbons in order of chronological award. Miniature medals and ribbons are worn with evening dress. Only with the full dress uniform is the full-size decoration worn. A few decorations, among them the highest military decoration in the United States, the Congressional Medal of Honor, are worn with full dress on ribbons around the neck. See also **service medal**.

decoy ship—See **Q-boat**.

decrypt—To convert encrypted text into its equivalent plain text by means of a cryptosystem.

decurion—1) In the early Roman Army, the officer in command of ten cavalrymen. 2) In Imperial Rome the officer in charge of a turma (q.v.) of 30 to 40 cavalrymen.

deep attack—Application of force or firepower deep behind enemy lines by means of air attacks, or long-range surface fire, or infiltration, or airborne operation. See also **follow-on forces attack**.

deep penetration—The action of a military force in advancing a considerable distance into enemy-held territory, either on the ground or in the air on a military mission. A force making a deep penetration may or may not maintain an overland line of communications with the main force from which the penetration was initiated. When overland communications are not maintained, the operation is a kind of raid (q.v.). Examples are Grierson's Raid of the US Civil War, and the long range penetration (q.v.) operations of the

Chindits (q.v.) In Burma during World War II.

deep-penetration bomb—A bomb so designed and fuzed as to penetrate a significant distance into a target before exploding.

defeat—Failure in combat, including one or a combination of the following conditions: inability to accomplish an assigned mission; suffering excessive casualties; loss of important terrain or resources. The consequences of defeat are increasingly serious at higher levels of combat. Thus, defeat in an engagement need not necessarily affect the outcome of a conflict, but defeat in a war could cause the collapse of a national government and irreparably damage a nation's social, political, and economic systems.

defeat in detail—n. The overcoming of the various parts or elements of a force one at a time, particularly when the parts are not close together. v. To overcome the separate parts of a force in turn.

defend—v. The process of defense (q.v.).

defender—One who defends.

defense—1) Resistance to attack; the opposite of offense. 2) Structures, tactics, or maneuvers whose purpose is to protect against attack. 3) Barring the way to the enemy or breaking his attack by the use of fire while taking advantage of the terrain in order to stop or delay him. The most important kinds of defense are: **active air defense**—Direct defensive action taken to destroy attacking enemy aircraft or missiles, or to nullify or reduce the effectiveness of such attack. **active defense**—The employment of limited offensive action and counterattacks to deny a contested area or position to the enemy. **area defense**—Defense of a general area, in contradistinction to defense of critically important points, as in "point defense." **cordon defense**—The siting of forces over the front line that have little or no depth and have insufficient reserve forces on hand. **deliberate defense**—A defense normally organized when out of contact with the enemy or when contact with the enemy is not imminent and there is time for organization. It normally includes an extensive fortified zone incorporating pillboxes, forts, and communications systems. **defense in depth**—The siting of mutually supporting defense positions in a zone extending back from the front line of contact with the enemy. It is designed to absorb and weaken progressively the force of attack, to prevent immediate observation of the defender's whole position by the enemy, and to allow the commander to maneuver his reserve. **fortified defense**—A comprehensive, coordinated defense system prepared by a defender having sufficient time to complete planned entrenchments, field fortifications, and obstacles in such a manner as to permit the most effective possible employment of defensive firepower. **hasty defense**—A defense organized while in contact with the enemy or when contact is imminent and there is little time to execute the organization. It is characterized by improvement of the natural defensive strength of the terrain by use of foxholes, emplacements, and obstacles; if occupied for a protracted period the hasty defense position can be improved to the status of prepared or fortified defense. **mobile defense**—That defense in which minimum forces are deployed forward to find out when an attack is impending and warn the main forces of it, to canalize the attacking forces into less favorable terrain, and impede and harass them so as to cause their disorganization. The preponderance of the combat power of the main, defending force is held conveniently in the rear and employed in vigorous offensive action to destroy the enemy at a decisive time and place. The mobile defense is a special case of the defense in depth. **passive defense**—Nonviolent or noncombat means of defense, such as the use of camouflage or the burying of wire telephone lines to protect them from hostile fire. **prepared defense**—A defense system prepared by a defender who has had sufficient time to organize the defensive position, but (due to lack of time or resources) with less than the strength of a fortified position.

defense area—See **area**.

defense attaché—See **attaché, military**.

defense classification—See **security classification**.

defense in depth—See **defense**.

defensive-offensive—A planned operation in which a military force in defensive posture is prepared to assume the offensive, in a counterattack or counteroffensive. The occasion for such a change in posture might be anticipated fatigue on the part of the attacker, the exposed position of his forces or lines of communication, or the disruption of his offensive by an ambush.

defensive operations—Operations to conduct defense (q.v.).

defensive position—1) Any area occupied and organized for defense. 2) A battle position. 3) A system of mutually supporting defensive areas or tactical localities of varying size, each having a definite assignment of troops and a mission. A defensive position may be fortified and otherwise organized to the extent that the situation requires or time allows.

defilade—n. Protection from hostile observation and fire provided by an obstacle such as a hill, ridge, or bank. v. To shield from enemy fire or observation by using natural or artificial obstacles.

defile—n. 1) A natural or artificial terrain feature that canalizes movement. Defiles frequently occur in mountain passes, woods, jungles, towns, river crossings, lake regions, and swamp areas. 2) A march in single file or a column of files. v. To march in single file or a column of files.

deflection—1) The turning aside of a moving object from its course by reason of an outside force being exerted upon it, for example, the force of a crosswind on the trajectory of an artillery shell or a falling bomb. 2) A setting on a bombsight that allows for deflection and permits sighting on the target or on an offset of the target. 3) In gunnery, the horizontal angle between the aiming point and the target. 4) The lead given a projectile from a weapon aimed at a moving target. 5) The direction in which a weapon is fired or is firing.

deflection error—The angle or distance measured between an imaginary line from the gun, drawn through the target or expected point of impact and an imaginary line drawn through the actual point of impact.

deflection pattern—Naval equivalent of a sheaf of fire (q.v.).

defoliate—To remove, or cause the removal of, the leaves of a tree or other plant, especially by the use of a chemical or biological spray or dust, known as a **defoliant**. The process and the result are **defoliation**.

degauss—To surround a ship with an electric coil through which a current can be passed that effectively neutralizes the ship's induced magnetic field and prevents it from detonating a magnetic mine. Degaussing was highly effective in reducing losses to mines in World War II. The name derives from the unit used to measure magnetic induction, the gauss, which was named for Carl Friedrich Gauss, the German scientist who developed it. See also **deperm**.

degradation—A reduction in strength, rank, value, or capability.

delayed action fuze—See **fuze**.

delaying action—A retrograde movement (q.v.) in which the defender inflicts maximum damage on an advancing enemy and delays him in order to gain time for his own withdrawal, but without becoming decisively engaged in combat or being outflanked. Also known simply as **delay**.

delaying operation—A tactic which involves a force under pressure trading space for time by slowing down the enemy's momentum by forcing him to deploy and maneuver, and inflicting maximum damage on him without becoming decisively engaged. The operation is usually undertaken by covering forces and other security attachments, and it is executed most effectively by highly mobile troops. Obstacles covered by fire are generally used to reinforce the delaying capability. Delaying operations are usually marked by deliberate, sequential withdrawal of the defender from one positon to another, defending each successive position only long enough to force the attacker to deploy and prepare his forces for a major attack, and then withdrawing to the next position before the defending forces are pinned down. A classic example was the German delay of Allied offensives in Italy in 1943 and 1944 in World War II.

deliberate attack—See **attack**.

deliberate defense—See **defense**.

Delta—Letter of NATO phonetic alphabet. See also **alphabet, phonetic**.

demibrigade—A military unit devised in

1793 by General Lazare Carnot for the French Army of the Revolutionary Period (1792–80), the demibrigade was a grouping of three battalions (q.v.), two of volunteers or conscripts and one of regular soldiers from the old Royal Army. The intention was to stiffen the untrained conscripts by the addition and example of professional soldiers. Ten years later demibrigades were designated as regiments, two of them being combined to form a brigade.

demiculverin—An artillery piece of the 16th and 17th centuries constructed to the same general design as a culverin (q.v.). Its length was eight and one-half feet, its weight 3,400 pounds, with a four-and-one-fifth-inch bore, a point blank range of 850 yards, and a maximum range of 5,000 yards. Its projectile weighed ten pounds.

demilitarized zone (DMZ)—A defined area in which the stationing or concentrating of military forces, or the retention or establishment of military installations of any description, is prohibited.

demilune—A crescent-shaped defensive structure, usually built at the entrance to a fort. The name, meaning "half moon," came to be applied to any structure designed to defend an entrance.

demisaker—See **minion**.

demobilization—The reversion of a person or a force from military to civilian status. Demobilization may also be used to mean the reversion of all sectors of a nation's economy from a wartime to a peacetime basis.

demolition—The destruction of structures, facilities, or material by use of fire, water, explosives, mechanical, or other means.

demonstration—1) An attack or show of force on a front where a decision is not sought, made with the aim of deceiving the enemy. 2) In an amphibious operation, an exhibition of force that may be a feint or a minor attack.

demotion—Reduction in rank or rating.

denial operation—Action to hinder or deny an enemy the use of space, personnel, or facilities. It may include destruction, removal, contamination, or erection of obstructions.

dependent—A person, usually a spouse, parent, or child, recognized by military regulations as having a claim upon an officer or an enlisted man or woman for support. The term is often used as an adjective, as in dependent housing, dependent travel.

deperm—Action to reduce the permanent magnetic field of a ship. Deperming is done during construction and before installation of degaussing gear. See also **degauss**.

deploy—1) In a strategic or tactical sense, to place forces in desired positions or areas of operations. 2) To change from an approach march, or a cruising approach, to a battle formation. 3) To extend the front of a unit, as in changing from column to line. 4) To extend a unit both laterally and in depth, or to increase intervals or distances, or both. The act of deploying is **deployment**.

depot—A facility for the receipt, classification, storage, accounting, issue, maintenance, procurement, manufacture, assembly, or salvage of supplies, or for the reception, processing, training, assignment, and forwarding of personnel replacements. It may be an installation or activity of the zone of the interior (q.v.) or area of operations.

depth bomb—See **bomb** and **depth charge**.

depth charge—Any charge designed for explosion under water, especially such a charge dropped or catapulted from a ship's deck. Also called a **depth bomb**.

deputy—An assistant or aide with authority to make decisions in the absence of the officer whose deputy he is.

desert—To leave one's post or ship without permission or orders, and without intention of returning. (Differs from absent without leave (q.v.) in terms of intent.) The person who so leaves is a **deserter**.

despatch— n. A naval message, usually briefly stated, particularly one sent by radio or other electronic means of communication. The US Army equivalent is spelled **dispatch.** v. To send a person to perform a task.

destroy—To damage a person, force, or object to the extent that it can no longer function in its normal or intended purpose and cannot be readily restored to its previous condition. See also **destruction** and **annihilate**.

destroyer—A fast warship smaller than a cruiser (q.v.), generally armed with three-inch or five-inch guns, surface-to-surface missles, torpedoes, and antisubmarine weapons. The destroyer is a versatile craft that can operate offensively—with strike forces, hunter-killer groups, and in support of amphibious assault operations—or defensively as escort for convoys and larger naval vessels. In order to perform these and other missions destroyers are built and armed in many different configurations. The US Naval designation for a destroyer is DD. The first destroyers, introduced in the 1890s to combat the threat of the motor torpedo boat, were called **torpedo boat destroyers**, hence the modern name.

destroyer-escort—A fast warship similar to but smaller than a destroyer. Like the destroyer, the destroyer-escort is versatile, but is particularly useful for convoy protection. The US Naval designation is DE.

destruction—1) The action of so damaging a person, force, or object, that it can no longer function in its normal or intended purpose, and cannot be readily restored to its previous condition. 2) The action considered as a military objective, or as a capability of a military force.

destruction fire—1) Fire delivered for the sole purpose of destroying objects, not enemy troops. 2) The field artillery technique for placing the center of impact of a gun on a target. See also **precision fire**.

detach—To separate a person or unit from the organization to which regularly assigned. In the US Army and Air Force this is a temporary separation. See also **detached duty**.

detached duty—A temporary duty assignment for a military individual.

detached unit—A unit serving away from the higher organization to which it is assigned.

detachment—1) A part of a military unit separated from its main organization for duty elsewhere. 2) A temporary military or naval unit formed from other units or parts of units. 3) The act of forming such a unit.

detail—1) n. A duty, normally noncombat, to which one or more troops or sailors are assigned, such as kitchen detail or clean-up detail. 2) The selection of people for such duty. 3) The people so selected. v. To assign to a specific duty.

detection—The act or action of discovering and locating that which is hidden or obscure, especially an enemy's equipment, installations, etc., or his electronic signals, as of radar or radio.

deter—To restrain an opponent, or potential opponent, from a course of action by displaying the harmful or disastrous consequences of retaliation that would be inflicted if he undertook that course of action. In a military sense, deter usually means to restrain from initiating a war or a major escalation of a war.

deterrence—The situation or action that deters an opponent or potential opponent from a course of action. Deterrence is used especially with respect to the restraints that prevent a nation from using nuclear weapons as a result of the fear of receiving a nuclear attack in response.

detonate—1) To cause the instantaneous combustion of an explosive material. 2) To explode.

detonator—A device that causes the explosion of the active ingredient in a round of ammunition. Basic to the development of effective detonators was an invention of Alfred Nobel, which used gunpowder or a fulminate to detonate a high explosive.

device—An insignia that identifies military personnel by service, unit, rank, specialty, experience, distinction, or the like—for instance, cap devices, shoulder patches, combat badges, etc.

DEW—Distant Early Warning line (q.v.).

diamond—A formation in which four units, or aircraft, proceed with one unit in advance, followed by a unit each to the left and the right, with the fourth unit further to the rear, directly behind the leader. When applied to aircraft, it is generally called a "diamond box."

diekplus—An ancient Greek naval maneuver in which a vessel, when about to ram an enemy galley, would suddenly back water on one side, causing its bow to sweep across and break the enemy's oars and his steering paddle, then turn forward again and proceed to ram.

diphosgene—A type of poison gas used in warfare.

direct fire—Gunfire delivered directly on a target that is visible to the person who aims the gun.

direct laying—The act of aligning the sight of a weapon with a target in direct fire (q.v.).

direct support—An action in which one force, especially artillery (**direct fire support**) or an air unit (**direct air support**), uses its weapons to provide assistance to another. An artillery unit in direct support of an infantry or armor maneuver unit is not usually attached to that unit (although it can be), but is expected to dedicate all of its fire support capability to the directly supported unit, unless given other orders by a higher commander. See also **general support**.

directing staff—In British usage, a group of officers who are members of the faculty of a military school or college or who are responsible for conducting and controlling a field exercise.

directive—Any oral or written communication that initiates, orders, or governs action, conduct, or procedure or that provides a plan of action in case of certain contingencies. Various kinds of orders and other publications when directive in nature are often called directives, including General Orders, Special Orders, and Court-Martial Orders.

director—A calculating machine or computer designed to convert commands, radar reports, and sensings into commands that will place fire accurately on a target. Usually used for naval and antiaircraft gun fire.

director fire—In naval gunfire and antiaircraft artillery fire the method of training and elevating the guns in a battery in accordance with signals transmitted by a director or master gun to the guns.

dirigible—A lighter-than-air craft with a rigid metal frame inside to strengthen and shape the gas bag. Dirigibles, called zeppelins (after their designer), were used by Germany in World War I to bomb England. The dramatic burning of the *Hindenburg* (1937), which was used for commercial transport, was one of several disasters which led to abandonment of the zeppelin as a military aircraft. See also **blimp**.

dirk—A short Scottish dagger, also used as a knife, carried by Scottish Highlanders and by British naval officers and midshipmen with nondress uniforms.

disappearing cupola—See **cupola**.

disappearing gun—An emplaced gun that emerges from cover and concealment, fires, and then, utilizing the force of recoil, is automatically or mechanically withdrawn from sight, usually behind a parapet. This was a typical coast-defense weapon of the early 20th century.

disarm—1) To remove or disassemble the detonating mechanism of an explosive device. 2) To relieve an armed person of all weapons.

disarmament—1) The voluntary or compulsory laying down of arms by a force or a nation. 2) The elimination of all weapons and war-making capability of a nation or group of nations by treaty or by compulsion. See also **arms control**.

disband—To disperse the members of an established unit or a force of any size and that had been formed for any purpose.

discharge—*v.* 1) To fire a weapon. 2) To release from active duty. *n.* 1) Release from active duty, and the document that proclaims it. In the US armed forces there are several types of discharge, including honorable discharge for one who has contributed to the good of the service; less than honorable; dishonorable; and various others, established by the Code of Military Justice.

discipline—*n.* The maintenance of order and esprit in a unit or formation by the adherence to rules and regulations of conduct. *v.* To impose punishment upon a person or a unit for an infraction of prescribed rules or regulations.

disengage—To break off combat with the enemy.

disinformation—An intelligence activity designed to mislead, deceive, or confuse an enemy by arranging for his intelligence collection activities to obtain false information.

dispatch—See **despatch**.

dispensary—A medical or dental facility for treating minor ailments and giving preliminary diagnoses. A dispensary may have some beds for patients not sick enough to be transported to a hospital, especially if there is no service hospital close by.

dispersion—1) The spreading or separating of troops, matériel, establishments, or activities usually concentrated in limited areas, with the purpose of reducing vulnerability to enemy action. 2) A scattered pattern of hits by bombs dropped under identical conditions or by projectiles fired from the same gun or group of guns with the same firing data. 3) In antiaircraft gunnery, the scattering of shots in range and deflection around the mean point of impact. As used in flak (q.v.) analysis, the term includes scattering due to all causes, with the mean point of impact assumed to be the target.

dispersion, cone of—The cone outlined by the flight paths of projectiles fired from a fixed installation.

dispersion error—The distance from the point of impact or burst of a round to the mean point of impact or burst. See also **probable error**.

displacement—1) In air intercept (by aircraft, not missiles), separation between target and interceptor tracks—that is, the path taken by an interceptor en route to target—established to position the interceptor so that sufficient maneuvering and acquisition space is provided. 2) The movement of artillery from one position to another.

disposition—1) Distribution of elements of a command within an area. 2) A prescribed arrangement of the stations to be occupied by the several formations and single ships of a fleet (or major subdivisions of a fleet) for any purpose, such as cruising approach, maintaining contact, or battle. 3) A prescribed arrangement of all tactical units composing a flight or group of aircraft. 4) The removal of a patient from a medical treatment facility by reason of return to duty, transfer to another treatment facility, death, or other termination of medical care. 5) The measures necessary to complete an assigned mission or staff action.

Distant Early Warning line (DEW)—A line of radar stations close to the 70th parallel, financed by the US government and operated in cooperation with the Canadian government for the purpose of detecting hostile aircraft or missiles en route to targets on the North American continent.

distributing point—A place or point at which supplies are distributed to troops in the field and to combat trains. See also **dump**.

distribution—1) The arrangement of troops for any purpose, such as a battle, march, or maneuver. 2) A planned pattern of projectiles about a targeted point. 3) A planned spread of fire to cover a desired frontage or depth. 4) An official delivery of anything, such as orders or supplies. 5) The functional phase of military logistics that embraces the act of dispensing matériel, facilities, and services. 6) The process of assigning military personnel to activities, units, or billets. 7) The list of people or offices to which an order or document is distributed.

district, military—Any of the several districts established throughout the US for administering and coordinating military reserve activity (like that of the Air National Guard), each district's boundaries usually being coincident with, or within, those of the state or District of Columbia with which identified.

district, naval—An administrative organization, including all of the US Navy's shore stations and activities, within one of the geographic areas into which the United States is divided by the Navy for administrative control of such installations.

dive bomber—See **bomber**.

dive bombing—The bombing of a target from a diving airplane, especially an airplane diving steeply.

diversion—1) An action, such as an attack

(**diversionary attack**) or increased activity, intended to occupy the attention of an enemy in an area other than the one in which a primary operation is planned. 2) A change in plans, especially a change in a planned route of march.

divided fire—Fire by a single ship at two or more targets at the same time.

division—1) An administrative and tactical ground formation containing units of the various combat branches of an army and large enough to operate independently. A modern division usually consists of three brigades (q.v.) or regiments (q.v.), usually containing three battalions (q.v.) each. The division dates from 1759, when the Duc de Broglie introduced a permanent formation of mixed bodies of infantry and artillery. In 1794 Lazare Carnot, the French minister of war, introduced a division of infantry, cavalry, and artillery, and the French Army adopted the formation as a standard. The British did not form divisions until 1807, and they did not become a standard organization of the US Army until 1917 (although divisions had been established in previous wars). 2) In the US Navy a tactical unit consisting of two or more vessels of the same type or similar types. 3) In the US Air Force, a tactical unit of two or more combat wings plus their service units. 4) An organizational part of an army headquarters that handles military matters of a particular nature, such as personnel, intelligence, plans and training, or supply and evacuation. 5) A number of personnel of a ship's complement grouped together for operational and administrative command.

division artillery—Artillery that is permanently an integral part of a division. For tactical purposes, all artillery placed under the command of a division commander is considered division artillery.

division slice—See **slice**.

division, square—A division (q.v.) in which the principal combat units are four infantry regiments or brigades. This was the standard division in most armies at the outset of World War I. By World War II most national armies had followed the German example and converted to triangular divisions. See also **division, triangular**.

division, triangular—The standard modern division, introduced by Germany in World War I, in which the principal combat maneuver units are three brigades or regiments.

divisional troops—Troops assigned to the headquarters staff of a division rather than to any of the combat or support components.

DMZ—See **demilitarized zone**.

doctrine—Fundamental principles and operational concepts by which the military forces or elements thereof guide their actions in military operations in support of national objectives.

Dog—Letter of old phonetic alphabet. See also **alphabet, phonetic**.

dog tag—Slang expression for one of two metal identification tags with the name, service number, blood type, religious preference, and tetanus record stamped on it, usually worn about the neck of a person in the military service.

dog watch—Naval term for one of the two-hour watches between 4 and 8 PM. Three watches are inserted in the usual four-hour watch schedule so that the starboard and port watches into which a ship's company is usually divided do not have duty at the same hours every night.

dogfight—An aerial battle, especially between a duel two fighter aircraft, involving considerable maneuvering and acrobatics.

doglock—An English variation of the Dutch snaphance gun. On the outside of the lock of the weapon, a dog catch, a device that caused ignition of the powder charge, was pivoted. It engaged the tail of the cock as a safety. The doglock and the snaphance were replaced by the flintlock (q.v.) in the second half of the 17th century. See also **snaphance lock**.

dolman—A jacket, usually elaborately decorated, often worn like a cape as part of a hussar's uniform.

donjon—French term for the keep, or fortified citadel, or main tower of a castle.

donrebusse—See **blunderbuss**.

Doodle Bug—Term used by the British for the German V-1 rocket in World War II.

dose rate—The rate at which a person receives nuclear radiation per unit of time, a measure of radiation intensity.

dosimetry—The measurement of radiation doses. It applies to both the devices used (dosimeters) and to the techniques.

double concentration—Two ships firing at the same target.

double envelopment—An attack directed around both of the enemy's flanks. The classic double envelopment was Hannibal's maneuver at the Battle of Cannae (216 B.C.).

double firing—A 16th-century method of firing explosive bombs or shells with wooden fuze plugs, in which the bomb or shell was inserted in the artillery piece with the fuze on the opposite side from the powder charge. The gunner lit the fuze with one hand and the touchhole of the gun or mortar with the other. This was a dangerous procedure that required precise timing, and sometimes resulted in disaster.

double redan—See **redan**.

double sap—See **sap**.

double shot—To fire two projectiles (balls, canister, grapeshot) at a time, instead of one, from a muzzle-loading piece.

double time—n. 1) A rate of march (while running) of 180 steps per minute, each step three feet in length. 2) The command to march at that rate. v. To run.

doughboy—A nickname applied to US soldiers, particularly infantrymen, in World War I.

Douhet theory—A thesis advanced by Italian Army General Giulio Douhet, in his book *Command of the Air* (1921), that future wars would be won by air forces, striking far behind national boundaries and fighting fronts on the ground. Air power would destroy the war-making capabilities of industrialized nations and so terrorize their populations that the will to continue the war would be eradicated long before a decision could be reached by embattled ground armies. Douhet visualized the air forces as first among a nation's armed forces, with armies and navies providing supporting elements for air power.

downgrade—To assign a lower defense classification to an item of classified matter.

draft—The compulsory induction of persons into military service. See also **conscription**.

draft board—A group of civilians responsible for local administration of Selective Service or other laws respecting compulsory military service.

draft dodger—One who seeks to evade the requirement for compulsory military service.

dragon (dragoon)—1) A short musket that could be hooked to a soldier's belt. It was carried by dragoons (q.v.), who probably took their name from it. 2) A standard, or square flag, derived from *draco*, which was a late Roman and Carolingian standard with a metal dragon's head and a trailing body of fabric.

dragon's teeth—Tank obstacles in the form of reinforced concrete tetrapods used as part of the German West Wall constructed before World War II. Long lines of dragon's teeth, four or five rows deep, stretched across all types of terrain and even through towns, and presented an effective delaying device against tanks when US troops approached the West Wall in 1944.

dragoon—A cavalryman, or mounted infantryman, beginning in the 16th century, who usually rode into battle, then dismounted to fight. The principal weapon of the dragoon was originally a short musket, or carbine. Beginning in the American Civil War, all US cavalrymen were dragoons, equipped with carbines; after that war they were usually equipped with the standard infantry rifle. The derivation of the term is in some dispute, but may have come from the short musket called a dragon (q.v.) or sometimes dragoon, carried by early dragoons.

drake—Name of certain light artillery weapons of the 17th and 18th centuries, mobile three-pounders.

D-ration—A US emergency ration with three four-ounce bars, each providing concentrated food—mostly chocolate—of 600 calories.

drawbridge—A bridge that can be raised or drawn aside either to prevent access to a castle surrounded by a moat, or to permit passage through a channel for vessels too high to pass underneath it.

dreadnought—The largest class of battleships, carrying at least eight large-caliber guns. Named for Britain's HMS *Dreadnought*, which was completed in 1906, these ships revolutionized naval warfare and caused all major navies to build comparable "all big gun" vessels. The *Dreadnought* displaced 21,845 tons and carried ten 12-inch 45 caliber guns, two to a turret. With four turbine-driven propellers she could make 21 knots.

dress—In formation, to straighten alignment, applied to a line of troops or sailors. On the command, "Dress right, dress!" each person turns his head and eyes to the right and extends his left arm to touch the shoulder of the person to his left, while aligning himself to the right. Dress is usually to the right, but can be to the left.

dress regulations—The rules in any military organization that indicate the times and places for wearing various types of uniform (dress, informal, work, etc.)., and the specific components of those uniforms.

dress uniform—A military uniform worn on formal occasions. The dress uniform may differ, in small details or completely, from the regular uniform. In many nations, including the United States, it is the only uniform on which full-size medals and decorations are worn.

dressing station—A place on or near a field of combat where medical personnel can give assistance to the wounded. Also called aid station.

Dreyse needle gun—See **needle gun**.

drill—1) Practice of military activities during training. 2) Practice specifically for combat (battle drill), including replication to the greatest possible extent of the sights, sounds, and sensations of battle, in order to accustom the troops to performing their duties under actual combat conditions. 3) Practice in administrative or ceremonial military movements (often called close-order drill, q.v.).

drill sergeant—A sergeant responsible for drilling the troops.

dromon—A fast, light galley used for coast guard duties in the ancient and medieval Byzantine navy. The name means "the runner."

drone—n. 1) A pilotless aircraft of conventional design, or of a modified conventional design, piloted by remote control guided and directed by a mother aircraft, or by a controller on the ground. 2) Any land, air, or sea vehicle that is controlled from a distance. *adj.* Subject to remote control.

drop zone—A specified area where airborne troops, equipment, and supplies are dropped by parachute, or upon which supplies and equipment may be delivered by free fall (that is, without parachutes). The NATO term is "drop" or "dropping zone".

drum magazine—See **magazine**.

drum major—An early British and US army noncommissioned officer whose responsibility was to train the regimental drummers. He usually led the field music (drums, bugles, and fifes) of the regimental band, and often served as bandmaster.

drummer—1) A person playing a drum in a military field music unit or band. 2) A person beating a drum to relay orders aboard ship or to members of a military force by beating certain patterns on the drum. Many drummers aboard ship were boys too young to serve as gunners.

drungarios—A senior Byzantine Army officer, equivalent of modern lieutenant colonel, who commanded a unit of from 300 to 400 men.

dual-capable forces—Forces capable of employing dual-capable weapons (such as conventional and nuclear) or of performing two different military functions.

dual- or multi-purpose weapon—A weapon that can be effectively applied in two or more basically different military functions or levels of conflict. The German 88mm gun of World War II is an example of a dual-purpose weapon, inasmuch as it served both antiaircraft and antitank functions. In fact this was a multi-purpose weapon, inasmuch as it could be, and often was, used as field artillery.

dud—Explosive munition that fails to explode as intended.

duel—A formal or informal combat encounter between two individuals, usually fought with matched or closely similar weapons. A duel can also take place between two self-contained weapons systems, as in a tank duel, an aircraft duel, or a warship duel. The emphases in a duel are on the attack (though simultaneous defensive maneuvers may be involved) and on the duality of the fight, in contrast to team or combined unit efforts at all other levels of military conflict.

duffel bag—See **bag, duffel**.

dugout—An underground shelter constructed to protect troops, ammunition, and matériel from gunfire, particularly from the effects of mid- or high-trajectory howitzer and mortar fire. In the US Civil War dugouts were called bombproofs (q.v.).

dum-dum bullet—A bullet that flattens excessively on impact, thereby causing extensive damage to flesh and bones. The use of this type of bullet is forbidden under international law. Named for a 19th-century British cartridge factory near Calcutta. Standard bullets can sometimes be modified to have a dum-dum effect.

dump—A temporary stock of supplies or a place of storage established in the field or afloat where military supplies are held temporarily. When supplies are issued from dumps, the latter are called distributing points (q.v.).

duty—The requirement of military men to perform in accordance with law and with the legal commands of lawfully appointed superiors.

duty officer—1) An officer on duty, or ready for duty, during periods when other personnel are off duty. 2) An officer charged with being personally present to take care of matters requiring instant and continuous attention. In the first sense, a duty officer in certain situations may simply be designated for duty if and when the occasion demands. In the second sense, a duty officer is one who looks after dispatches. He is said to "have the duty." This is also a naval term for being on watch and not free to leave a ship or station. See also **officer of the day**.

duty roster—A listing of the times at which people are scheduled for various duties.

duty station—The station or base to which a person is assigned for regular duty.

dynamite gun—A small cannon hurling a dynamite charge by means of compressed air. Adapted briefly by the US Navy in the late 19th century. A dynamite field gun was used by US Army troops at the siege of Santiago, Cuba, in 1898.

E

eagle (Roman, Napoleonic)—The national emblem of the Roman Republic (and later the Empire). A golden or gilded eagle on a pole was carried by a senior soldier wherever the cohort went—in garrison, in the field, and in battle. Loss of the eagle in battle was considered a disgrace to the cohort. In the French Napoleonic Army the national flag, the tricolor, was flown on a staff surmounted by a gilded eagle, much like that of the Roman cohort.

eagle boat—A 200-foot, 615-ton craft built by the Ford Motor Company for convoy escort and similar service in World War I. Twelve of these vessels—the first ships to be built on an assembly line—saw service in the war, primarily as submarine chasers. The cost was approximately one-quarter that of a destroyer. The concept was less than successful, however, and the eagle boat was phased out of service by the 1930s.

early warning system—An electronic or visual system of surveillance whose purpose is to detect and report the approach of airborne enemy weapons or weapons carriers as much in advance as possible of their arrival on target. See also **Distant Early Warning line**.

earmarked for assignment—North Atlantic Treaty Organization (NATO) designation for forces assigned to the operational command or control of a (NATO, SEATO, CENTO) commander at some future date, either in peaceful circumstances or in the event of war. The term earmarked for assignment was originated in 1952 by one of the author's of this book.

earnest gun—A breech-loading (q.v.) rifle with a fixed chamber closed by a movable breechblock on the left side of the gun.

earthworks—Any field fortification con- structed primarily of earth, including trenches, walls, barriers, or dugouts (q.v.). See also **entrenchments**.

Easy—Letter of old phonetic alphabet. See also **alphabet, phonetic**.

E-boat—See **motor torpedo boat**.

echelon—An arrangement of troop units, ships, or aircraft, that suggests sequence, steps, or a scale. Echelon may refer to: 1) A subdivision of a headquarters, as the forward and the rear echelons. 2) A level of command. For example, a battalion is a lower echelon than a regiment, and a division a higher echelon than a regiment. 3) A portion of a command with a principal combat mission, that is, the attack echelon, the support echelon, and the rear echelon. 4) A division of forces, used primarily by the Soviet Union, with a first echelon, consisting of from about half to two-thirds (or more) of a unit or formation, and a second echelon, which will attack behind or through the first, usually at a predetermined time or in a predetermined situation to a predetermined objective. See also **first echelon** and **second echelon**. 5) Positioning of forces, usually in the advance to attack, in such a way that elements of a laterally distributed force are also staged sequentially to the rear, either from the right or the left. This was the kind of echelonment employed by Epaminondas, most notably at Leuctra (371 B.C.), and Frederick the Great, particularly at Leuthen (1757). The effect of such echelonment is a refusal of the flank which is echeloned to the rear. See also **refuse a flank**.

Echo—Letter of NATO phonetic alphabet. See also **alphabet, phonetic**.

ECM—See **electronic countermeasures**.

economic potential—The total capacity of a nation to produce goods and services, particularly in relation to the national capability to wage war.

economic warfare—Aggressive use of economic leverage to achieve national objectives in peacetime or wartime.

economy of force—A principle of war embodying the thesis that in order to be able to employ adequate forces in the most important portions of an operational area, forces employed elsewhere should be kept at minimum strengths. Thus, to achieve mass, or concentration, at the decisive point or points, economy of force—sometimes called "economy of forces"—should be employed elsewhere.

EDC—See **European Defense Community**.

EEI—See **Essential Elements of Information**.

effective—*adj.* of troops, fit for service. Of fire, accomplishing its purpose. *n.* A soldier fit for service.

effective range—The maximum distance at which a weapon may be expected to fire accurately.

effectiveness—See **combat effectiveness**.

eighty-eight (88)—A gun of 88mm (or three and one-half inch) caliber used by the German military in World War II. Initially designed by the Krupp Munitions Works as an antiaircraft weapon, the 88 was also used as an antitank weapon and as a field artillery piece. With its long range—as an antiaircraft weapon its effective range was 34,770 feet, as an antitank or artillery gun its maximum range was 16,600 yards—the 88 was equally effective against Allied bombers and Allied tanks because its high muzzle velocity enabled the Germans to place the weapon in a commanding position at a considerable distance from its targets.

élan—A high-spirited morale usually associated with exceptionally self-confident and elite units.

elbow-gauntlet—A gauntlet of plate armor, reaching to the elbow. Developed in Asia, it was adopted in the West in the 16th century.

electrician—US Naval warrant officer.

electrician's mate—US Navy enlisted specialty rating, first, second, and third class.

electronic barrage—See **barrage**.

electronic countermeasures (ECM)—A major subdivision of electronic warfare involving actions taken to prevent or reduce the effectiveness of enemy equipment and tactics that employ or are affected by electromagnetic radiation. Also, to exploit through electronic means the enemy's use of such radiations.

electronic deception—The deliberate radiation, reradiation, alteration, absorption, or reflection of electromagnetic radiations so as to mislead an enemy in the interpretation of data received by means of electronic equipment.

electronic intelligence (ELINT)—The technical and intelligence information gleaned from monitoring the electronic transmissions of an enemy or potential enemy.

electronic jamming—The deliberate radiation, reradiation, or reflection of electromagnetic signals with the object of preventing or impairing the use of electronic devices by the enemy. See also **jam**.

electronic warfare (EW)—The military employment of electronics to prevent or reduce an enemy's effective use of radiated electromagnetic energy, and to insure one's own effective use of radiated electromagnetic energy.

element—1) Any subdivision of a military organization that may function apart from the whole or be considered a separate entity. 2) A component of an organization, as in command element or service element.

elephant dugout—A large dugout (*q.v.*) that is protected by steel ribs or girders. The large semi-cylindrical, corrugated pieces of thin steel sheeting so used for this purpose in World War I, sometimes in dugouts, sometimes for surface protection, were called elephant iron.

elevate—To raise the muzzle of a weapon, or the aim of the gunner, in order to deliver the projectile on or near a target.

elevating arc—1) That portion of the elevating mechanism of a gun in the shape of a segment of a circle, with ratchets along the circumference which are engaged by a worm screw, which rotates when the gunner turns the elevating handle or wheel to raise or lower the muzzle of the gun. 2) An arc attached to the base of the breech of some 19th-century cannon, parallel to the ratchets and graduated in degrees and parts of a degree, to measure elevation. See also **gunner's quadrant**.

elevation—In gunnery, the inclination or angle of the axis of the cannon or gun above ground level to counteract the effect of the force of gravity. The higher the elevation, or angle of the gun (up to 45°), the longer the projectile's range. If the gun can be elevated above 45°, the range decreases as the elevation increases above 45°. See also **site, angle of** and **quadrant elevation**.

ELINT—See **electronic intelligence**.

elite corps—A body of troops designated or regarded as elite (see **elite troops**). Examples include the Roman Praetorian Guard, Napoleon's "Old Guard" (q.v.) and the Confederate Stonewall Brigade and the Union Iron Brigade of the US Civil War.

elite troops—Troops that, because of special training or some other distinctive and above-average quality, are used for special missions, or as a nucleus for troops of lesser quality. In some nations elite troops have been considered socially superior to the rest of the armed forces.

embarkation—The loading of troops with their supplies and equipment into ships or aircraft.

embattle—1) To become engaged in battle. 2) To arrange in order of battle; to draw up, as troops for battle; to prepare or arm for battle. 3) To furnish with battlements.

embrasure—An opening built into a wall or a parapet through which guns can be fired, with minimum exposure of the gunners.

emplacement—A prepared position (usually in a field fortification) for housing and operating one or more weapons or pieces of equipment, offering protection against hostile fire or bombardment.

encamp—To settle down in a spot or locality and prepare for the shelter and subsistence of a force, generally on a temporary basis.

encampment—A place, usually temporary, where a military force resides. It is generally a collection of temporary structures, often tents, with portable facilities which provide for the subsistence of the people and the administration of the force.

enceinte—A wall that surrounds a fortress. An enceinte may be a single, solid wall or a series of more or less concentric defensive works.

encipher—To convert, by means of a cipher system, a plain-text message into language unintelligible to anyone except one possessing the key, in order to prevent an enemy from understanding the message if intercepted.

encircle—To surround. The process of surrounding with a military force is **encirclement**. The purpose of an encirclement is to cut off the encircled force from sources of supply and support and thereby influence it to surrender. Encirclement is often achieved by two or more units approaching the enemy from different directions.

encounter—n. A minor battle. v. To come against, meet; to engage in conflict with.

encounter battle—A battle that occurs when two hostile forces meet unexpectedly, without preparation or reconnaissance. See also **meeting engagement**.

encrypt—To convert a plain-text message into unintelligible form (a cryptogram) by means of a cryptosystem. See also **encipher**.

endurance—The time an aircraft can continue flying or a ground vehicle or a ship can continue operating under specified conditions, such as without refueling.

enemy—One who is hostile to another. A foe, adversary, antagonist.

enemy capabilities—Those courses of action of which the enemy is capable and which, if adopted, will affect the accomplishment of one's own mission. The term includes not only the general courses of action open to the enemy, such as attack, defense, and withdrawal, but also particular courses of action. In assessing enemy capabilities all known factors affecting military operations, including time, space, weather, terrain, and the strength and disposition of enemy forces are considered. In strategic thinking, the capabilities of a nation represent the courses of action, in peace or war, that a nation has the power to undertake in order to accomplish its national objectives.

Enfield rifle—Any of a number of rifles, or earlier rifle-muskets, originally manufactured at the royal small arms factory in Enfield, England. Enfield rifles were used by both British and US soldiers in World War I. The magazine held six cartridges, the barrel was 26 inches long, and the weapon's total length was 46 inches. The Enfield rifle-musket was a mid-19th century weapon.

enfilade—1) A position, usually on a flank, that permits a force to fire down the length of a hostile line, thus increasing the possibility of hitting a target because range dispersion will not matter. 2) Firing of a gun or guns along the entire line of men or works, from one end to the other, equivalent to rake (q.v.) in naval battle.

engage—To enter into conflict, or join battle; to fight.

engagement—A battle, or a subdivision of a battle, between elements of larger forces. See also **conflict levels of intensity.**

engagement, naval—A battle or engagement fought at sea.

engineer, military—A person in military service whose duties involve the construction, repair, operation, or maintenance of structures, defensive positions, and equipment required for combat and noncombat purposes. In the US Army, the Corps of Engineers performs these duties, which range from the construction of fortification, the building of roads, and the laying and removal of mines to the maintenance of inland waterways and harbors. The Civil Engineer Corps of the US Navy performs similar functions and is also charged with the operation and maintenance of the propulsion machinery of US naval vessels.

Engineers, Amphibious—A sub-branch of the US Army Corps of Engineers, established in World War II and active in all Allied landings. Amphibious Engineers were specially trained in the techniques of amphibious operations, including the operation of landing craft and organizing and controlling beach activities. In the Pacific Theater all of these tasks were performed by naval personnel.

engineman—US Navy specialist rating, first, second, third class, and chief.

English lock—A type of flintlock (q.v.).

enlist—1) To enroll oneself voluntarily in military service. 2) To enroll another person in military service.

enlisted personnel—Those in the lower ranks of the armed services who do not hold rank by either commission or warrant. Includes privates, corporals, and sergeants (E1–E8), and equivalent ranks in other services. The British equivalent term is "other ranks."

enlistment—1) A voluntary entrance into military service, as distinguished from induction through conscription. 2) A period of time, contractual or prescribed by law, an enlisted person serves between enrollment and discharge. 3) The act of enrolling in military service.

ensign—1) In the US and British Royal Navy, the lowest commissioned rank, corresponding to second lieutenant in the other services. 2) A flag, especially the national flag, displayed on ships or at naval stations.

entrench—To dig trenches and construct hastily thrown-up fieldworks in order to strengthen a force in position. The fieldworks so constructed are **entrenchments.**

entrenching tool—A small spade or pick, often collapsible, that is normally carried in the pack of infantrymen or other soldiers for use in digging trenches or building earthworks.

envelop—The act of maneuvering all or part of a military force in such a manner as to attack the enemy's flank, or pass around the enemy's flank for the purpose of attacking lines of communication or the enemy rear.

envelope—The three-dimensional space that is within range, altitude, and deflection reach of a weapon, particularly an air defense weapon.

envelopment—A maneuver in which the main attack is directed against the flank (close-in envelopment) or in the rear (wide envelopment) of the enemy's main forces. A wide envelopment is often called a turning movement (q.v.). The term vertical envelopment (q.v.) is often used to describe an airborne operation in the enemy's rear area.

epaulement—A wall or other fortification built out from a bastion to protect the flank. The term comes from the French word for shoulder, *epaule*.

epaulette—A decorative device worn on the shoulder, now usually with dress uniforms. At one time commonly worn by both army and naval officers, epaulettes derived from the knot of ribbons that was used to hold a baldric (q.v.) at the shoulder. They consist of a strap fastened near the neck, widening and usually ending in a circular piece at the end of the shoulder, to which gold fringe is attached. Devices indicating rank or service are attached to the top. The shoulder board worn by US and other naval officers is a derivation.

epibatae—Greek soldiers who fought aboard ship, like modern marines.

epitagma—A Greek cavalry formation of 4,096 men, usually deployed in battle with one-half on each wing of the line of battle.

equerry—An officer in a royal or noble household whose primary function is managing the care of horses, but who may have other duties as well.

equipage—All the equipment, including transportation means but not weapons, required by a military unit or a ship in order to support the life and operations of its personnel.

equipment—All articles needed to outfit an individual or organization. The term refers to clothing, tools, utensils, vehicles, other similar items, and may include weapons. The specific details of the equipment authorized for a unit in the US Army are prescribed by tables of allowances, and by tables of organization and equipment—TO&Es (q.v.), a type of authorization in which weapons are included.

equites—The collective name for the knights or noble horsemen who made the cavalry arm in Roman armies before the military reforms initiated by Consul Marius in 105 B.C. As a result of the Marian reforms, the equites disappeared, replaced for the most part by allied and mercenary cavalry, recruited mainly from Thrace and Africa.

equivalent airspeed—See **airspeed**.

escalade—The climbing of a wall or rampart with ladders, especially by an attacking force.

escalation—An increase in scope or violence of a conflict, deliberate or unpremeditated. It is usually a sequential process of two or more steps of increasing volatility in which one side responds to a step just taken by the opponent.

escarp—In fortification, the surface of a moat or ditch next to a rampart, the surface toward the enemy being called the counterscarp.

escarpment—1) A steep slope created by cutting away ground about a fortified position, in order to render it less accessible to the enemy. 2) A natural terrain feature that because of precipitous cliffs or bluffs, imposes similar limitations upon an attacker.

escort—One or more people, units, or vessels assigned to accompany people, ships, or cargo, for protection or courtesy.

esmeril—A 2½-foot-long artillery piece of the 16th century, with a one-inch bore and a weight of 200 pounds. It had a maximum range of 750 yards and a point-blank range of 200.

espadon—1) A gigantic two-handed double-edged sword used by Swiss and German foot soldiers in the 15th century. With a blade over 50 inches long, the espadon, or *zweihander* (which means two-handed), hung from a scabbard down the back like the two-handed samurai sword. In order to avoid slicing the shoulder when the sword was drawn, part of the blade, near the hilt, was usually covered with leather. 2) Any double-edged weapon.

espignole—A kind of blunderbuss (q.v.) that was loaded with several balls.

espionage—Actions directed toward the acquisition of information through clandestine and covert operations.

espontoon—A 17th-century half-pike, about three feet in length. Also called "spontoon" (q.v.).

espringal—A war engine used in pregunpowder times to throw arrows, bolts, rocks, or other missiles. The espringal derived its propulsive force from the principle of tension. Also called "springal." See also **catapult** and **ballista**.

esprit de corps—A feeling of pride in a unit and of pervasive enthusiasm and confidence among its members. See also **élan**.

essedarius—An ancient Roman gladiator who fought in a heavy chariot called an **esseda**. The esseda, used by the Gauls, the Belgae, and the Britons, was open in front instead of behind. The warrior could run forward along the pole to which the horses were harnessed to seek an advantageous position or angle for the discharge of his missiles.

essential elements of information (EEI)—The critical items of information about the enemy and his situation that must be analyzed in relation to other available information and intelligence so that a logical command decision can be reached.

establishment—1) An organization or installation, together with its personnel and equipment, that forms an operating entity. 2) The table setting out the authorized numbers of men and major items of equipment in a unit or formation.

estacade—A barricade of piles driven into the sea, a river, or a morass to halt an enemy's advance.

estec—A long, narrow thrusting sword, called a **tuck** in Elizabethan times.

estimate, intelligence—A systematic appraisal of available intelligence relating to a specific military situation or condition. The capabilities, limitations, vulnerabilities, and probable intentions (or probable reactions) of an enemy or potential enemy in a given situation must be evaluated in order to determine the courses of action an enemy might take and the probable order of their adoption.

estimate of the situation—A comprehensive review of the circumstances bearing on a military situation. From such an estimate, a commander decides the course of action to take to accomplish his mission. Sometimes called an **appreciation**.

estos—A German-Bohemian thrusting sword with a long thin, rigid blade, designed to penetrate the joints of armor. It was a forerunner of the rapier.

estramaeon—1) A two-edged sword used in the 16th and 17th centuries. 2) A blow with the edge of a sword.

European Defense Community (EDC)—French proposal of the early 1950s to create an integrated European army with a single defense ministry and logistical system.

euthytonon—A light, portable ballista (q.v.). Alexander the Great had 150 euthytonons and 25 palintonons (q.v.) in his army, which he used as field artillery.

evacuate—1) To remove all people from a position, a town, or a locality. 2) To move ill or injured people, generally from a medical facility near a combat area to one that is in a noncombat zone.

evacuation, axis of—A route or series of routes along which sick or wounded troops, prisoners of war, captured equipment, and civilian refugees (and in some instances unneeded troops and equipment) may be transported away from the area of active combat. Such a route is particularly important in a planned withdrawal or delaying action (q.v.).

evacuation point—Medical treatment facilities behind the combat front from which the sick and wounded are evacuated.

evaluation—Appraisal of an item of intelligence for its credibility, reliability, pertinence, and accuracy.

evasion and escape—The procedures and operations whereby military personnel, escaped prisoners of war, and other selected individuals are enabled to escape from an enemy-held or hostile area to areas under friendly control.

evolutions—The movements of troops as they change position. All such movements as marching, countermarching, changing front, forming line, and defiling are evolutions.

EW—See **electronic warfare**.

exchange of prisoners—The trading of one or more prisoners captured by one side in a conflict for prisoners taken by the other. In ancient times prisoners commonly were retained by the victors as slaves. Before the Christian era, however, the exchange of prisoners and the payment of ransom had become a customary practice. Enslavement of captured soldiers, however, was not eradicated until the 17th century.

executive officer—1) The officer second in command of a military unit. 2) An assistant or deputy to a commanding officer responsible for seeing that command policies as applied to the details of day-to-day operations are carried out. The executive officer in the lower echelons of a military organization combines the equivalent duties of the deputy and the chief of staff in higher echelons.

exercise—Any practice operation—large or small—for the purpose of increasing the proficiency of military personnel and of units in the performance of their tasks. In an exercise, simulated enemy action is used as a training device. An exercise may involve a unit or an entire military force of any branch, and may be conducted at sea or in the field. Exercises may also be simulated on a map, computer, or other device. Those conducted in the field, as opposed to in garrison, are usually called **field exercises**. See also **command post exercise**.

expedition—1) A journey made by a military force from its base to an area where combat is anticipated. 2) A force making such a trip, including all its equipment and means of transportation.

expeditionary force—An armed force orga-

nized to accomplish a specific objective in a foreign country. Examples are the British Expeditionary Force (BEF) and the American Expeditionary Force (AEF) in World War I.

exploitation—1) Taking full advantage of success in battle by following up initial gains. 2) Taking full advantage of any information that has come to hand for tactical or strategic purposes. 3) An offensive operation, usually one following a successful attack that is designed to disorganize the enemy in depth.

explosive—A solid, liquid, or gaseous substance that when ignited or detonated expands instantly, or breaks down chemically or physically in a violent manner.

exposure dose—A measurement of radiation in relation to its ability to produce ionization. The unit of measurement of the exposure dose is the roentgen.

expugnable—Capable of being conquered or taken by assault; susceptible to the use of force.

extend—To widen the space between files of a formation or units on a front in order to occupy a greater area of ground.

extended order—The deployment of troops in a tactical formation with more than the minimum of space needed for their weapons between ranks and files.

exterior lines—A situation of a military force, or group of military forces, which are divided into two or more widely separated elements facing a more centrally disposed force, which is said to operate on interior lines (q.v.). Operations by the divided forces are convergent, and are said to be on "exterior lines."

external ballistics—See **ballistics**.

F

Fabian tactics—A type of warfare in which combat is avoided and instead a policy of delay and harassment is directed at the enemy. The name derives from Quintus Fabius Maximus, who defended Rome against Hannibal in 217 B.C. with such a campaign, never joining battle, and thereby earning for himself the name *Cunctator* (Delayer).

fabric armor—See **armor**.

face—*n*. One side of a formation of troops or a fortified position. *v*. To present the front of a formation or object in a given direction. See also **about face, left face, right face**.

facility—Part or the whole of a building, base, or other structure or group of structures, including associated attachments or equipment, which houses and supports any agency or operating entity related to the armed forces. The entity itself may be considered a facility, independent of, or in combination with, its quarters.

facings—1) The collar, cuffs, and trimmings of a uniform coat. 2) Changes in direction or front of a military unit. See **face**. 3) The movements of troops by turning on their heels to the right, left, right about, etc.

falarica—An incendiary defensive weapon of classical times. It consisted of a long iron javelin with a wooden shaft, bound with tow and smeared with inflammable pitch or resin. The falarica was set on fire and thrown against an attacker's siegeworks.

falarique—A combustible dart or arrow about three feet long, used in the Middle Ages for setting fire to a ship or a wooden structure.

falarira—Medieval weapon, similar to a ballista (*q.v.*), which threw darts wrapped with inflammable materials.

falchion—A sword with a short, broad, thick, heavy blade that widened toward the point and had a single convex cutting edge. It was used in Western Europe in the 12th and 13th centuries, primarily as a slicing weapon.

falcon (falconet)—An iron or brass cannon used on ships and ashore from the 15th through the 17th centuries. Weighing about 800 pounds, the falcon was six feet long, with a bore of about two and one-half inches, and fired a three-pound ball. Its maximum range was about 2,500 yards, its point-blank range about 400. A smaller version, the falconet, was 3.7 feet long and weighed 500 pounds. With a bore of two inches, it could fire a one-pound ball as far as 1,500 yards, a point-blank range of about 280 yards.

fall in—*v*. to assume an assigned position in a rank or formation. *n*. The command to assume such a position.

fall out—*v*. To break ranks on command. *n*. The command to break ranks.

fallout—The falling to earth of radioactive particles from a nuclear cloud; also applied to the particles themselves.

fallout-safe height of burst—The height of burst at or above which no militarily significant fallout will be produced as a result of a nuclear-weapon detonation.

fascines—A tied bundle of wooden sticks or branches, either carried with a unit or gathered up in an area, used to build fortifications, strengthen earthworks, fill ditches, or otherwise as needed.

fatigue—Noncombat work, generally of a

manual or menial kind, such as cleaning barracks. **Fatigue duty** is any labor other than use of arms performed by a soldier, and is assigned to troops both to accomplish the job and to keep them occupied.

fatigue, combat—Intense weariness or mental illness caused by continued exposure to hostile fire sufficiently severe to render a soldier incapable of continuing in active combat. Also called **battle fatigue**. In World War I it was called shell shock.

fatigue duty—See **fatigue**.

fatigue party—A detachment of soldiers detailed on noncombat-related duty.

fatigues (fatigue uniform)—Outer clothing, usually made of denim or another coarse, durable material, worn by soldiers on fatigue duty (q.v.). Slang for work clothes.

fauchard—A 16th-century weapon that looked like a very large razor blade attached to the end of a staff.

FBS (Forward-Based Systems)—Short-range nuclear weapons systems that can reach superpower targets by virtue of their location in forward areas. For example, US quick-reaction alert aircraft in Western Europe can quickly strike at targets in Eastern Europe and the Soviet Union.

FDC—See **fire direction center**.

FEBA—See **Forward Edge of the Battle Area**.

feint—A pretended attack designed to distract, divert, or deceive the enemy, particularly to focus his attention on a place or an area distant from the site where decisive action is planned.

fencible—A soldier enlisted for the defense of the country, and not liable to be sent abroad.

Ferguson rifle—A rifled infantry firearm with a screw-plug breech action invented by Major Patrick Ferguson, a Scottish officer, during the American Revolution. The Ferguson rifle, probably the most effective military firearm of the period, could fire six shots a minute to an extreme range of 300 yards. Ferguson was killed in the Battle of King's

Mountain (9 October 1780), and his rifle was never adopted by the British Army.

ferret—An aircraft, ship, or vehicle specially equipped for the detection, location, recording, and analyzing of electromagnetic radiation.

feu de joie—French phrase meaning "fire of joy," describing a firing of muskets one after another, closely timed to make a continuous noise, in celebration. The term was also used for a bonfire.

field—1) A place where a battle is fought. 2) Away from garrison or home, usually for combat, training, or field exercises. As in "in the field."

field army—An administrative and tactical organization composed of a headquarters, certain organic, or permanently assigned, army troops, service support troops, two or more corps and independent divisions, and other units. See also **army**.

field artillery—Those mobile cannon (guns and howitzers), rockets, and surface-to-surface missiles usually employed to provide indirect fire support to maneuver forces (infantry and armor). This includes most mobile artillery pieces, although coast artillery did include some large mobile weapons. It does not include air defense weapons. It does not usually include mortars, which are usually infantry weapons, although in some armies heavy mortars are field artillery. See also **artillery**.

field duty—Service or duty performed away from garrison, in training, or in operations against an actual or potential enemy.

field exercise—See **exercise**.

field force—1) A combined arms ground force operating under actual or assumed combat circumstances. 2) The units, general headquarters, installations, and equipment that comprise, or are intended to comprise, the forces in a theater of operation.

field fortifications—Emplacements and shelters of temporary construction built during combat, as opposed to more elaborate permanent fortifications (q.v.) built prior to hostilities. Field fortifications can be simple, and can be constructed with reasonable ease by units requiring no more than minor engineer supervision and equipment, or they can be built by engineers and can approximate the strength and complexity of permanent fortifications.

field glasses—A small compact binocular telescope. Military field glasses are usually four to seven power in magnification.

field grade—The officer ranks of major, lieutenant colonel, and colonel.

field marshal—In some armies, a senior rank, usually held only by the head of large military organizations such as a group of armies or the entire national army. In the US Army the equivalent rank is general of the army.

field music—A small group of fifes, drums, trumpets, bugles, bagpipes, or other instruments and their players who play for marching, regimental calls, or certain ceremonies.

field officer—A senior army or marine corps officer with the rank of major, lieutenant colonel, or colonel.

field of fire—The area that a weapon or a group of weapons may cover effectively with fire from its position.

field kitchen—See **field ration**.

field radio—A sending and receiving set for communications between elements of a military force in the field. With power supplied by a hand-cranked generator, radio was first used by combat forces in World War I. By the end of World War II radio had largely replaced the field telephone (q.v.) as a means of communication in mobile warfare when security was not of primary importance and laying telephone lines was difficult.

field ration—Food required for one man for one day of field operations. The meals are either prepared in advance and carried by a soldier, usually in prepackaged form, or prepared in a central **field kitchen** and delivered to the troops in containers.

field staff—A staff carried by gunners in the field to hold lighted matches for discharging cannon.

field telephone—A portable, battery-powered telephone designed for field or combat use. The field telephone became the common means of communication and control early in the 20th century.

field training—Training conducted under field conditions, as opposed to in garrison (q.v.).

fieldpiece—A mobile cannon, towed or self-propelled.

fieldwork—A temporary construction built for defense by an army in the field. Most often used in the plural, **fieldworks** are entrenchments and other temporary fortifications thrown up by an army in the field, either for defense against attack or to cover an attack. All nonpermanent fortifications are fieldworks. See also **earthworks**.

fife—A high-pitched flutelike instrument, with no keys, about 14 inches long, with six fingerholes. In military music it is used primarily with drums. In the 16th century it was used in Switzerland, and later adopted by English military units. A **fifer** is one who plays a fife.

fifth column—A subversive organization that operates in secret inside a nation prior to or in conjunction with a military attack from without. Its members work by infiltrating legitimate organizations and agencies, and by exploiting all opportunities to weaken the legal government politically and militarily. The expression was first used in 1936, during the Spanish Civil War, by General Emilio Mola, commander of four columns of Nationalist forces that were attacking Madrid. Mola referred to the Nationalist underground within the city as his "fifth column."

fifty-mission cap—A service cap worn without the spring stiffener and grommet (q.v.). Wearing caps in this manner was customary in World War II among flying officers of the US Army and Navy, originally because of the need to wear earphones over the cap.

fight—To engage an enemy in combat (q.v.).

fighter aircraft—Highly maneuverable aircraft that are relatively light in weight, have a crew of one or two, and are designed primarily for air-to-air combat. Also called fighter-interceptors. **fighter bomber**— Fighter aircraft are designed both for air-to-air

combat and for close support of ground forces. They are likely to be assigned to units that have air defense and ground attack roles.
fighter, all-weather—A fighter aircraft equipped with radar devices and other special equipment which enable it to intercept its target when visibility is reduced by darkness or weather conditions.

fighter bomber—See **fighter aircraft** and **bomber**.

fighter cover—The maintenance of a number of fighter aircraft over a specified area or military force for the purpose of repelling hostile air activities.

fighter sweep—An offensive mission by fighter aircraft to seek out and destroy enemy aircraft or targets of opportunity in an allotted area of operations.

fighting patrol—See **patrol, fighting**.

file—A column of individual troops or of vehicles traveling one behind the other.

file off—To leave a formation, generally consisting of several files, by marching away in single file.

final protective fire—See **fire, final protective**.

fire—n. 1) The discharge of one or more firearms. Often used collectively, as in, "The fire came from the left." 2) The command to discharge one's weapon. The earliest guns, in the 14th and 15th centuries, were discharged by applying fire to a primary charge connected to the propellant in the gun. Thus the origin of the term "fire" for discharge of hand guns and cannon. v. To discharge a weapon.

fire and maneuver—The concept of combining movement with firepower to attack an enemy force in position under the most favorable conditions. Movement or maneuver cannot be effectively performed in combat without fire superiority (q.v.).

fire arrow—A small iron dart to which was attached a match impregnated with powder and sulphur, used to set fire to the sails of ships.

fire base—The weapons available to a unit commander for achieving fire superiority,

and to maintain such superiority while the unit's maneuvering elements are closing with the enemy.

fire, concentrated—Fire from a number of weapons, directed at a single point or small area.

fire, continuous—1) Fire conducted at a normal rate without interruption. 2) In artillery and naval gunfire support, loading and firing as rapidly as possible, consistent with accuracy, within the prescribed rate of fire for the weapon.

fire control—Control over the direction, volume, and time of fire of guns or rockets by a systematic process which may include the use of certain electrical, electronic, optical, or mechanical systems; a fire-control system.

fire controlman—A US Navy enlisted specialty, chief, and first, second, and third class.

fire, coordination of—The planning and execution of fire so that targets are adequately covered by a suitable weapon or group of weapons.

fire cross—A signal in ancient Scotland indicating that the people should take up arms.

fire direction—The process of controlling the fire of an artillery unit primarily at the battalion and group level, although fire direction can be conducted at the battery level. This involves fire control as well as the calculation and transmission of data necessary for placing effective fire on targets.

fire direction center (FDC)—That element of an artillery command post, consisting of gunnery and communication personnel and equipment, by means of which the commander exercises fire direction (q.v.) and fire control (q.v.). Receiving target intelligence and requests for fire, the fire direction center translates them into appropriate firing data and fire commands.

fire discipline—Control over the firepower resources of a unit to assure that ammunition is not wasted and that information is not given to the enemy.

fire fight—The exchange of fire between two opposing units, usually at platoon or company level.

fire, final protective—A prearranged barrier of fire designed to impede enemy movement across defensive lines or areas. See also **barrage**.

fire for effect—1) Fire delivered after the center of impact or burst is within the desired distance of the target or adjusting/ranging point. 2) Term in a fire message to indicate that the adjustment (or ranging) is satisfactory and fire for effect is desired.

fire for record—The process of formally testing an individual or unit for proficiency in the use of a weapon. Depending upon the score achieved, an individual firing a small arm in the US Army is qualified as marksman, sharpshooter, or expert.

fire, Greek—See **Greek fire**.

fire, massed—The process of bringing to bear in a relatively small target area the concentrated fire of a number of artillery batteries or battalions simultaneously or in a short period of time.

fire mission—1) A specific target assignment given to a fire unit, such as an artillery battery. 2) An order used to alert personnel at the guns that the message following is a call for fire.

fire pan—The receptacle that held the primer in early gunpowder weapons.

fire plan—A tactical plan for coordinating the use of the weapons of a unit or formation so that their fire will be effective.

fire position—A position from which fire is opened.

fire, prearranged—Gunfire delivered on known or suspected targets in accordance with a planned schedule on either a time or an on-call basis. Such a schedule is usually prepared with an attack plan. See also **scheduled fire**.

fire, searching—Fire distributed by successive changes in the elevation and deflection settings of a gun. The purpose of such fire is essentially for harassment and interdiction.

fire storm—A windstorm induced by an immense fire, generally in an urban area, which causes warm air to rise above the burning area in such volume that cooler air is rapidly drawn in toward the burning area from all directions, increasing the intensity and destructiveness of the fire. Often the result of an intensive aerial bombardment.

fire superiority—The battle application of a greater amount of effective firepower than is applied by the opponent. The side with fire superiority has enhanced maneuver capability.

fire support—In US Army usage, the assistance or protection given ground forces in direct contact with the enemy by the firepower of ground or naval guns or by aircraft engaging in close air support.

fire support coordination—The planning and executing of the application of firepower to targets in such a way as to use available weapons most efficiently.

fire support coordination line—A line established by the appropriate ground commander to ensure coordination of fire that is not under his control but which may affect tactical operations. The establishment of the fire support coordination line is normally coordinated with the appropriate tactical air commander and other supporting elements. The fire support coordination line should follow well-defined terrain features.

fire support ship—A ship whose mission is to support amphibious operations or land operations with weapon fire.

fire support team (FIST)—In the US Army, a small group of artillerymen, under a lieutenant, one of whom is attached to each company-level maneuver unit (infantry and armor) for the purpose of locating targets, adjusting indirect fire of supporting artillery and mortars, and coordinating these and all other forms of fire support (including naval gunfire and air support), as well as planning for such support with the supported commander. The FIST includes forward observers, communications personnel, and one or more drivers for the FIST vehicle or vehicles. See also **forward observer**.

firearm—An individual weapon, such as a musket, rifle, or pistol, using gunpowder to propel a projectile.

fireball—1) The luminous sphere of hot gases

which forms a few millionths of a second after detonation of a nuclear weapon and immediately starts expanding and cooling. 2) Before the era of nuclear weapons, a ball filled with powder or other combustibles to be thrown among enemy troops.

firelock—A term formerly used to describe a flintlock (q.v.) musket.

firepower—1) The capability of delivering fire. 2) The fire itself, or the quantity or effectiveness of fire delivered. 3) Any explosive or missile that wreaks damage upon the target against which directed. 4) The amount of fire that may be delivered by a position, unit, or weapon system measured in various units (of firepower), including volume and weight.

firepower score—A quantitative measure of the effectiveness of a weaon in comparison to other weapons.

fireship—A vessel loaded with combustible material designed to be set afire and let drift so as to destroy docked or anchored enemy ships.

firing chart—Map, photo map, or grid sheet showing the relative horizontal and vertical positions of batteries, base points, base point lines, checkpoints, targets, and other details needed in preparing artillery firing data.

firing line—The line on which troops or weapons systems are placed to deliver fire against a target.

firing range—See **range**.

firing squad—1) A detachment of troops assigned to execute a condemned person by gunfire. 2) A detachment that fires the final salute at a military funeral.

firing step—In trench warfare, a small step, board, or ledge about a foot from the bottom of a trench on which men could stand to fire or look out.

firing tables—A compilation of detailed data on the firing characteristics of a type of weapon, based upon exhaustive firing tests at a proving ground. For example: "Firing Tables for Howitzer 105mm, M2 and M2A1, Firing Shell H.E. M1." Typically includes such information as: the range in yards or meters for every mil or fraction of a degree

of elevation, the fork (q.v.), and probable error (q.v.) for range, deflection, etc.

first call—A bugle call alerting troops for roll call or for ceremonies or drill.

first echelon—Under the Soviet concept of committing forces by echelon (q.v.), the leading element of a Soviet or Warsaw Pact unit or formation when advancing to combat. In World War II first echelons of Soviet forces were usually about one-half to two-thirds the strength of a command. The remainder was in the second echelon (q.v.), although it could sometimes be in a reserve. (On rare occasions, in a narrow sector, there were three echelons.) Complicating this doctrine, from the standpoint of both the Soviet Army and its possible opponents, is the fact that there can be echelonment of the units or formations within each level of command; that is, a regiment could have a first and second echelon, and both of these could be in the first echelon of a division. Similarly, a division could have both a first and second echelon, and both of these could be in the first echelon of an army, and so on up through front (or army group) and theater. There is some evidence that the Soviet Army makes less use of the echelonment concept in the later 20th century than it did in World War II.

first graze—The point where a bullet following a ballistic trajectory will, if not deflected, first strike the ground. See also **point blank**.

first lieutenant—1) A commissioned rank or person in the US Air Force, Army, or Marine Corps, immediately above second lieutenant. 2) In the Navy, an officer in the complement of a ship or naval station who is responsible for maintenance and repair.

first line—1) In a formation of troops, vehicles, or units composed of two or more lines, the line most forward, or closest to the enemy, not including skirmish lines (q.v.), screens, or covering forces. 2) A slang expression, which, when qualified, denotes the weapon, weapon system, technique, or capability that will best serve a given purpose or at a given time. For example, "The F-16 Falcon is a first-line fighter aircraft."

first strike—The first offensive move of a war, generally associated with use of nuclear weapons.

FIST—See **fire support team**.

fix bayonets—An order to attach bayonets to rifles.

fixed ammunition—See **ammunition**.

fixed artillery—See **artillery**.

fixed gun—A machine gun attached rigidly to the fuselage of an aircraft and aimed by moving the aircraft.

fixed-wing aircraft—A conventional jet or propeller aircraft, as distinguished from a rotary wing aircraft (q.v.) or helicopter (q.v.).

flag—A banner of distinctive size, color, or design, which serves as a symbol, a marker, or a decoration.

flag bag—The chest or locker aboard ship in which signal flags are kept.

flag bridge—The bridge on a warship where the admiral observes or directs a battle. On an aircraft carrier it is the first bridge above the flight deck.

flag captain—The officer of whatever rank in command of a flagship.

flag country—The area of a flagship that is used as quarters and for other purposes by the flag officer.

flag lieutenant—An aide to an admiral.

flag mast—A disciplinary procedure in the US Navy corresponding for officers to the captain's mast (q.v.) for enlisted men.

flag officer—A naval officer with a rank higher than captain. Flag officers are entitled to fly a navy blue flag showing rank by means of the number of stars on the flag, one to five. See also **admiral**.

flag of truce—A white flag carried or displayed to an enemy, as an invitation to conference or parley. A white flag can also be a signal of surrender.

flag plot—On a flagship the room where the admiral has the equipment and information for controlling his command.

flag rank—The ranks of naval officers above the rank of captain. See also **flag officer**.

flag signals—Signals representing information or orders that are transmitted by flags flown from a staff or mast; the flags' number, shape, or color has a meaning known to both sender and recipients. The only means of communicating in the days of sail and before the invention of the wireless, flag signals have now largely been replaced by short-range radio and are used only when visibility is very good, when secrecy is paramount, and there is danger of being overheard on radio. Flag signals may also be flown to indicate unclassified information as to the status or operation of a vessel.

flags—Naval slang for a signalman, referring to the device of crossed flags he wears on his rating badge.

flagship—A ship carrying the admiral or other senior officer in command of a fleet, squadron, or other naval organization and flying his flag. A flagship is usually, but not always, the largest of the ships.

flak—1) Antiaircraft fire directed at enemy aircraft. 2)The shell bursts from such fire. Originating in World War II, the word was adopted from the German acronym for *Flieger Abwehr Kanone*, meaning aircraft defense gun.

flak jacket—A jacket or vest of heavy fabric containing metal or other bulletproof plates, designed especially for protection against flak. The usual type of flak jacket covers the chest, abdomen, back, and genitals, leaving the arms and legs free.

flak tower—A towerlike structure as many as ten stories high, in which one or more antiaircraft weapons are mounted. Such a tower might also house the crew and store ammunition.

flame thrower—A weapon that projects and ignites incendiary fuel. First used in modern warfare in 1915 during World War I. See also **Greek fire**.

flanchards—Medieval defensive armor to protect the flanks of a horse in combat. The name first appears in a French text of 1302 as *flanchieres*.

flank—n. One of the sides—right or left—of a military formation or position. v. To place one's force in position to threaten or attack the opponent's flank. An **open flank** is a flank without adequate protection. A **protected flank** is one resting on an obstacle not easily attacked by an enemy. When a force has one flank open and one protected, the open flank is the **tactical flank**. A **strategic flank** is one that, if enveloped, would permit an enemy to threaten a force's line of communications.

flank attack—An attack directed at the flank of an operational force. See also **attack**.

flank company—In the British Army of the American Revolution period, the light infantry and grenadier companies in each battalion that were composed of the battalion's best soldiers. The other companies were called **battalion companies**. Flank companies from the various battalions were often combined to fight in special units.

flank guard—A security element deployed to protect the flank of a moving or stationary force.

flanker—1) A detachment of troops assigned to guard the flank of a marching column. 2) pl. Troops belonging to a flank company (q.v.). 3) Formerly used of riflemen and all light troops because of their position on the flanks of a force. 4) A fortification jutting out so as to command the flank of an attacking body.

flanking position—A position that threatens the security of an advancing enemy by attacking on the advancing force's flank.

flare—A pyrotechnic device that burns with a brilliant, in some cases colored, light, either stationary or shot into the air, used to signal, illuminate a target or an area, or identify those who fire the device.

flare gun—A Very pistol (q.v.) or other type of small gun used for shooting flares. Used particularly to indicate the position of a ship or unit in distress.

flash—The flame that issues from the muzzle of any firearm or piece of ordnance when it is fired.

flash hider—A device on the muzzle of a gun

that hides the flame or flash when the gun is fired. Also known as a flash suppressor.

flash in the pan—A firing of the primer of a firearm without igniting the main charge.

flash message—A category of precedences reserved for initial enemy contact messages or operational messages of extreme urgency. Brevity is mandatory.

flash ranging—Finding the position of an enemy gun by observing its flash.

flash suppressor—See **flash hider**.

flask, powder—A bottle or similar vessel in which gunpowder or shot was carried before the introduction of the cartridge.

flattop—Slang for an aircraft carrier. The name has been credited to Lieutenant Commander Robert E. Dixon on 8 May 1942, when, after US naval aircraft sank the Japanese carrier *Shoho*, he radioed back to his carrier, "Scratch one flattop!"

flèche—From the French word for arrow, a fortification of earth shaped like a V, with the point toward the potential attacker.

flèchette—A small metal device shaped like a dart; flèchettes are clustered and fired in a shell that explodes above a target to scatter the projectiles broadly. Larger dartlike steel missiles called flèchettes de guerre were dropped from airplanes in World War I.

fleet—1) An organization of ships, aircraft, marine forces, and shore-based naval activities under the command of an officer who may exercise operational as well as administrative control. 2) A less formal organization of a group of warships under a single command. 3) All the naval ships belonging to a nation.

Fleet Air Arm—1) The air arm of the British Royal Navy, separated from the Royal Air Force in 1939. 2) Generically, any naval air arm.

fleet ballistic missile—A ballistic missile carried on and launched from naval vessels.

fleet ballistic missile submarine—A nuclear-powered submarine designed to deliver ballistic missile attacks against either submerged or surfaced targets. Designated in the US Navy as SSBN.

fleet in being—A naval force that avoids action but is large enough and so positioned that an opposing naval force must station a considerable number of ships to watch it and be prepared for sudden movements.

Fleet Marine Force—A US Marine Corps organization serving with an integral part of a US fleet. It was originally set up in December 1933 as a brigade-sized unit and a part of the US fleet. A balanced force with land, air, and support elements, the Fleet Marine Force has the primary mission of capturing bases for the US Navy.

Fleet Reserve—In the US Navy the reserve organization composed of retired officers and enlisted men.

flight—1) In US Naval and Marine Corps usage, a specified group of aircraft usually engaged in a common mission. 2) The basic tactical unit in the US Air Force, consisting of four or more aircraft in two or more elements, or flight subunits. 3) A single aircraft airborne on a nonoperational mission. 4) a trip by air from one point to another.

flight cap—Same as overseas cap (q.v.). See also **fifty-mission cap**.

flight deck—1) The upper deck of an aircraft carrier, from which aircraft take off and on which they land. 2) An elevated compartment in some aircraft where the crew operates the aircraft.

flight leader—A pilot in command of a flight (q.v.) of aircraft.

flight lieutenant—A commissioned rank in the Royal Air Force and the Royal Canadian Air Force, corresponding to that of captain in the US Air Force.

flight nurse—An Air Force nurse who cares for patients aboard a medical air evacuation aircraft.

flight officer—In the US Army Air Forces in World War II, a rank corresponding to warrant officer, junior grade.

flight sergeant—An enlisted rank or person in the Royal Air Force or Royal Canadian Air Force, corresponding to master sergeant in the US Air Force.

flint—A very hard, fine-grained quartz. Because it produces sparks when struck with steel, it was used in early firearms to ignite gunpowder. See also **flintlock**.

flintlock—A small-arms ignition system invented by the French gunsmith Le Bourgeoys in 1615. It consisted of a flint held in a vise at the end of a cock. When the trigger was released, the flint struck against a piece of steel, creating sparks. The priming powder was ignited, which in turn ignited the powder that fired the weapon. Flintlock muskets apparently were first issued to two companies of Cromwell's New Model Army in 1645. The flintlock musket was less accurate and had a slower rate of fire than the model with the improved matchlock, but was more reliable and less affected by weather. It was standard in European armies by 1699. The snaphance lock (q.v.) was generally considered a type of flintlock, as were the English lock, the dog lock, the Scandinavian snaplock, and the miquelet lock.

flissa—An Oriental sword with a straight handle that with the blade forms a Latin cross.

floating battery—A group of guns placed on any floating platform, usually protected with plate armor. Such batteries were used by the Allies against the Russian forts on the Black Sea in the Crimean War, and in the American Civil War.

floating bridge—See **bridge**.

floating reserves—In an amphibious operation, reserve troops that remain embarked until needed.

FLOT—See **forward line of own troops**.

flotilla—1) An administrative or tactical organization consisting of two or more squadrons of destroyers or smaller ships, together with such additional ships as may be assigned. (Flotillas are not currently used by the US Navy.) 2) A group of ships, usually small ones, more or less operating together formally.

flying army—In the 19th century, a large body of cavalry and infantry capable of rapid movement, and sometimes kept in motion because their mission was to cover the approaches to an area or to protect a frontier. Also called a flying camp (q.v.).

flying artillery—In the 19th century, artillery trained to move rapidly. This was usually horse artillery in which each man had his own horse, but it could be horsedrawn artillery with the men mounted on caissons (q.v.) and limbers (q.v.).

flying camp—1) In the 19th century a body of troops prepared for rapid movement from one place to another. 2) A term used in the period of the American Revolution for a mobile, strategic reserve.

flying column—A term frequently used in the 19th century for a rapidly moving force, usually cavalry, operating behind enemy lines or sent to seize a key objective.

Flying Fortress—Nickname for a four-engine World War II bomber of very long range. Manufactured by Boeing, its Air Force designation was B-17; the Navy designation was PB-1W.

flying officer—1) A commissioned rank in the Royal Air Force and the Royal Canadian Air Force corresponding to first lieutenant in the US Air Force. 2) In a general sense, an officer who flies or who holds an aeronautical rating.

flying pig—In World War I, a large projectile fired by a German trench mortar.

flying sap—See **sap**.

"Flying Squadron"—A force of fast US Navy armored cruisers and smaller vessels, under the command of Commodore W.S. Schley, held in place off the US Atlantic Coast early in the Spanish-American War in fear of a possible attack against coastal cities by a Spanish naval squadron under Admiral Pascual Cervera.

FO—Forward observer (q.v.).

fodder—Coarse food, such as hay, for cattle, horses, and sheep. A sufficient supply of fodder was always a logistical challenge for armies dependent upon horse-drawn transportation.

foe—Enemy.

fog of war—The situation reflecting the inherent inability of commanders and staffs to know and understand everything they want to know about the activities, organization, and plans of the enemy, and about the details of combat activities of their own forces. The fog of war can be reduced, but not eliminated, by good intelligence information about the enemy, by good communications between commanders and subordinates, as well as between adjacent commanders, and by close staff surveillance of subordinate activities. The concept was first articulated by Clausewitz.

fogy, (fogey, fogie)—An allowance paid to a military person for a specified period of service. Because such allowances are given periodically, a career serviceman may have several fogies. If, for example, a lieutenant gets a $10.00 per month fogy for every five years of service, after he has served ten years at the rank of lieutenant his monthly base pay will be increased by $20.00

follow-on-forces—A NATO term for Warsaw Pact forces expected, in the event of war, to arrive on the battlefield to reinforce Pact first echelon forces (q.v.) already engaged against NATO forward defense (q.v.) forces. See also **second echelon**.

follow-on-forces attack (FOFA)—A NATO concept for employing long-range weapons (aircraft and highly accurate surface-to-surface missiles) to disrupt the planned and anticipated arrival of Warsaw Pact follow-on-forces. (q.v.). See also **deep attack**.

follow-up, amphibious—The landing of reinforcements and stores after the assault and assault follow-on echelons have been landed.

foot—A term once commonly used for infantry or foot soldiers; that is, troops who do not fight mounted on horses, tanks, or other means of transportation.

foot guards—An elite infantry unit, with a mission of providing security for a sovereign.

forage—To seek and take food and other necessary items from civilian sources in the area where a military force is stationed or operating.

forage cap—A uniform cap with a low, flat crown and a visor. Usually softer and with a lower crown than a kepi (q.v.). Forage caps were common in the 19th century, and were worn by both sides in the US Civil War.

foray—n. A raid, usually by a portion of a combat force, on a target other than the main objective. v. To make such a raid.

force—n. 1) An aggregation of military personnel, weapons systems, vehicles, ships, and necessary support. 2) Superior strength applied in order to overcome resistance. v. To apply superior strength, mental or physical, in order to overcome resistance.

force de frappe—The independent air striking force with nuclear capability established by General Charles de Gaulle when he was President of France.

force multiplier—A factor, tangible or intangible, that increases the combat value of a force. Clausewitz's famous statement that "defense is the stronger form of combat" indicates that he considered defensive posture such a multiplying factor. In the US Army force multipliers are considered to include (in addition to defensive posture) such factors as terrain, training, leadership, and morale.

force ratio—The ratio of the strengths of two opposing forces. Strength is indicated by such measures as number of personnel and firepower score.

forced march—A march performed at an unusually fast pace, generally for a long distance, in order to reach an objective in less time than usual. See also **march**.

Foreign Legion, French—A French military organization of mercenaries of many nationalities who serve under French officers. The Foreign Legion was established in 1832 and served primarily in the French colonies in North Africa, but also in France's possessions in Indochina. After most French colonies became independent, the Legion was based in Corsica, though it has overseas units in former colonies.

Foreign Legion, Spanish—A Spanish military organization, similar to the French Foreign Legion, established early in the 20th century. It served mainly in the Spanish colonies of North Africa.

fork—Technical artillery term signifying, for a given elevation setting, that angle in elevation that will move the center of impact four probable errors (q.v.) on the ground.

form—To assemble in any formation of ranks and files.

form on—To connect one individual or several to an object or a formation so as to lengthen a line.

formation—1) An ordered arrangement of troops, ships, vehicles, or aircraft for a specific purpose. 2) In US military terminology, a combat organization of division size or larger.

formers—Round pieces of wood of slightly less than the diameter of the bore of a gun, around which could be rolled the paper or other material of a cartridge in separate-loading ammunition (q.v.).

fort—1) A permanent defensive structure often including living quarters for a sizable force. 2) A fortification (q.v.). 3) The term used to designate a permanent US Army installation or garrison, such as Fort Myer, Fort Leavenworth, etc.

fortification—1) The military art of strengthening a place, position, or area in order to increase the defensive power of troops against attack. 2) A military defensive work. 3) The act of fortifying. The general classification of fortifications is into permanent fortifications (q.v.) and field fortifications (q.v.). Throughout the ages fortifications have been an important, frequently employed asset in warfare. Modern warfare, especially since the Industrial Revolution, radically changed the conduct of warfare by making possible the development of weapons of greatly increased power, requiring men and armies to resort more frequently than ever before to fortifications. In World War I machine guns, shrapnel, and high-explosive shells drove the armies into extensive trench systems. For the first time in history (although this development had been presaged to some extent in the American Civil War), armies fought a war in which there were no flanks. Victory could be gained only by costly frontal attacks or by exhausting the enemy's ability or will to make war. The appearance of the first tanks in 1916, and, later, the introduction of "Hutier tactics" (q.v.) by the Germans, restored a measure of mobility to the battlefield, but combat in World War I was characteristically a combat of trenches and artillery. During the interwar years, many European nations, having analyzed the military experience of World War I, erected permanent fortifications along their frontiers, primarily for strategic purposes. The best remembered of these works are the French Maginot Line and the German West Wall (or Siegfried Line). Both permanent fortifications (q.v.) and field fortifications (q.v.) were important factors in all World War II land campaigns and in all major wars since the Second World War.

fortified defense—See **defense**.

fortified line—A series of permanent defensive positions of great depth and hardness, as in the French Maginot or the German Siegfried lines. Such a line employs concrete or steel-protected gun emplacements, preplanned supporting fire coverage, heavily protected troops and supplies, and coordinated series of obstacles to stop combat vehicles. Also called a **fortified zone** because of its depth.

fortify—To strengthen a position by means of forts, entrenchments, batteries, or constructions of any kind that improve the possibility of defense against attack.

fortress—A fortified place, especially a large and permanent military stronghold, often including a town. A fort.

forward air controller—An air officer, stationed in an advance ground position or airborne, who directs the activities of aircraft engaged in close air support of ground troops.

forward area—An area at or near the front, in or near a combat area, or near the enemy. See also **area**.

Forward-Based Systems—See **FBS**.

forward command post—See **command post**.

forward defense—A NATO concept that, in the event of an invasion of Western Europe by the Warsaw Pact, NATO forces will defend close to the "Iron Curtain" (the frontier between East and West Germany). This requires a high level of readiness and of defensive effectiveness on the part of NATO forces. The reason for this concept is partly political, partly military. Politically it is important that the West German government and people be assured that NATO will endeavor to save as much of West Germany as possible from invasion. Militarily, West Germany is so narrow that some military theorists believe there is not enough space for traditional defense in depth. Other theorists believe that the forward defense concept has the weaknesses of a linear or cordon defense (qq.v.).

forward echelon—That echelon of a headquarters or of a unit that is located in or sent ahead to a forward area, distinguished especially from a rear echelon.

forward edge of the battle area (FEBA)—That line or zone where the forward elements of two opposing armies are engaged in direct combat. Virtually synonymous with FLOT (q.v.).

forward line of own troops (FLOT)—The front line of a force engaged in combat. See also **FEBA**.

forward observer (FO)—An individual located at or near the front line whose mission is to adjust the fire of heavy weapons on hostile targets, using telephone or radio to communicate with the weapons, which are usually to the rear in defilade (q.v.). A forward observer is usually an artillery officer attached to an infantry or armor unit for the purpose of adjusting supporting artillery fire to help the supported unit accomplish its mission. See also **fire support team**.

foss (fosse)—A moat or a ditch, particularly a moat around a fort.

fougasse—A crude type of antipersonnel mine that projects a charge of broken stone or scrap metal or both. This weapon has been employed since the early days of gunpowder.

fourragère—A unit decoration given by France and Belgium, consisting of braided cords of distinctive color and design, usually worn on the right shoulder, as opposed to the aiguillette (q.v.), which is worn on the left.

fox—A pre-19th-century term for sword.

Fox—Letter of old phonetic alphabet. See also **alphabet, phonetic**.

foxer—A World War II device for counteracting the German acoustic torpedo, which could be detonated by the noise of a ship's propellers. The foxer was towed behind a ship and made a noise louder than the propellers, so that it would detonate the acoustic torpedo at a safe distance from the ship.

foxhole—A hole in the earth, shallow or deep, for a single soldier, usually dug hastily for protection against enemy fire.

Foxtrot—Letter of NATO phonetic alphabet. See also **alphabet, phonetic**.

fragmentary order—An abbreviated form of an operation order, usually issued on a day-to-day basis, which eliminates the need for restating information contained in a basic operation order. It may be issued in sections. Sometimes called **frag order**.

fragmentation—The scattering of pieces of metal casing from a grenade, bomb, or shell, generally by explosion above a target area. See also **shell fragment**.

fragmentation bomb—See **bomb**.

fragmentation grenade—See **grenade**.

fraise—From the French word meaning ruff or ruffle. A fraise is a barrier consisting of a series of pointed stakes stuck into the ground at an angle, or a similar construction of barbed wire.

francisca—A single-bladed axe with a particularly heavy head set on the haft at an angle. Used by the Franks from the 6th to 8th centuries as a throwing axe, as well as in hand-to-hand combat, it disappeared when the Franks changed to mounted combat. See also **axe**.

franc-tireur—A French guerrilla, particularly of the period of the Franco-Prussian War.

free company—A group of medieval mercenaries, notably English, who hired themselves out to any European ruler who needed tough, skilled, fighting men. See also **company**.

free war game—A war game in which there is no predetermined scenario and in which the controller freely applies his own judg-ment and military experience or insight to make assessments. See also **war game**.

French 75—A 75mm field gun developed in France in 1897 and widely used during and after World War I. The finest field gun of its time, it was adopted by armies throughout the world. It was the principal light artillery piece of the American Expeditionary Force in World War I. In France the weapon was called a **soixante-quinze**.

friction—Interactions among individuals, units, and formations, manifested by delays, confusion, misunderstandings, and generally reduced tempo of operations in comparison with the ideal. The larger the force, the greater the friction and the performance degradation resulting therefrom. Friction can be reduced, but not entirely eliminated, by intensive training, careful planning, and close surveillance of operations by commanders and staffs. The concept was elucidated by Carl von Clausewitz in his classic book, On War.

frigate—A class of fast, light warships. The name in ancient times was used for a variety of light, open craft, with oars or sails, in the Mediterranean. Various types of fast ships were called frigates by the Spanish, Portuguese, and the French in the 16th, 17th, and 18th centuries. At the time of the Seven Years' War (1756–1763) the frigate was standardized as a three-masted, fully rigged (square-rigged) vessel with from 24 to 50 guns, most of them on one deck but with some on the poop and forecastle. In the 19th century, steam frigates were developed, and they usually carried fewer guns than frigates under sail. The designation has been used by the US Navy since World War II to designate a class of fast ships weighing 5,000 to 7,000 tons, which can operate independently or with others to counter submarine, air, and surface threats. US frigates normally carry three-inch or five-inch dual-purpose guns and surface-to-surface missiles, as well as advanced antisubmarine warfare weapons. The US designation is DL.

frill—An ornamental appendage to the shirt usually worn with regimentals (q.v.), attached by a hook and eye to the uniform coat. Enlisted men wore frills not attached to the coat.

frizzen—The piece of steel arched over the

priming pan of a snaphance (q.v.) against which the flint was struck to ignite the powder. Also called **battery**.

frogman—A man, usually in the navy, equipped with a wet suit, swim fins, and breathing apparatus, who performs a variety of underwater tasks.

front—1) In general, the combat area. 2) The line of contact between two opposite forces. 3) The side of a force or a position facing the enemy. 4) Translation of a Russian term for army group.

front line—1) The line formed by the foremost units in a combat situation. 2) In a more general sense, the line or area of contact with the enemy, the front (q.v.). See also **FEBA** and **FLOT**.

frontal attack—1) An attack that strikes the enemy along his front. 2) In air interception, an attack by an interceptor aircraft in which the attacker crosses the heading of the other aircraft at an angle greater than 135 degrees. See also **attack**.

full charge—The maximum charge when the two or more propelling charges are available for guns with semifixed or separate-loading ammunition (qq.v.)

functional bombing—Bombing of a special class of key targets, like dams or marshaling yards, within an industrial complex or transportation system, as opposed to bombing the entire system.

funda—1) A Roman sling. See also **funditor**. 2) A crude catapult, used by Visigoths and Franks in sieges after the fall of Rome.

funditor—In the ancient Roman Army, a lightly armed skirmisher who carried a sling rather than bow and arrow or a javelin. See also **veles**.

furlough—An authorized absence of enlisted personnel from a duty station, usually for a period longer than a weekend. Now officially called **leave**.

fusil—1) A 17th-century light flintlock musket (q.v.). 2) A steel that strikes fire out of a flint. 3) A tinderbox. 4) The piece of steel that covered the pan in early firearms.

fusilier—Originally, a soldier armed with the fusil (q.v.). Fusiliers were originally used as artillery guards. Fusilier units, armed with modern weapons, still exist in the British Army, the Royal Welch Fusiliers, for example.

fusillade—n. The simultaneous or near-simultaneous discharge of many firearms in combat or in a military exercise. v. To shoot down by a simultaneous discharge of firearms.

fustibal—A sling on the end of a four-foot-long staff. Fustibals were used during the Hussite campaigns as auxiliary weapons and were used in the 17th century to hurl grenades.

fuze—An instrument for detonating an explosive device. Originally, fuzes were strings of hemp or other slow-burning combustible material that were lighted at the end farther from the powder. Mechanical or electrical mechanisms have replaced the older type of fuze, and are of a number of types, including: **impact**, which triggers the explosion when the projectile hits the ground or the target; **proximity**, which detonates at a set distance from the target; **delayed action**, which detonates at a set time after impact; **time**, a clock-like mechanism that causes a projectile to explode after a period of time to produce an air burst over the target; **base**, a fuze located in the base of a projectile or a bomb.

fyrd—The militia in England before the Norman Conquest.

G

gabion—A cylinder of wicker or other material, sometimes closed at one end. It was filled with dirt or rocks and used for building field fortifications. The gabion has been to some extent superseded by the sandbag. A work made with gabions is called a **gabionade**.

gallease—A heavily armed three-masted galley used in the Mediterranean in the 16th and 17th centuries.

galleon—A large, three-masted sailing vessel, with two or more decks. Numerous countries, but most notably Spain, built galleons in the 15th and 16th centuries for use as both warships and merchantmen.

gallery—1) In a fort, a passageway that is covered overhead and protected by walls or parti-walls on the sides. 2) A mine tunnel or shaft. 3) A portable Roman structure covered with hides and used to conceal and protect men moving forward to attack a besieged fortification.

galley—1) A long, open, or partially decked vessel propelled by oars and sometimes by one or more sails, used particularly in the Mediterranean for centuries. The ancient galleys were often rowed by slaves and had as many as five banks of oars (quinquereme, q.v.). They were used militarily both for fighting naval battles and to carry troops for land battles. In medieval times galleys became much larger. They were gradually abandoned after the development of bigger ships propelled entirely by sail. See also **bireme** and **trireme**. 2) A place aboard ship where food is prepared.

galliot—A light, fast war galley, equipped with both sails and oars, that was used in the Mediterranean region.

galloper—A mounted messenger.

game, war—See **war game**.

gantlet—1) A military punishment in which the culprit ran between two rows of men, who hit him with sticks, clubs, or other devices as he passed. 2) Variant spelling of **gauntlet**.

gap marker—A device used to mark a minefield gap, an area (usually a lane) cleared of mines. Because gap markers must describe a line or path, they are referenced to a landmark or intermediate marker to indicate the boundary of the gap.

Garand rifle—A semiautomatic rifle designed by the American small-arms designer John C. Garand. Adopted in 1936, the Garand served as the basic US infantry weapon in World War II and the Korean War. Officially designated the M-1, it is a .30 caliber gas-operated weapon, weighing about nine and one-half pounds and firing an eight-round clip from a 24-inch barrel with four rifled grooves.

Gardner gun—A .45 caliber machine gun named for its designer, William Gardner of Toledo, Ohio. The 1879 model had twin air- or water-cooled barrels. A hand crank on the right side of the action loaded first one barrel and then the other; it also fired and ejected the shell case with one revolution.

garrison—n. 1) All units assigned to a base or area for defense, development, operation, and maintenance of facilities. 2) A military post. v. To assign troops or units to, or move troops into, a post.

garrison ration—In the US Army, the standard peacetime ration for all persons entitled to it. See also **ration**.

gas grenade—See **grenade**.

gas, nerve—Any chemical agent whose physiological effect is a paralysis or derangement of the nerve centers.

gas, poison—Any chemical substance that can produce toxic or irritant effects. Gas was first used in modern warfare during World War I, on the Western Front on April 22, 1915, when the German forces released chlorine against the Allied positions in the Ypres salient, with lethal results. Subsequently, both sides made widespread use of a variety of poison gases. The use of chemical weapons was banned by international agreement in Article 171 of the Versailles Treaty and by similar prohibitions in other treaties signed at the end of World War I. These prohibitions were reinforced by the Treaty of Washington (1922) and by a protocol of 1925 that was signed by more than 40 states. See also **chemical warfare; Lewisite; mustard gas; phosgene**.

gas, war—See **gas, poison**.

gas mask—A mask worn over the face to protect the eyes from toxic gases and to remove gases from the air as it is inhaled.

Gatling gun—A multibarreled machine gun designed by Richard J. Gatling. Patented in 1862, it was the first effective machine gun, and was adopted by the Union Army in the Civil War. Manually operated by a hand crank, but with automatic loading and ejection of cartridges, the Gatling would now be called a semiautomatic weapon.

gauntlet—A glove, particularly a medieval armor glove, with a wide wrist to give extra protection. See also **gantlet**.

gear—1) The personal belongings of a sailor, corresponding to the soldier's kit (q.v.). 2) The equipment required for a particular job, such as degaussing gear, flight gear.

gendarme—1) A member of a national organization of military police, called the **gendarmerie**, responsible for internal law enforcement and security. 2) A member of a royal French cavalry corps that existed in various forms from the 15th century to 1789.

general—1) A senior rank in most national ground and air forces. A **general officer** is one holding a rank above colonel, and entitled to display a flag (red in the US Army) with from one to four stars, depending upon whether the rank is **brigadier general** (one star), **major general** (two stars), **lieutenant general** (three stars), or **general** (four stars). The rank of four-star general with title of "General of the Army" has been conferred three times in the United States since the Civil War: in 1866 when it was given to Ulysses S. Grant, in 1869 to William T. Sherman, and 1888 to Philip H. Sheridan. In 1944, during World War II, the title "General of the Army of the United States" (five stars) was given to Henry H. "Hap" Arnold (changed to General of the Air Force when an independent air force was established in 1946), Dwight D. Eisenhower, Douglas MacArthur, and George C. Marshall, and in 1950 it was conferred on Omar Bradley. This rank is the equivalent of field marshal in other armies. From 1799 to 1802 the US Army had an official rank of **general of the armies** (presumably four stars), which was also the equivalent of field marshal but was never awarded. Re-established in 1919, this rank was conferred on John J. Pershing, who wore four stars.

general court martial—See **court martial**.

general mobilization—See **mobilization**.

general officer—See **general**.

general of the armies—See **general**.

general orders—1) Permanent instructions, issued in order form, that apply to all members of a command, as compared with special orders, which affect only individuals or small groups. General orders are usually concerned with matters of policy or administration and include such things as citations of units, awards, important appointments, and other matters involving information or orders that are general in application and permanent in character. 2) A series of permanent guard orders that govern the duties of a sentry or post.

general quarters—A condition of readiness of a warship's crew when action is imminent. All battle stations are fully manned and alert; ammunition is ready for instant loading; guns and guided missile launchers may be loaded.

general reserve—A reserve of troops and

units under the control of the commander of a force for use when and where he decides.

general staff—A selected group of military generalists whose function is to assist a nation's military leadership—or a general commanding a field force of combined arms elements—in planning, controlling, directing, coordinating, and supervising the activities of all subordinate military elements to achieve an assigned objective. The leader or leadership makes decisions and gives commands; the general staff's responsibility is to provide all possible support to assure that the decisions and commands are timely, sound, and effective.

general support—Support that is given a whole force rather than to any particular subdivision thereof. It differs from direct support (q.v.), which is provided to a single, specified unit.

general support artillery—Artillery that executes the fire directed by the commander of the unit to which it organically belongs or is attached. It fires in support of the operation as a whole rather than in support of a particular subordinated unit.

general support-reinforcing—A tactical artillery mission. General support-reinforcing artillery has the primary mission of supporting the force as a whole and a secondary mission of providing reinforcing fire for another designated artillery unit.

general war—Armed conflict between major powers in which the belligerents' total resources are employed, and the national survival of a principal belligerent is in jeopardy.

generalissimo—1) Commander in chief of all of a nation's armed forces. The rank, which is the Italian superlative of general, has long been used, but was first used in its modern sense by Benito Mussolini and subsequently adopted by some others, notably by Stalin of the Soviet Union and Chiang Kai-shek of China. 2) Commander of a combined force of allied military forces.

Geneva Convention—One of a series of international agreements signed at Geneva, Switzerland, in 1864, 1906, 1909, 1929, and 1949, establishing rules for the treatment of prisoners of war and of wartime sick or wounded.

Geneva Cross—The symbol used by the Red Cross and by medical services of armed forces of most nations to indicate their neutrality. The red cross, reversing the national heraldic emblem of Switzerland, was adopted at the 1864 Geneva Convention that established rules for the treatment of prisoners and the sick and wounded.

genitors—Spanish light cavalry, armed with light lance or javelin (q.v.), introduced in the wars with the Moors about the 11th century. The genitors continued to serve in Spanish armies through the Italian Wars of the 16th century. In weapons and tactics they imitated the Moorish light cavalry.

George—Letter of old phonetic alphabet. See also **alphabet, phonetic**.

GI—From Government Issue; the nickname "GI" was given to US soldiers in World War II, because all of their equipment was provided by the government.

GI Bill—The Veterans' Benefit Act of 1944, which provided for federal funding of education for veterans, among other entitlements.

gig—The small boat used by the commander of a vessel for transport to and from the main ship. Called the "captain's gig."

glacis—From French *glace*, meaning ice, it is a cleared slope extending outward from the base of a fortification so that attacking enemy troops are within sight of the defenders.

gladius—A two-foot sword with a two-inch-wide blade adopted by Roman armies about 250 B.C. The hastati and principes (qq.v.) wielded the gladius at close quarters after throwing their javelins. It was primarily a thrusting weapon, and could also be used as a hand axe.

glide bomb—v. 1) To bomb a target by releasing one or more bombs when the bombing plane is gliding. 2) In a less exact sense, to dive-bomb at a shallow angle. n. A bomb fitted with airfoils to provide lift, carried and released in the direction of a target by an airplane.

glide path—1) The flight path of an aircraft

or winged missile as it glides downward, the line of which forms an angle with the longitudinal axis of the aircraft or missile. 2) The line to be followed by an aircraft as it descends for landing. Also called **glide slope**.

glider—A fixed-wing airplane having no internal source of power, and constructed so as to glide and soar. Military gliders, used to carry troops or supplies, are normally towed aloft by a powered airplane. Released from the tow, they have limited range, but are given lift by their airfoils, and thrust by the force of gravity combined with momentum and the reaction of aerodynamic forces upon the lifting surfaces.

G-1, G-2, G-3, G-4—The four main sections of an army staff in the US Army; the organization was adopted in 1917 from the French example. G-1 is concerned with personnel; G-2 with intelligence; G-3 with plans, training, and operations; G-4 with supply. There is sometimes a G-5, which has had such diverse assignments as war plans and civil affairs.

Gold—Letter of NATO phonetic alphabet. See also **alphabet, phonetic**.

gorget—A piece of armor developed in the 15th century consisting of several articulated plates hinged to fit around the neck and fasten with a clasp. Gorgets were worn with buff coats after armor fell into disuse. The term was later used for a badge of rank worn at the throat.

Grande Armée—The name applied to the principal field army of Emperor Napoleon Bonaparte of France. Formed in 1804, when Napoleon was contemplating an invasion of England, it was first employed as the Imperial Army in the Ulm–Austerlitz Campaign (1805). The Grande Armée of 1805–1806 numbered some 200,000 men, two-thirds of them veterans; it was the best trained, equipped, and led of Napoleon's armies, and was the instrument of his greatest victories, including Austerlitz, Jena, Auerstadt, and Friedland. The nucleus of the Grande Armée was largely destroyed in the 1812 Russian Campaign.

grande batterie—A concept devised by Napoleon for massing artillery to support his main attack. Hitting the enemy's front line with concentrated artillery fire would cut a breach and permit the infantry to advance.

Grand Fleet—The designation of the British battle fleet that had the mission of protecting the British Isles against the German High Seas Fleet during World War I. The great fleet, which in 1914 included the 24 battleships and battle cruisers of the First, Second, and Third Home Fleets, was commanded by Admiral Sir John Rushworth Jellicoe during 1914–1916 and defeated the German fleet in the Battle of Jutland (31 May 1916), the greatest naval battle of the war. Following Jutland, Jellicoe was appointed First Sea Lord of the Admiralty; he was succeeded in command of the Grand Fleet by Admiral Sir David Beatty, who had commanded the British battle cruisers at Jutland.

grand strategy—An imprecise term, used by many military theorists of the late 19th and early 20th centuries, generally signifying strategic concepts at a level above single-theater and campaign strategy, and comprising a mixture of military strategy and national strategy. See also **strategy; strategy, military; strategy, national**.

grand tactics—An imprecise term used by many military theorists of the late 19th and early 20th centuries in a largely unsuccessful effort to make a clear distinction between battlefield tactics (q.v.) and military strategy (q.v.). Conceptually akin to the more modern concept of operations (q.v.).

grapeshot—A type of ammunition composed of a number of iron balls about twice the size of a musket ball or of canister shot, (q.v.), placed in layers—usually three separated by metal plates—inside a shell and fired from cannon. Generally used at short range against attacking troops. Grapeshot, or grape, was used in both land and naval warfare in the 18th and 19th centuries.

Gras rifle—See **Chassepot**.

graves registration—Supervision and execution of matters pertaining to the identification, removal, and burial of the dead, and the collection and processing of their effects.

gravity drop—The vertical deflection of a bomb or projectile caused by the action of gravity. See also **trajectory**.

gray propaganda—Propaganda that does not specifically identify its source or its origins. See also **propaganda**.

graze—1) Richochet. 2) As an artillery sensing in time fire it is the term used to indicate a burst on impact and not in the air.

grazing fire—Fire that is approximately parallel to the ground and does not rise above the height of a man standing.

Great White Fleet—The 16-battleship battle fleet of the United States which, at the direction of President Theodore Roosevelt, made a round-the-world cruise in 1907–1909. It was called "the Great White Fleet" because all of the warships were painted white.

greave—Armor for the lower leg, ranging from a shin guard to a type that covered the leg from the knee to the bottom of the heel.

Greek fire—An incendiary mixture, probably including sulfur, pitch, niter, and petroleum with othre substances, used by the Byzantine Greeks from at least the 7th century A.D. in both land and naval warfare. Greek fire is the Crusader term for what by the Byzantines was called "liquid fire." Although it is alleged that liquid fire was invented by an engineer named Kallinokos, it seems likely that he merely refined an existing weapon.

grenade—A hollow, explosive-filled projectile hurled at the enemy by hand or by some mechanical device—launcher, gun, or mortar—designed and fuzed to explode on impact or when triggered by a self-contained timer. In the 19th century crude pots of clay filled with wet quicklime and other combustible substances were flung at an enemy. Such devices were used until the end of the 13th century. Explosive containers using gunpowder were used in siege operations in the 16th century. In 1591 Maurice of Nassau invented a cast-iron grenade. Grenades became a permanent part of the equipment of the French army in 1667. They fell into disfavor, although not entirely out of use, but became popular again in the Crimean War. Grenades of various kinds are in modern arsenals. Grenades are classified by means of delivery or propulsion and by their effects. Types of grenades classified by means of delivery or propulsion include: **hand grenades** and **rifle grenades** (See also **grenade launcher**). Types of grenades classified by effects include: the **antitank grenade**, designed for use against tanks and armored vehicles. The high-explosive, antitank (HEAT)

rifle grenade consists of a body assembly, stabilizer, and fin assembly. The body contains about 12 ounces of high explosive in the form of a shaped charge and a base detonator. A point-initiating fuze is attached to the nose of the grenade. Upon impact, this fuze, which has a small shaped charge, initiates the base detonator, which, in turn, explodes the main charge. At right angles, the jet from this grenade will penetrate upwards of eight inches or armor plate and is effective up to 65 degrees obliquity. The **fragmentation grenade**. This typical hand grenade, used by the US Army in both World Wars I and II, weighed about 22 ounces. Its serrated cast-iron body originally contained two ounces of black powder. However, in 1943 the powder was replaced by TNT, which shattered the case more effectively and scattered fragments over a circular area of about ten yards. The **gas grenade**, filled with tear gas or other disabling chemicals, is used primarily by police agencies for riot control. The **illuminting grenade**, fired by a rifle, consists of a cylindrical body and a stabilizer assembly. The body contains one-half pound of illuminant, a base-igniting fuze, and a quickmatch. Upon impact, the illuminant burns for 55 seconds with 80,000 candlepower, illuminating an area 240 yards in diameter. The **incendiary grenade** contains thermite (TH) which burns 30 seconds at 4,300° F., and is used to set fire to enemy matériel. The **offensive grenade** is a cardboard cylinder with metal ends, containing six or seven ounces of TNT. It creates an explosive blast and is most effective in a confined area like a trench or a pillbox. It gets its effect from blast rather than fragmentation. The **smoke grenade** produces smoke in various colors, which is used to conceal troop movements, for signaling or for other purposes. See also **grenade, "jam pot"** and **grenade, "potato masher."**

grenade, "jam-pot"—An early form of the fragmentation grenade (q.v.), the "jam-pot" was a makeshift grenade used by the British Army in World War I; it was a tin can containing an explosive and scraps of iron.

grenade launcher—Any of several devices for projecting grenades. In World War II a launching device that could be fitted to the muzzles of standard rifles fired hand grenades from 100 to 200 yards. One of the best-known types was the Japanese "knee mortar" of World War II.

grenade, "potato masher"—A German anti-personnel grenade used in both world wars, so called because it was shaped like an old-fashioned potato masher, with a long handle attached to a cylinder containing the explosive.

grenadier—Originally a soldier whose primary function was to throw grenades. At the defense of Ratisbon in 1632, the name was first used for soldiers who threw grenades from the walls, for which they received extra pay. By the mid-17th century special companies were being formed of especially strong men, who were employed to throw grenades. Although the use of grenades declined during the 18th century, the elite units of grenadiers remained in many armies, and—as in the British Army—often became elite attack troops.

grid (military grid)—Two sets of parallel lines, labeled by consecutive letters or numbers, intersecting at right angles and forming squares. The grid is superimposed on maps, charts, and other representations of the earth's surface in an accurate and consistent manner to permit precise identification of ground locations, facilitating the computation of direction and distance to other points.

grommet—A ring of metal or other stiff material placed inside a visored cap to keep it taut.

ground alert—A status in which aircraft on the ground are fully serviced and armed with combat crews in readiness to take off within a specified short period of time after receipt of a mission order.

ground control—1) The guidance given an aircraft or missile in flight by a person on the ground. 2) The electronic equipment or network used in exercising this control. 3) A system of accurate measurements used to determine the distances and directions or differences in elevation between points on the earth. The system provides a basis for mapmaking.

ground crew—A team of mechanics and technicians who maintain and service aircraft, and aircraft armament.

ground fire—Small arms ground-to-air fire directed against aircraft.

ground force—1) A force made up of ground troop units as opposed to air or naval units. 2) A force belonging to the ground echelon of an air organization.

ground troops—Troops that are generally employed in ground warfare, such as infantry and artillery (or US Marines), as opposed to air or naval forces.

ground zero—The point on the surface of land or water, at or just below the center of burst of a nuclear weapon.

group—1) A flexible administrative and tactical unit composed of two or more battalions or two or more squadrons or two or more ships. 2) A number of ships or aircraft, normally a subdivision of a force, assigned for a specific purpose. See also **task group**, **groupment**.

group captain—A commissioned rank, title, or person in the British Royal Air Force, equivalent to that of colonel in the US Air Force.

groupment—A temporary task force, usually of artillery units, consisting of two or more battalions under the command either of the senior commander or—if a direct-support groupment—of the organic direct-support battalion commander.

guard—n. 1) A military person, usually but not always enlisted, who is assigned the duty of protecting a place or its occupants against intruders. 2) A body of persons detailed to escort a person, or people, or material items requiring safeguard; such guards may perform exhibition drills or carry out other ceremonial functions. 3) Member of an elite body of troops assigned permanently to the protection of a monarch or other national leader. 4) A person who supervises and controls prisoners. v. 1) To perform protective functions. 2) To monitor (usually a radio frequency). An **honor guard** is detailed to greet or accompany distinguished persons at a ceremony or, at a funeral, to accompany the caskets of military personnel. Also known as **guard of honor**.

guard mail—Mail that is carried by or accompanied by an armed guard.

guard of honor—See **guard**.

Guardia Civil—Elite gendarmerie of Spain.

guard-mount—A ceremony, usually held daily, at which the assigned guard, usually a temporary unit with personnel from other permanent units, for an installation, garrison, or unit is relieved by a new guard. In anachronistic terminology, though still occasionally used, it was called "guard-mounting."

guardship—A ship detailed for a specific duty for the purpose of enabling other ships in company to assume a lower degree of readiness. A guardship, for example, may monitor various radio frequencies on which communications for the company may be sent in order that the other ships may maintain radio silence.

guardsman—1) One who acts as a guard. 2) A member of the US National Guard. 3) A soldier in a national unit of guards, as in the British Army a member of a regiment of the royal household guards.

guarrel—See **crossbow** and **quarrel**.

guerre de course—The interruption of an enemy's seaborne commerce by the destruction of its merchant shipping. Such naval warfare is usually carried on by fast cruisers, capable of fighting small enemy warships, but able to avoid enemy capital ships by speed, maneuver, or stealth (as in a submarine).

guerrilla—Spanish for "little war." *adj.* Pertaining to irregular warfare. *n.* A participant in fighting not directly connected with a formal military organization or operation. See also **partisan**.

guerrilla warfare—Military and paramilitary operations conducted in enemy-held or hostile territory by irregular, predominantly indigenous forces.

guidance—1) The action of pointing out a course to be followed. 2) The action of providing information and advice required by another to carry out an operation. 3) The action of guiding a moving object along a course.

guide—1) A local individual, usually civilian, attached to an army to provide information and assistance in respect to local areas. 2) A soldier (or other military person) at the extreme right-front or left-front of a unit marching in close order, whose duty is to set the direction of the unit, in accordance with orders given by the commander.

guided missile—An unmanned missile that moves above the surface of the earth, whose trajectory or flight plan is capable of being altered either by a preset or self-reacting device within the missile or by radio command outside the missile. Guided missiles are divided by their ranges into long range (more than 5,500 kilometers), mid-range (601–5,500 kilometers), and short range (less than 601 kilometers).

guidon—A small flag or pennant, originally carried by a guide (q.v.) to mark the end of a line along which a unit was to form. Guidons are used to identify company-sized units in the US Army, having different colors for different branches of the service.

gun—1) A mechanism consisting essentially of a barrel, receiver, breech mechanism, and carriage (or stock or handle, depending on the type of gun), using controlled explosives to shoot projectiles or signal flares. 2) A cannon, as distinguished from a small arm. 3) In precise technical definition a gun is a cannon with a relatively long barrel (usually 25 calibers or more), high velocity, and a consequent flat trajectory, as distinguished from a howitzer or a mortar.

gun, assault—See **assault gun**.

gun camera—1) A camera mounted on a gun stock. 2) A camera that is operated by the same mechanism that fires the guns of an aircraft and is used to record the effect of fire on the target.

gun captain—A petty officer, normally a gunner's mate, in charge of a gun crew on a warship.

gun carriage—A platform, fixed or mobile, that provides a secure base for a gun, including towed or immobile framework, truck, tank, or other vehicle. Carriages usually are considered to include the elevating and traversing mechanisms of the gun.

gun crew—Personnel who load and fire an artillery piece or a naval gun.

gun deck—The deck of a wooden warship on

which most of the vessel's guns were placed, or where the gunroom (q.v.) was.

gun, evening—The ritual firing of a gun on army posts at sunset. See also **retreat**.

gun pit—An excavation in which a field-piece is placed for protection and, sometimes, concealment.

gun shield—See **shield**.

gun tackle—The blocks and tackle attached to the inside of a ship's sides, during the period from about 1650 to about 1850, to haul a gun to and from the gunport.

gunboat—Any of a number of types of small craft carrying mounted cannon. Gunboats are normally used in shallow coastal waters or rivers.

guncotton—A cellulose nitrate used in smokeless powder.

gun-howitzer—A cannon that has characteristics of both a gun and a howitzer (qq.v.).

gunlaying—The aiming of a large gun at a target by setting azimuth, range, and elevation. One who does this may be called a gun-layer, or, more usually, a gunner.

gunnage—The number of guns on a warship.

gunner—1) One who aims or fires a gun. 2) In a US Field Artillery gun crew the second in command and the one responsible for aiming the piece in deflection by means of the gun sight. Usually performed by a corporal. 3) In British parlance, any artilleryman. 4) In the US Navy, a warrant officer concerned with maintenance and firing of a ship's guns. The commissioned officer senior to him is a **chief gunner**. Enlisted personnel junior to him are the several ranks of gunner's mates (q.v.), who are commonly dubbed "guns."

gunner, master—See **master gunner**.

gunner's mate—A petty officer rank in the US Navy for those specializing in ordnance.

gunner's quadrant—See **quadrant, gunner's**.

gunnery—The science concerned with the use and performance of guns and projectiles. It overlaps the science of ordnance (q.v.).

gunnery officer—A naval officer in charge of the ship's guns, their operation and maintenance, and the crews that maintain and fire them.

gunnery sergeant—In the Marine Corps, a noncommissioned rating between staff sergeant and first sergeant.

gunpointer—A naval enlisted man who aims and fires a manually operated ship's gun.

gunport—An opening in the side of a ship, the wing of an airplane, a pillbox, or other structure, through which a gun can be fired.

gunpowder—1) An explosive combination of sulfur, saltpeter, and charcoal. 2) Any explosive mixture used as a propelling charge in a gun. Gunpowder was apparently used in some form in China as early as the 12th century. It seems to have been independently developed in Western Europe in the 14th century, and its introduction as a generally employed weapon propellant revolutionized warfare. It is widely accepted that the first use of gunpowder weapons in battle was by the English at the Battle of Crécy in 1346. See also **corned gunpowder**.

gunpower—The total weight of metal fired by a warship's major battery in a single broadside.

gunroom—An apartment at the after-end of the lower gun deck, occupied by the gunner, or, in the British Navy, as a mess room by the lieutenants.

gunship—An aircraft (usually a helicopter) used as a weapons platform.

gunsight lamp—An optical device that enables a gunner to take aim at an aircraft flying out of the sun.

gunsmith—One who makes or repairs firearms.

gunster—An early term for gunner.

gunstone—A stone chipped to spherical shape, used as cannon shot during the 15th and 16th centuries. After the invention of iron balls, stones became less frequently used.

Gurkha—A soldier from Nepal in the British or Indian Army. Gurkhas are renowned for their courage and battlefield tenacity.

hackbut (hackenbusch)—British name for a 15th-century matchlock (q.v.) gun, known as harquebus in England, in France as arquebus (q.v.), in Germany as hackenbusch, and by variations of all these terms.

Hague Conventions—Seventeen conventions establishing laws for land and naval warfare, peace, and neutrality, adopted as the result of international conferences held at The Hague, Holland, in 1899 and 1907.

hailshot—Grapeshot (q.v.).

halberd (halbert)—A long-handled axe used as an infantry weapon during the late Middle Ages. The halberd is a pike to which a broad axe blade is affixed just below the point. Halberds in various forms were used from at least the 14th century, and the long weapon enabled a foot soldier to cleave the armor of either mounted or foot warriors. With the abandonment of armor the halberd became a ceremonial weapon or a badge of rank.

Hale rocket—A stickless rocket developed by William Hale in 1840. Hale controlled the flight of his rockets by placing three metal vanes in the exhaust nozzle, which caused the rocket to rotate in flight. The Hale rocket was adopted by the United States, Great Britain, Austria, and Russia, and a similar one was manufactured in France. Some types of Hale rockets were used into the 1890s.

half-armor—Armor covering the torso and arms only, used mainly by cavalrymen during the 16th and 17th centuries, when armor was disappearing from the battlefield.

half face—To turn half the distance between one's position and a right or left face; such a facing by troops in line makes the line oblique.

half-moon (demi-lune)—An outwork in the shape of a half-moon, built to protect the entrance to a fort.

half pike—A short pike (q.v.) used from the 16th century on; it was used by naval boarding parties and carried by infantry officers as a badge of rank.

half stripe—Quarter-inch braid, usually of gold, that in combination with one or more full stripes on uniform sleeves or shoulder boards indicates ranks of certain naval officers. For instance, in the US Navy one full stripe and one half stripe designate a lieutenant, junior grade; two full stripes and one half stripe designate a lieutenant commander.

half-track—A vehicle with front wheels and reticulated tracks like a tank instead of rear wheels. Half-tracks were developed in the 1930s and are used both as personnel carriers (usually armored) and as weapons platforms.

Hall rifle—A flintlock (q.v.) weapon, the Hall rifle was the first breech-loading rifle formally adopted for military use. It was invented by John H. Hall in 1811 and adopted by the US Army six years later. The lock mechanism and chamber were in a pivoted breechlock that could be rotated up for loading, and closed and locked in position for firing. The first military weapon in America with standardized parts, the Hall rifle remained the Army's primary infantry weapon and (as a carbine) cavalry weapon for about 35 years. About 1840 it was modified as a percussion weapon.

hand grenade—See **grenade**.

handgun—A firearm that is held and fired with one hand; synonymous with pistol (q.v.).

hang fire—A delay in the discharge of a round of ammunition after it has been struck by the firing pin or subjected to other igniting action. See also **misfire**.

hangar—A building used to house aircraft.

hanger—1) A short cutting sword, with a curved blade and a knuckleguard, carried by infantrymen of European armies during the 18th and 19th centuries. The naval cutlass was developed from it. 2) A device for attaching a sword or dagger to a belt.

harass— To disturb an opponent by any of a variety of means, such as artillery fire, raids, patrols, etc.

harassing fire—See **harassment**.

harassment—A function in which the primary objective is to disrupt the activities of a unit, of installations, or of a ship, rather than to inflict serious casualties or damage. **Harassing fire** is fire designed to disturb the enemy's troops, to curtail movement, and, by threat of losses, to lower morale. Harassment can also be carried out by air attacks, usually called air harassment.

harbor defense—The defense of a harbor or anchorage and its water approaches against naval surface or submarine attack, enemy mine-laying operations, and sabotage. However, the defense of a harbor against guided or dropped missiles while such missiles are airborne is considered a part of air defense.

hardened site—An installation (which may or may not be a fortification) constructed to withstand the blast and associated effects of a nuclear attack and likely to be protected against a chemical, biological, or radiological attack.

hardstand—See **apron**.

harpax/harpago—An innovation in naval warfare introduced by Octavian's admiral, Agrippa, it was a pole, with a hook on the end, that was shot from a catapult into the side of an opposing ship, where the hook would hold it fast. Attached to the end of the harpago was a rope, which could be winched in to bring the two vessels together and facilitate boarding. This, the precursor of the whaling harpoon, was first used effectively

by Agrippa at the Battle of Naulochos (36 B.C.) and again at Actium (31 B.C.).

harquebus—See **arquebus**.

harquebusier—See **arquebusier**.

hash mark—See **service stripe**.

hastatus—(From the Latin, meaning one armed with a spear or javelin). The hastati (young but experienced soldiers) who carried two javelins (pilum, q.v.) about seven feet long, a shield, and a short broadbladed sword (gladius, q.v.). They were the principal manpower component of the Roman legion (q.v.). After throwing their javelins the hastati would close and fight with their swords.

hasty defense—See **defense**.

haubergeon—A short coat of mail (q.v.), generally sleeveless. Also known as the lesser hauberk, it was a shortened version of the hauberk (q.v.).

hauberk—A tunic or shirt or coat of mail. Although haubergeon (q.v.) was sometimes used for short tunics, hauberk seems to have included all types.

havelock—A covering of cloth that fits over a cap and has a flap to protect the back of the neck from the sun. The havelock is named for Sir Henry Havelock, a British general who served in India in the mid-18th century. The name was adopted by the US Navy for a rain cover for the WAVES uniform hat.

haversack—A one-strapped canvas bag worn over a shoulder to carry supplies on a hike or a march.

havildar—A noncommissioned rank in the Indian Army corresponding to sergeant. The most senior holders of this rank were **havildar-majors**.

headquarters—1) The command echelon of a unit or other organization consisting of a commander and his staff. 2) The geographical center of authority of any organization, that is, the building, base, or other place where the commander and his staff are located.

headquarters company—A military unit that supports the headquarters of a battalion or larger unit. In the US Army all enlisted members of a headquarters are assigned for administration, food, and housing to a headquarters company (**headquarters battery** for an artillery unit, **headquarters troop** for a cavalry unit).

heavier-than-air (HTA)—Designation of airplanes and helicopters to distinguish them from blimps and other lighter-than-air craft.

heavy—*adj.* Used to denote a large weapon, ship, tank, or unit. For instance, infantry is normally categorized as light or heavy; artillery as light, medium, and heavy (and sometimes superheavy).

heavy artillery—See **artillery**.

heavy cavalry—See **cavalry**.

heavy cruiser—See **cruiser**.

heavy force—A military organization equipped with weighty and cumbersome weapons and other matériel which provide powerful combat capability, but which limit mobility and transportability.

heavy infantry—See **infantry**.

heavy tank—See **tank**.

heavy weapons—US Army infantry term for such crew-served ordnance as mortars, infantry cannon, heavy machine guns, and recoilless rifles.

hedgehog—1) A portable obstacle, made of crossed poles laced with barbed wire, in the general shape of an hour glass. 2) A beach obstacle, usually made of steel bars or channel iron, imbedded in concrete and used to interfere with beach landings. 3) An antisubmarine device with 16 charges that could be fired simultaneously. 4) A close formation of pikemen with their long spears that pointed out in all directions like the spines of a hedgehog and were used against attacking cavalry. 5) In World War II, a fortified strongpoint or grouping of fortifications capable of all-around defense.

hedron—US Navy designation of the headquarters of an air squadron, including administration and support activities.

height of burst— The vertical distance of an exploding projectile from the earth's surface or from a target. For a nuclear weapon, the optimum height of burst is that at which it will produce destruction over the maximum possible area.

helicopter—An aircraft that derives lift over an approximately vertical axis from revolving engine-driven wings or blades. A helicopter possesses the capability of taking off and landing without a run, and of hovering at a fixed altitude above a given point on the ground or water surface. Helicopters are used for carrying airborne units in landing operations, for submarine search-and-strike purposes, for reconnaissance, for artillery spotting, for liaison, for evacuating the wounded, and as weapons platforms.

heliograph—A device with a movable mirror that flashes the light of the sun in order to send messages in a code of long and short flashes (as in the Morse Code).

Hellcat—Single-engine US Navy World War II fighter, designated F6F and manufactured by Grumman.

helm—1) The device that steers a ship; thus the word also denotes command of a ship or an organization. 2) A steel helmet introduced about 1200 to wear over the steel or iron skullcap. The helm was cylindrical, or nearly so, and evolved from a flat-topped piece to a rounded helmet long enough to attach to the breastplate. It was particularly popular for wearing in tournaments.

helmet—Any special piece of headgear designed to protect the wearer from such hazards as blows, shell fragments, crash injury, and cold. Helmets are usually made of steel, leather, or some other form of armor, and are lined or padded.

Henry rifle—A breech-loading repeating rifle designed by B. Tyler Henry of the New Haven Arms Co. in the mid-19th century. It had a tubular magazine that held 15 rimfire .44 (11.2 mm) cartridges. A toggle-link lever action moved the cartridges into the firing chamber seriatim, and they could be fired quite rapidly—about 15 shots in 11 seconds with a sustained rate of fire of about 20 rounds per minute, including reloading, over a period of six minutes. The Henry rifle was a forerunner of the Winchester (*q.v.*).

herald, military—An official in a feudal army whose function was to issue proclamations by the commander and carry important messages.

heraldry, military use of—The design and granting of a right to wear a device or coat-of-arms, which indicates the origin or ancestry of a unit or organization and identifies membership in that unit or organization. Heraldry was developed to identify noble and royal families, and thus feudal armies and their commanders. It survives in modern military organizations in flags and pennants, badges, shoulder patches, etc.

hersillon—A strong beam with spikes protruding from all sides, thrown across a breach in a fortification to prevent attackers from passing through.

H-hour—A term customarily used to refer to the precise time at which some specified operation—an attack, amphibious assault, or movement—will begin. See also **D-day**.

Higgins boat—A landing craft designed by Andrew J. Higgins in 1939 and used in amphibious landings in World War II. Its principal features were shallow draft and a bow ramp by means of which troops and vehicles could be disembarked rapidly on a beach or in shallow water close to the beach. It was the prototype for most subsequent landing craft.

high altitude bombing—Horizontal bombing from an aircraft flying at an altitude over 15,000 feet.

high angle fire—Fire delivered at elevations greater than the elevation of maximum range. In high-angle fire the range decreases as the angle of elevation is increased. Should not be confused with indirect fire (q.v.), which is entirely different. See also **elevation** and **low angle fire**.

high burst adjustment—A registration of an artillery piece upon an arbitrary point in the sky when, for any reason, it is not possible to register on a base point (q.v.). A single gun is registered by time fire at an elevation and time fuze setting which will assure that the rounds (all at the same settings) will burst in the air, and in the field of view of observers with survey instruments at positions precisely located with respect to the guns by survey. Each observed burst is then plotted by trian-

gulation; the center of impact is located on the firing chart; and this becomes the base point until there is an opportunity to register on an identifiable point on the ground. The range elevation for the guns to hit this arbitrary base point is calculated by deducting algebraically the value of the angle of site and complementary angle of site from the quadrant elevation (q.v.).

high explosive—Any powerful, non-nuclear explosive characterized by extremely rapid detonation and having a powerful disruptive or shattering effect. Distinguished from a low explosive like gunpowder, which was used for exploding or detonating shells or devices before the pre-World War I adaptation of TNT as the standard military explosive.

high explosive antitank (HEAT)—See **ammunition, armor-piercing**

high explosive (HE) bomb—See **bomb**.

high frequency direction finder (HF/DF)—A radio receiver that is equipped to indicate the direction from which a broadcast is being sent. The point at which the direction lines from two or more receivers in different locations intersect identifies, within a reasonable degree of accuracy, the location of the transmitting radio. "**Huff Duff**," as HF/DF was commonly called, was extremely important in World War II in directing antisubmarine units to the areas where German U-boats were operating.

high-intensity conflict—See **conflict levels of intensity**.

hilt—The end of a sword by which it is held. Hilts have been designed in many shapes and sizes. Most distinctive is the **basket hilt**, which includes a guard, more or less in the shape of a basket, that provides cover and protection for the swordsman's hand.

Hindenburg Line—A highly organized defensive zone prepared by the German Army in northern France in early 1917 to shorten the front line and permit it to be held by fewer troops. Behind a lightly held outpost line heavily sown with machine guns lay two successive defensive positions, heavily fortified. Behind these again lay the German reserves concentrated and prepared for counterattack. Each successive defensive line was so spaced in depth that, should one be taken, the attacker's artillery would have to move to another, more forward fire position before taking action against the next zone. Between the original line and the new zone, the countryside was devastated: towns and villages were razed, forests leveled, water sources contaminated, and roads destroyed. See also **Siegfried Line**.

hipparch—A commander of cavalry in ancient Greek armies.

hit—n. A blow or impact on a target by a projectile or hand-held weapon. v. To strike with a hand-held weapon or a projectile.

hitch—In the US armed forces, a term popularly used for a period of voluntary enlistment.

hoarding (fortification)—Wooden fencing, generally used to support the sides of trenches.

hobelier—A light horseman of the Middle Ages who was used for reconnaissance, carrying intelligence, harassing troops on a march, intercepting convoys, and pursuing a routed enemy.

hoistman—The member of a ship's gun crew who directs the supply of ammunition from a magazine below decks to the gun compartment.

holding attack—See **attack** and **secondary attack**.

hollow charge—An explosive charge designed to penetrate armor or concrete. It uses the Munroe effect, whereby a cavity in a high-explosive charge focuses the effect of the detonation in a narrow cone of highly penetrating flame and blast. Hollow charges were first used extensively in armor-piercing projectiles in World War II. Also called a **shaped charge**.

hollow square—See **square**.

holster (saddle, belt)—A leather case shaped to hold a pistol. An item of uniform for all military persons armed with pistols.

holy water sprinkler—A mace with protruding spikes and a long handle used in the 16th and 17th centuries. Such a mace was sometimes called a morning star (q.v.).

home defense—See **home guard**.

home guard—An organization of volunteers or militia set up to provide defense for an area or a nation from which regular military forces have been sent away or are hard-pressed and diverted by battle engagement. In World War I and World War II, for example, many states of the United States, particularly those on the coasts, formed home guards to replace National Guard units that had been called to active military service.

honor guard—See **guard**.

honors—Recognition, in the form of an award, a ceremony, a written commendation, or other device, of an outstanding achievement or performance, or of the presence or arrival of a person of particular distinction.

honors of war—A measure taken by a victor in recognition of merit in a vanquished enemy, such as permitting him to march out of a town or position he has surrendered, armed and with colors flying. Because British General Clinton denied honors of war to the army of General Benjamin Lincoln that surrendered at Charleston in 1779, General Washington denied honors of war to General Cornwallis's army when it surrendered at Yorktown in 1781.

hoplite—A well-disciplined Greek infantryman who fought in the phalanx formation. The hoplite carried an 8-to-10-foot pike and a short sword. He wore a helmet, breastplate, and greaves (q.v.), and he carried a round shield, called a **hoplon**. In the Macedonian army of Phillip II and Alexander the Great, most hoplites were armed with a longer spear, the sarissa (q.v.), and a larger shield than were the Greek hoplites.

hoplon—See **hoplite**.

horizontal bombing—Bombing from an aircraft flying at a constant altitude above a target.

horizontal error—The error—in range, or deflection, or the error that is expressed in radius of distance from target—that a weapon may be normally expected to exceed. The horizontal error of weapons making a nearly vertical approach to the target is often described in terms of circular error probable (CEP, q.v.) in radius distance. Horizontal error of weapons producing an elliptical dispersion pattern (that is, the dispersion pattern of shell fragments as they strike a target area) is expressed in terms of probable error (q.v.), which can be expressed as an angle (degrees or mils) or as a distance at different ranges.

hornbow—See **bow**.

hornwork—A small outwork protruding from a major defensive work or wall, consisting of two half bastions, or demi-bastions, joined by a curtain. Hornworks were designed to protect the approach to a gate or postern entrance in the major work.

horse— Collectively, cavalry or mounted foot soldiers.

horse artillery—Artillery designed to support cavalry, distinguished from horse-drawn artillery by the fact that all personnel were mounted on horses to enhance mobility.

horse guards—1) A mounted formation of guards (q.v.). 2) Designation of a formation of mounted guards in the British Army.

horse holder—1) A soldier designated to hold the horses of a small unit of dragoons or cavalry who have dismounted to fight on foot. 2) A soldier designated to hold the horse of an officer while he is performing dismounted duties.

horsetail—A Turkish standard, denoting a pasha's rank. Commanders were distinguished by the number of horsetails carried before them.

hospital—A medical facility fully equipped for treatment of the sick and wounded. The US Army maintains both general hospitals, comparable to civilian hospitals, and field hospitals, which are set up temporarily in or near a combat zone.

hospital apprentice—See **hospital corpsman**.

hospital corpsman—In the US armed forces, an enlisted man in a medical unit, including, in the Navy, **hospital apprentices** (unrated personnel, corresponding to apprentice seamen) and the various pharmacists' mates and other medical ratings.

Hospitaler—A member of a military religious order, the Knights Hospitalers (Knights of St. John), founded among European crusaders in 12th-century Palestine. Later, the Hospitalers were the defenders of Rhodes and then Malta against Turkish attacks.

hospital ship—A ship equipped as a hospital, completely unarmed, usually painted white and marked with prominent red crosses or other indications of its function, and illuminated at night for ready identification. Hospital ships are used particularly during amphibious operations, and also for evacuating casualties to medical facilities farther away from combat areas.

hostage—A person seized by a belligerent, or sometimes given by one belligerent to another, and held to insure that certain terms or agreements will be kept.

hostilities—1) pl. Acts of battle or of fighting; acts of warfare. 2) The state or situation under which armed conflict takes place or can be expected to occur. Hostilities may begin or cease without causing the beginning or end of a war.

Hotchkiss gun—Originally a five-barrel 37mm revolving cannon designed in 1871 by an American, B. B. Hotchkiss, and used by the British Royal Navy. Later, any gun made by the Hotchkiss Company. Also, a small, single-barreled (8mm) machine gun made by the Hotchkiss Company early in the 19th century and used as a British cavalry weapon into the 1930s.

Hotel—Letter of NATO phonetic alphabet. See also **alphabet, phonetic**.

hot shot—Ammunition, especially cannon balls, heated before firing and generally used against combustible ships or structures. In order to avoid detonation of the propelling charge by the heat of the ball, wet clay or straw was inserted between the powder and the shot. The first use is unknown, but hot shot appears to have been used at the siege of Cherbourg in 1418. Incendiary shells replaced hot shot in the 19th century.

houfnice—The basic Hussite field gun of the 15th century, with a caliber of 150–250mm and a chamber of smaller diameter. The important, innovative feature of this gun was that it was mounted in a two-wheeled carriage with a trail, which made laying the guns much easier for both range and direction. The name comes from the Czech term *houf* meaning a group of men.

house carls—A member of the bodyguard or household troops of a Danish or early English king or noble.

household cavalry—Mounted units of royal guards.

household troops—Troops whose duty is to guard the security of a sovereign and his domestic establishment. See also **guards**.

How—Letter of old phonetic alphabet. See also **alphabet, phonetic.**

howitzer—1) A cannon that combines certain characteristics of guns and mortars. The howitzer is lighter than a gun of comparable caliber, and heavier than a mortar, and delivers projectiles with medium velocities, either by low or high trajectories. 2) A cannon with a tube length 20 to 30 calibers. See also **gun** and **cannon.**

"Huff Duff"—See **high frequency direction finder (HF/DF).**

hung bomb—A bomb that accidentally clings to the bomb rack after the releasing action has been taken.

hunter-killer operations/group—Offensive antisubmarine operations combining searching, tracking, and attacking capabilities of air, surface, and subsurface forces in coordinated action to locate and destroy submarines at sea. Hunter-killer operations involve groups of ships consisting of one or more antisubmarine carriers and a number of escort vessels.

hurdles—Obstacles, usually movable and made of lumber or logs, that can be readily emplaced to block roads and defiles (q.v.).

hussar—A light cavalry soldier with special uniforms and traditions. The original hussars were formed by King Matthias Corvinus of Hungary in 1458, and they performed with considerable success against the Turks. Similar units, generally uniformed in Hungarian-type uniforms with bright color and frogging, were formed and named hussars in other European armies. Like other cavalry units, hussars have been converted from horse cavalry to tank units.

hut—A temporary building, usually small, for housing troops or storing arms. A collection of such buildings is a **hutment.**

Hutier tactics—The name given by the Allies to German infantry tactics that were characterized by a short, sharp concentration of artillery fire, followed immediately by infantry assaults, both guns and troops being brought into position at the last possible moment to ensure surprise. Heavy concentrations of smoke shells fired on enemy strongpoints permitted the attacking infantry and light guns to bypass them. The term "Hutier tactics" was given to this type of offensive because it was first employed on the Russian Front in 1917 by a German army under General Oscar von Hutier, who later commanded an army employing these tactics in the German Somme Offensive of 1918. It was called **infiltration tactics** by the Germans.

hypaspist—An ancient Macedonian infantryman. Hypaspists originally carried the shield or weapons of a heavily armed hoplite (q.v.). Philip of Macedon reduced the weight of their equipment and trained them as highly mobile troops. He used them skillfully, notably at the Battle of Chaeronea in 338 B.C., when they attacked on his army's right, then withdrew, pursued by the Athenians, who thus created a gap through which Philip's son, Alexander, drove, to strike the Thebans in the rear.

hypervelocity armor-piercing (HVAP)—See **ammunition, armor-piercing.**

ICBM—See **intercontinental ballistic missile**.

Identification Friend or Foe (IFF)—1) A system using electronic transmissions to which equipment carried by friendly forces automatically responds, for example, by emitting pulses. It is a method of determining the friendly or unfriendly character of aircraft and ships using electronic detection equipment and associated Identification Friend or Foe systems. 2) A general term applied to electronic equipment used to identify hostile or friendly aircraft or ships; any of several systems used in such identification. Before and during much of World War II, identification depended upon the use of predesignated signals for challenge and reply.

IFF—See **Identification Friend or Foe**.

IFV—See **infantry fighting vehicle**.

illuminating grenade—See **grenade**.

illuminating shell—A shell designed to become a flare when exploded well above a target area, so as to light it for bombing or identification. The flare can be made more effective if combined with a parachute which is released upon explosion. Also called a **star shell**.

immediate message—See **message precedence**.

Immortals—An elite corps of bodyguards in the ancient Persian armies of Achaemenian Persia, the dynasty that ruled ancient Persia from 559 B.C. to 330 B.C.

"Immunes"—A brigade of 10,000 enlisted US volunteers in the Spanish-American War whose immunity to tropical diseases (particularly malaria) allowed them to be used in Cuba, where large numbers of nonbattle casualties had been sustained because of communicable diseases of local origin. The "Immunes" were organized in ten infantry regiments, the 1st through the 10th Volunteer Infantry, six of white troops and four of black.

impact, mean point of—The point whose coordinates are the arithmetic means of the coordinates of the separate points of impact of a finite number of projectiles fired or released at the same aiming point under a given set of conditions. In bombing, it is the mean, or geometrical, center of a bombfall pattern, excluding gross errors. See also **center of impact**.

impact area—An area having designated boundaries within the limits of which it is intended that all ordnance, including projectiles, bombs or other missiles fired by artillery or launched from aircraft, will make contact with the ground.

impact fuze—See **fuze**.

impedimenta—Supply train; the term comes from the Latin word for baggage.

impi—A Zulu military formation, usually consisting of several thousand men.

implosion weapon—A weapon in which a quantity of fissionable material, less than a critical mass at ordinary pressure, has its volume suddenly reduced by compression—a step accomplished by using normal chemical explosives around the fissionable material—so that it becomes supercritical, producing a nuclear explosion.

impressment—The forcing of service in an army or navy through kidnapping or the involuntarily seizure of citizens and transporting them to an army unit or a ship. This practice was common in England, France, Prussia, and other nations into the 19th century, to the extent that some units were predominantly composed of impressed men who usually fought without enthusiasm. Press gangs roamed seedy, poorly lit areas, kidnapping drunks and other unsuspecting men; others were sent from jails to military units. One of the causes of the War of 1812 was the British practice of sending boarding parties aboard US vessels and impressing men they claimed were British, US citizens among them, to fight in the Napoleonic wars.

inactive status—The opposite of active duty. A reservist who is not performing full-time military duty with an organized, fully commissioned unit is in an inactive status. An installation or organization that is not actively functioning or not in use by a military organization is in inactive status.

in being—Organized, manned, equipped (though not necessarily at full strength), and ready to operate. Used in reference to any military force.

incendiary—n. Any chemical compound that causes fire. adj. Causing fire (incendiary attack, incendiary bomb).

incendiary bomb—See **bomb**.

incendiary grenade—See **grenade**.

independent unit or formation—A military organization that is not an organic part of a larger unit or formation. An independent unit may be attached temporarily to another, usually larger, unit, but is available for transfer from one to another as the senior commander considers desirable or necessary.

India—Letter of NATO phonetic alphabet. See also **alphabet, phonetic**.

indicated airspeed—See **airspeed**.

indirect fire—Fire delivered on a target which cannot be seen by the aimer. Should not be confused with high-angle fire (q.v.), which has a totally different meaning.

induct—To conscript, or to take a person into an armed service without voluntary action on his part. In the United States this has been

performed through the Selective Service System.

industrial mobilization—The transformation of industry from peacetime activities to the industrial program that supports national military objectives. It includes the mobilization of materials, labor, capital, production facilities, and contributory items and services essential to the wartime program.

infantry—A branch of an army in which soldiers are organized, trained, and equipped to fight on foot. Infantry units are designated **light** or **heavy,** according to the equipment they carry and the speed with which they can move. In modern armies many if not most infantry units are **mechanized**—that is, they move in vehicles and sometimes fight from them. Light infantry became standard parts of armies in the 17th century. Predecessors were the psiloi (q.v.) of the Greek and Macedonian armies and skirmishers (q.v.) in Roman armies and those of the Middle Ages.

infantry division—A combined arms ground force of division (q.v.) type and size, the principal combat components usually being three infantry regiments or brigades. It may or may not be motorized.

infantry fighting vehicle (IFV)—An armored vehicle, tracked or wheeled, primarily designed to transport infantry into combat and permit a majority of the transported personnel to engage the enemy from the fully enclosed vehicle.

infiltrate—To carry out infiltration.

infiltration—The movement through or into an area or territory occupied by enemy troops or organizations. The movement is made, either by small groups or by individuals, at extended or irregular intervals. Maximum use is made of deception, to avoid detection.

infiltration tactics—See **Hutier tactics**.

infirmary—A medical facility smaller and having less equipment than a hospital. Patients may be given emergency treatment and kept for minor problems or brief periods of recuperation and treatment.

influence fuze—See **proximity fuze**.

information officer—An officer whose duty is to provide authorized information about a command to the nonmilitary public.

infrastructure—A term used in NATO that is generally applicable to all fixed and permanent installations, fabrications, or facilities for the support and control of military forces. Derived from a French railroad term.

initial issue—First combat supply, placed at the disposal of units of all arms of all services in peacetime. In principle it enables these units to fulfill their first mission without further supply. It can be transported by the organic transport of the unit.

initial point—1) Any designated place at which a column or element thereof is formed by the successive arrival of its various subdivisions, and comes under the control of the commander ordering the move. 2) A well-defined point, easily distinguished visibly or electronically, used as a starting point for the bomb run on a target. The first point at which a moving target is located on a plotting board.

inshore patrol—A naval defense patrol operating generally within a defensive coastal area and comprising all elements of harbor defenses, the coastal lookout system, patrol craft, supporting bases, aircraft, and Coast Guard stations.

insignia—Any distinctive device that identifies a person or object as to nationality, organization, office, rank, or branch of service. Examples of insignia are badges of rank, badges designating branch of service, shoulder patches, chevrons, aircraft and vehicle markings.

inspection—The examination of personnel, organizations, activities, or installations to determine their effectiveness and economy of operation, adequacy of facilities, readiness to perform assigned missions, or compliance with directives; also, the examination of matériel to determine quality, quantity, or compliance with standards. An inspection of personnel or an installation by a visiting dignitary is often a mark of respect, and is included in "honors" (q.v.).

inspector general—1) An officer whose duty is to investigate and report on the condition and noncombat performance of a military or-

ganization. 2) In some armies (as the German) the senior officer.

installation—A military facility in a fixed or relatively fixed location, together with its buildings, building equipment, and subsidiary facilities such as piers, spurs, access roads, and beacons. See also **base, post, camp, station**.

instrument bombing—Bombing by the use of radar or other instruments to locate the target and direct and release the bombs, without visual reference to the ground.

insurgency—A condition resulting from a revolt or insurrection against a constituted government that falls short of civil war. See also **conflict levels of intensity**.

intelligence—The product resulting from the collection, evaluation, analysis, integration, and interpretation of all available information that concerns one or more aspects of foreign nations, their areas of operations, and their military establishments, and which is immediately or potentially significant to military and diplomatic planning and operations.

intelligence estimate—See **estimate, intelligence**.

intention—An aim or design (as distinct from capability) to execute a specific course of action.

intercept—v. 1) To interfere with, and if possible prevent, the arrival of a hostile military force at its destination. 2) To overhear, without permission, communications between hostile or potentially hostile electronic stations. 3) To make contact, visual or by radar, of a friendly aircraft with an unidentified one. This includes **close controlled interception**, in which the interceptor is continuously controlled by a surface or air station; and **broadcast controlled interception**, in which the interceptor is given the area of interception by a surface or air station but effects interception without further control.

interceptor—A manned aircraft used for identification and engagement of airborne objects. Usually an interceptor is characterized by a high rate of climb, high speed, high ceiling, effective armament, and relatively short endurance.

intercontinental ballistic missile (ICBM)—A ballistic missile with a range over 3,000 nautical miles. See also **missile, ballistic**

interdict—To prevent or hinder enemy use of an area or route for supply, communications, or movement. Interdiction may be accomplished by gunfire, by aerial bombing, or by otherwise destroying significant parts of the enemy's route or attacking his combat or support forces on the route.

interdiction fire—Fire placed on an area or point for interdiction purposes. See also **interdict**.

interior guards—The guards who police the entrances of, and key points or areas in, a base or other installation.

interior lines—Advantageous lines of communication resulting from a situation wherein a force, owing to its position with respect to the enemy forces, or because of superior lateral communications, enjoys advantages of time and space in the employment of its units against the enemy. Offensive operations on interior lines are usually conducted from a central locality against enemy forces advancing along convergent lines.

intermediate range ballistic missile (IRBM)—A ballistic missile with a range of up to 1,500 nautical miles. See also **missile, ballistic**.

intermediate-range bomber—See **bomber**.

internal ballistics—See **ballistics**.

internal security—The state of law and order prevailing within a nation that protects it from subversion by a hostile force.

interoperability—The capability of different national forces in an alliance to work together in a combat situation. This is in part a function of weapons and equipment standardization and in part a function of common or compatible doctrines. Particularly used with respect to NATO.

interrogation—The procedure or activity of questioning prisoners of war.

interrupted screw—Construction of breechblock and breech receiver of a gun so that the breechblock can be swung into position and

achieve, by a one-quarter or one-eighth turn of the mechanism, full obturation through the close fit of breechblock and receiver.

interval—1) The space between adjacent groups of ships or boats measured in any direction between the corresponding ships or boats in each group. 2) The space between adjacent individuals, vehicles, or units in a formation that are placed side by side, measured abreast. 3) The space between adjacent aircraft measured from front to rear in units of time or in-flight distance.

intervisibility—The existence of a visually unimpeded, direct line of sight between two geographically identifiable points.

in the clear—A message transmitted in the language in which it was written, as opposed to being sent in code.

intruder—1) A hostile or unidentified aircraft that appears in a defended airspace. 2) An aircraft sent on an intruder mission.

intruder operation—An offensive operation by day or night over enemy territory with the primary objective of destroying enemy aircraft in the vicinity of their bases.

invade—To cross the borders of a nation by force or with a military force for the purpose of seizing territory.

invalid—A soldier who for physical reasons, particularly as the result of enemy action, is assigned to limited, noncombat duty.

invasion—An intrusion of military forces into the territory of another nation.

invest—To deploy troops or ships around an enemy objective in such a fashion as to cut all lines of communication. To besiege (q.v.).

investment—The deployment of troops or ships around an enemy objective, cutting it off from communication with its own forces.

IRBM—See **intermediate range ballistic missile**.

ironclad—A naval vessel, usually built of wood, and covered with plates of iron (up to four and one-half inches thick) for protection against enemy shot and shell. Ironclad floating batteries were used by French and British forces in the Crimean War. The most famous ironclads were the *Monitor* and the *Merrimack*, which fought each other in the Chesapeake Bay in March 1862, in the Battle of Hampton Roads, during the American Civil War. A number of other ironclads were built before the end of the 19th century when iron plates gave way to hulls built of steel.

Iron Cross—A German military decoration, in the shape of a Maltese Cross of cast iron edged with silver. The *Eisernes Kreuz* was instituted in 1813 as a reward for distinguished service in the War of Liberation against Napoleon, was revived at the time of the Franco-Prussian War (1870–1871), and was again issued for service in World Wars I and II.

iron ration—An emergency ration consisting of concentrated food.

"Ironsides"—Nickname for highly disciplined cavalry units of the Parliament Army during the English Civil War. The Ironsides were cuirassiers (q.v.), first recruited and trained by Oliver Cromwell, beginning in late 1642. Cromwell stressed training, unit cohesion, and morale; he would have only "honest and religious" men in his units, and saw to it that they were paid regularly, well equipped, and adequately clothed. These troopers were more than a match for the excellent but undisciplined cavalry of the Royalist Army and became the cavalry backbone of the New Model Army with which the Parliament won the war. See also **"Old Ironsides."**

irregular troops—Armed individuals or groups who are not members of regular forces. They are also known simply as **irregulars**.

irregular warfare—Warfare conducted by armed groups of guerrillas, partisans, or other nonregulars usually within enemy-held territory.

Item—Letter of old phonetic alphabet. See also **alphabet, phonetic**.

J

jack—1) A type of body armor worn by the common soldier and by sailors until it was replaced by the buff coat at the end of the 16th century. It consisted of two layers of linen or canvas, with pieces of iron or horn sewed between them. 2) A small flag, usually flown on the bow of a naval vessel, indicating its nationality, or used otherwise as a signal flag. The jack used by the US Navy consists of the union of the national ensign (and is therefore called a "union jack"). It is flown from a small staff called a **jack staff** when a naval vessel is at anchor or moored.

jack boots—Cavalry boots of thick, firm leather, specially hardened, sometimes lined with plates of iron.

jacket—1) British term for the outer casing of a shell or a bullet. 2) Old British military slang expression for promotion to the rank of senior captain of artillery, because of the distinctive uniform jacket worn by artillery officers of that rank.

jack man—One wearing a jack (q.v.); a horse soldier.

jack staff—See **jack**.

jaculator—In the Roman Army, a lightly armed soldier whose weapon was the javelin. Derived from the Latin word for throw. See also **veles**.

jaeger or **jäger**—Rifleman. The first jaegers (literally huntsmen) were light infantrymen, organized by Frederick the Great during the Seven Years' War from foresters and gameskeepers. They were expert marksmen, but armed with hunting rifles whose slow rate of fire put them at a disadvantage against regular troops. The term "jaeger" persisted in the German and Austrian armies, designating both lightly armed and hunter-type units of various kinds, for example, the Panzer Jaeger (or tank destroyer) units and the Jaeger division, or light division, of World War II.

jaeger rifle—A flintlock (q.v.) hunting rifle, with a grooved barrel, developed in Germany in the 17th century. (Also known as a **jäger**, the German term for hunter). The spin thus given to the bullet gave it a great advantage in accuracy and range over the smoothbore muskets then the standard infantry weapon. It was, however, much heavier and more cumbersome than the musket, and took longer to reload after firing. This heavy weapon was modified in North America, mainly by German craftsmen in Pennsylvania, becoming a lighter, longer-barreled weapon in the 18th century, and thus was the forerunner of the grooved, or rifled, weapons that gradually replaced the smoothbores (q.v.) in the early decades of the 19th century. See also **rifle** and **Kentucky rifle**.

jam—1) To make the transmission of a radio unintelligible; to make a radar or radio set ineffective, either by use of countertransmissions or by the use of a confusion reflector. 2) Of a firearm, to stick or become inoperative because of improper loading, ejection, or the like.

janissary—1) A Turkish guard, one of an elite body organized in the 14th century and maintained until 1826 as the permanent element of the army of the Ottoman Empire. The first janissaries were former Christians who had been captured in childhood and brought up to become fanatical Moslems. They were highly disciplined infantry that evolved into a formidable military and political force and a privileged social class but then deteriorated. In June 1826 the janissaries revolted against Sultan Mahmud, whose attempt to modernize and Westernize the army was correctly seen by the janissaries as a threat to their existence. They were massacred by loyal troops and the mobs of Istanbul. 2) Term for a loyal follower, generally used in the plural.

javelin—A light spear—soldiers often carried two or more—that was common in ancient armies. Javelins were usually thrown simultaneously by a massed or formed group against a similar enemy group, although individuals achieved considerable skill at hitting individual targets. After the phalanx appeared, javelins were used by lightly armed auxiliary troops, to shake the enemy's morale, open gaps, and harass flanks.

jeep—1) A rugged four-wheel drive quarter-ton-capacity vehicle first widely used by the US Army in World War II. The name derives from the pronunciation of the initial letters of General Purpose, G.P. 2) In the US armored forces of World War II the term was applied to the one-half-ton command-car vehicle; in those units and forces the one-quarter-ton vehicle was called a **peep**. 3) A smaller version of other military items, including a small airplane used for reconnaissance and liaison, an amphibious truck, and an escort carrier.

jemadar—A rank for native officers in the Indian Army corresponding to lieutenant in the US Army.

Jig—Letter of old phonetic alphabet.See also **alphabet**, **phonetic**.

jihad—A holy war of Moslems against infidels.

joint—In the United States armed forces connoting participation by two or more services, as "joint operation" or "joint command." See also **combined**.

Joint Chiefs of Staff (JCS)—A body of senior officers within the Department of Defense consisting of the Chief of Staff, United States Army, the Chief of Naval Operations, the Chief of Staff, United States Air Force, the Commandant of the United States Marine Corps, and a Chairman, all of whom serve as the principal military advisers to the president, the National Security Council, and the Secretary of Defense. The JCS was formally authorized by the National Security Act of 1947, but had actually been in existence since 1942, when the group was established by action of the president under statutory war

powers. There have been minor changes in the organization and functions of the JCS by legislation in 1949, 1953, and 1958.

joint servicing—An administrative function performed by a jointly staffed and financed group whose activity is in support of two or more military services.

joint staff—1) The staff of a specified or unified command, or of a joint task force, which includes members from the several US armed services comprising the force. A joint staff is created to ensure that the commander understands the tactics, techniques, capabilities, needs, and limitations of the force's component parts. Positions on the staff may be divided so that service representation and influence generally reflect the composition of the force. 2) The staff of the Joint Chiefs of Staff as provided for under the National Security Act of 1947, and as amended by later legislation.

joint strategic capabilities plan—A short-range current capabilities plan of the US Defense Department, prepared by the JCS, that translates US national objectives and policies for the next fiscal year into terms of military objectives and strategic concepts and that defines military tasks for cold, limited, and general war which reflect actual US military capabilities. Referred to as JSCP (or jay-scap).

joint strategic objectives plan—A mid-range objectives plan of the US Department of Defense, prepared by the JCS, that translates US national objectives and policies for a time frame of from five to eight years in the future into terms of military objectives and strategic concepts. The plan also defines basic undertakings for cold, limited, and general war that may be accomplished with the objective force levels. Referred to as JSOP (or jay-sop).

jointure of command—A US Navy term for the transfer of command of an amphibious landing force from the Navy commander to the ground force commander at a certain point in the landing operation.

joust—1) A peacetime competition, usually among knights in medieval times, in the form of serious but usually nonlethal combat, usually with lances when the opponents were mounted; a tilting match; a tournament. 2) Any combat suggestive of such a competition.

judge advocate general—1) The senior legal officer in one of the US military services. He supervises the administration of military justice and such other legal duties as may be required within his service. 2) The legal officer on the staff of a high-level military command. An officer serving as prosecutor in a court martial, or serving as legal adviser to a commander, is a **judge advocate**.

Judge Advocate General's Corps—A corps of qualified attorneys who are also military officers and who serve under the judge advocate general of their service, assisting him in the supervision of military justice. Most assignments as staff judge advocates or judge advocate generals in higher level commands are held by members of this corps.

Juliet—Letter of NATO phonetic alphabet. See also **alphabet**, **phonetic**.

jumpmaster—An officer or noncommissioned officer who commands parachute troops in an aircraft, and controls their jumping and the dropping of their equipment from an aircraft.

jump of a gun—1) The motion of a small-arms weapon during the firing period, caused by the force of recoil (also known as "kick"). 2) Any jerky movement of a land or naval cannon, or their carriages, due to play in the mechanism, nonuniform application of moving power, or the like.

junction point—See **coordinating point**.

jungle warfare—Warfare in an environment of dense, intermingled growth which is distinguished from combat in other environments only by those special considerations relating to visibility, mobility, supply, climate, and the like.

junior officer—1) An officer junior to another by virtue of his rank or his date of rank. 2) In general, an officer of the rank of captain and below in the Army, Air Force, or Marine Corps, and the rank of lieutenant senior grade and below in the navy. In the US Army those who hold this rank are also generally called "company grade (q.v.) officers."

junta—A council or board, usually composed of a group of military officers who govern a country, particularly a group which has taken power in a coup d'état.

jus ad bellum—See **jus belli**.

jus belli—Law of war, or the body of doctrine, handed down by tradition, that specifies the circumstances under which war may be resorted to (**jus ad bellum**) and the practices which should not be engaged in during war (**jus in bello**). The last mentioned are often called **laws of war** or **rules of war**.

jus in bello—See **jus belli** and **laws of war**.

justice, military—The legal system governing the military forces of a nation.

just war—A concept developed by Christian thinkers in the Middle Ages which affirms that war can be considered just under, and only under, specified conditions. These conditions, as generally stated, are (1) that the war have a just cause—that is, that it be intended to remedy or prevent an injustice; (2) that no peaceful means can adequately accomplish the same end; (3) that the suffering that is likely to result from the war not outweigh the injustices it is intended to remedy; (4) that a legitimate political authority have authorized the war after giving proper consideration to the requirements of a just war; and (5) that the political authority sanction the war for just motives only.

K

kamikaze—A specially trained Japanese pilot in World War II; also, his airplane when filled with explosives. The pilot's mission was to fly his plane suicidally into an Allied target, usually a warship. The pilots were given minimal training in the handling of the aircraft and prepared ritualistically for the sacrifice before takeoff. Their sudden appearance in the last months of the war was an act of Japanese desperation. Although many were shot down, and others aborted, a number succeeded in evading antiaircraft protection and causing considerable damage to US vessels. The name means in Japanese "divine wind."

kampfgruppe—German designation of a combat team or task force (q.v.).

K-day—1) In World War II, the day set for a strike by a US Navy carrier's aircraft. It corresponds to the land forces' designation of D-day, with which it may or may not coincide. 2) The date for introduction of a convoy system on any particular convoy lane.

keelhauling—A punishment used in sailing navies, in some cases into the 19th century. The offender, weighted with iron or lead, and with ropes attached, was dropped from one yardarm of a vessel, hauled beneath the keel of the vessel, and then hoisted to the opposite yardarm. The punishment was usually fatal.

keep—The strongest area of a fortification or a castle, designed to be held in the final stages of a siege or an attack. The keep, also called a "donjon," was usually centrally located in the fortification. See also **citadel**.

Kentucky rifle—A long-barreled muzzle-loading rifle first developed in Pennsylvania in the 18th century by German immigrants and used for hunting, particularly in the wilderness areas of western Pennsylvania, Virginia, Ohio, Kentucky, and Tennessee. The Kentucky rifle proved an effective weapon in the French and Indian War, the American Revolution, and the War of 1812. Kentucky rifles were frequently elaborately decorated, with maplewood stocks, and the barrels were sometimes octagonal, like the jaeger rifle (q.v.) upon which they were modeled. See also **rifle** and **Pennsylvania rifle**.

kepi—A uniform cap worn particularly by French soldiers. The kepi has a visor and rises directly from the head to a circular, flat top, sometimes higher in back than in front. Such caps were worn by soldiers of both sides in the US Civil War. See also **forage cap**.

khaki—Often in the plural, a uniform of tan cotton khaki cloth worn by US servicemen and others in warm seasons and warm climates. The original khaki, from the Persian word for dust, is a cotton cloth of a light yellowish brown. Khaki has come to mean the color, and modern uniforms are made in a variety of materials of that color.

kick—See **jump of a gun**.

killed in action (KIA)—A battle casualty who is killed outright, or who dies as a result of wounds or other injuries before reaching any medical treatment facility.

killing area/zone—An area in which a commander plans to force the enemy to concentrate so as to destroy him with conventional or nuclear weapons.

kill potential—The potential of a weapon for killing people or destroying aircraft or weapons.

kill probability (kp)—The probability of a

given percentage of deaths resulting from a shell, bomb, or warhead, or the like exploding under a set of given conditions.

Kilo—Letter of NATO phonetic alphabet. See also **alphabet, phonetic.**

King—Letter of old phonetic alphabet. See also **alphabet, phonetic.**

kit—1) The personal belongings of a soldier. 2) The parts and instructions for use with a piece of equipment or performing some function, as in repair kit, mess kit.

kit bag—A small bag used by soldiers for carrying personal articles.

kiwi—1) A nonflying air officer. 2) A student pilot who has not flown solo. The name is taken from a New Zealand flightless bird.

klibanion—A corselet of lamellar armor (q.v.) worn by Byzantine soldiers, generally waist length with short sleeves or sleeveless.

knapsack—A case or bag, usually of canvas or leather, worn on the back to carry supplies and equipment, especially on a hike or march. In the US Army a knapsack is generally called a pack.

knee mortar—A 50mm Japanese mortar used in World War II for firing grenades and flares. The name was given it by US forces because of the mistaken belief that the base plate was concave in shape so that the mortar could rest on and be fired from the knee. Efforts to test this with captured mortars resulted in a number of broken American legs. Actually the concavity facilitated rapid use of the mortar from a solid base on the ground or on a fallen tree trunk.

knight—An aristocrat or nobleman of medieval times who served in warfare as a mounted man-at-arms, usually as a leader or officer of a feudal contingent in the service of his liege landlord. A knight was trained from youth as a page and a squire, and attained a privileged position in service to his lord.

K-9 Corps—The organization established in the US services in World War II for the purpose of using dogs for various tasks, such as guard and patrol.

knot (symbol of rank)—1) An epaulet in the form of a braided knot, usually gold braid in the ceremonial uniform of an officer. 2) A tassel of knotted leather or gold braid on an officer's sword or saber guard.

kontos—A Byzantine lance, originating with the Sarmatians and the Alans. Cavalrymen carried a 12-foot kontos, or kontarion, and infantrymen one of the same length or longer.

Krag/Krag-Jorgensen rifle—A rifle, invented by two Norwegians, Krag and Jorgensen, that was used by the armies of Denmark and Norway and, in a modified version, as the standard rifle of the US Army from 1892 to 1903. It was a .30 caliber, single-shot, bolt-action weapon with a five-round magazine.

K-ration—A packaged emergency ration for use under combat conditions, developed for US forces in World War II. Each ration contained three concentrated meals, separately packaged, including bread or biscuits, canned meat, a powdered beverage, and a confection.

Kriegsakademie—Literally war academy, this school was founded by General Gerhard von Scharnhorst in 1810 for the instruction of selected officers of the Prussian Army for duties with the General Staff. Originally called the School for Young Officers, it was renamed Kriegsakademie in 1859. It continued as the senior military educational institution of the German Army after the unification of Germany in 1871.

kriegspiel—Literally, war game, a game used for instruction and planning, employing a large map or a contoured model and markers or small figures to represent the opposing forces. Introduced into the Prussian Army about 1821.

kris—A Malayan sword with a straight handle and a wavy, double-edged blade.

Krupp gun—Any cannon produced by the factories of the Krupp family, including most of the weapons of the Prussian and German armies in the 19th and 20th centuries.

kukri—A short, heavy-bladed sword of Nepal; the hand weapon of Britain's and India's Gurkha troops (q.v.).

kyklos—A defensive naval tactic in ancient

warfare in which fleets formed a circle with all rams pointing outward. It was used by the Greeks in fighting the Persian fleet at Artemisium (480 B.C.), and by the Peloponnesians against the Athenians at Rhium.

L

laager—n. An encampment for defense, with wagons or armored vehicles encircling it for protection. *v.* To form such an encampment. Laager is an Afrikaaner word.

Lafayette Escadrille—1) A volunteer squadron made up principally of American fliers who fought for France in World War I before the United States entered the war in 1917. In 1918 it was transferred to the American service and became the 103rd Pursuit Squadron. 2) In World War II, a fighter squadron of French pilots under American command and equipped with American aircraft, which fought in North Africa.

lamellar armor—Armor originally developed in eastern Asia, composed of small plates of stiff laminated fabric laced together. By the end of the 14th century such armor was being worn in Scandinavia, where it seems to have been introduced from Russia. See also **armor.**

Lancaster rifle—A naval gun invented by the Englishman Charles William Lancaster. The first rifle cannon to be used in naval warfare, it was introduced by the British in the Crimean War. It had a smooth, elliptical bore that was twisted slightly, and fired a projectile that was also slightly elliptical in cross-section.

lance—A long, metal-tipped spear used primarily by mounted soldiers. Used by many ancient peoples, including the Assyrians, Egyptians, Greeks, and Romans, lances remained common weapons until the introduction of gunpowder.

lance corporal—1) An enlisted rank in the US Marine Corps above a private and below a corporal; comparable to private first class in the US Army. 2) A British Army private

acting as a corporal. A British corporal acting as a sergeant is a **lance sergeant**.

lancer—A cavalryman armed with a lance.

lance rest—A bracket on the right side of an armor breastplate, or a leather pocket on the saddle, into which the butt end of a lance fitted when the rider charged, preventing the lance from being driven backward when the tip hit a target.

lance sergeant—See **lance corporal.**

Lanchester Equations (Also known as **Lanchester Laws** or **Lanchester's Equations** or **Lanchester's Laws**)—A concept of military combat power and attrition developed by British aeronautical engineer Frederick William Lanchester in 1914, to show the relationship between force strengths and combat outcomes, particularly combat losses, in standard circumstances. Lanchester Equations have been used since World War II as the basis for the calculation of attrition or casualties in most ground combat models (q.v.) and simulations (q.v.). Lanchester's **Linear Law** applies when the location of the hostile force or forces is known only generally. Lanchester's **Square Law** applies when the location of the hostile force is known precisely.

land mine—An explosive device placed on or slightly beneath the surface of the ground and usually designed to explode when pressure, particularly that of a passing person or vehicle, is applied.

landing area—1) That part of the objective area within which an amphibious force initiates a landing operation. It includes the beach, approaches to the beach, transport areas, fire support areas, air occupied by close supporting aircraft, and the land through which an advance inland to the initial objective is made. 2) An **airborne landing area** is the general area used for landing troops and matériel by either air delivery or air landing. This area includes one or more drop zones or landing strips. 3) Any specially prepared or selected surface of land, water, or deck designated or used for the takeoff and landing of aircraft.

landing attack—An attack against enemy defenses by troops landed from ships, aircraft, boats, or amphibious vehicles. See also **amphibious operation**.

landing beach—A beach upon which a landing force disembarks (usually in an assault landing).

landing craft—Any of various craft employed in amphibious operations, specifically designed for carrying troops, supplies, and equipment and for beaching, unloading, and withdrawing. See also **Higgins boat**.

landing force—A task organization of troop units, aviation and ground, assigned to an amphibious assault. It is the highest command echelon in the amphibious operation.

landing operation—See **amphibious operation**.

landing ship—Any of various types of assault ship designed for long sea voyages and for rapid unloading over and onto a beach. It includes various special types, including Landing Ship, Dock (LSD) Landing Ship, Tank (LST), and Landing Ship Infantry (LSI)

landing site—A continuous segment of coastline over which troops, equipment, and supplies can be landed by surface means. It includes one or more landing beaches.

landing team, regimental—See **regimental landing team**.

landing vehicle, tracked—A lightly armored amphibious vehicle designed for amphibious assaults and operations inland.

landsknecht—One of a body of mercenary foot soldiers formed by Emperor Maximilian I at the end of the 15th century as a counter to the Swiss infantry. Organized, equipped, and trained as spearmen, just like the Swiss, they were nearly as sought for by European monarchs for the armies of the 16th century

as the Swiss themselves. The name means "man of the plains" in German.

landsturm—A German Army reserve category of men who are older or not fully fit. The term, meaning literally "territorial assault force," was originally applied to a general levy of militia. Composition of the landsturm as part of reserve organizations has differed in different historical times.

landwehr—The army reserve category in German and Austria, and some other countries, consisting of those who have relatively recently completed compulsory service. Meaning literally "territorial defense force," landwehr originally referred to the entire militia.

lane marker—A marker used to mark a minefield lane.

lanyard—1) A short piece of line used for fastening items on ships. 2) The cord or line on which a sailor carries a boatswain's pipe, or a knife, or similar object. 3) A cord or thong attached to a pistol butt, for looping about the neck to prevent loss of the pistol. 4) A cord, wire, or cable for firing certain cannon, pyrotechnics, and so forth. 5) Any of certain foreign decorations worn over the shoulder.

late medieval concentric castles—See **concentric castles**.

lateral communications—See **communications, lateral**.

latrine—A communal toilet in use by a barracks, camp, or the like.

launch—n. 1) A small boat, powered by sail or motor, used as a tender. 2) The largest boat belonging to and carried by a warship. 3) The transition of a missile from static repose to dynamic flight. v. To set in motion an attack, an aircraft, a ship, a torpedo, a rocket, a missile, or other item that once released or given an impulse will proceed by momentum or under its own power.

launcher—1) A structural device, erected on a **launch pad**, or carried on a truck, designed to support and hold a missile in position for firing. 2) The catapult that launches an aircraft from the deck of a warship.

law, military—The laws governing the operation of the military establishment and the conduct of its members. See also **justice, military**.

laws of war—Rules of behavior for states and individuals engaged in war. These rules have become "laws" by virtue of having been codified and agreed upon by treaties and by the general conventions and declarations of Paris (1856), Geneva (1864, 1906, 1909, 1929, 1949), St. Petersburg (1868), The Hague (1899, 1907, 1923), and London (1909, 1930). The first important codification of the laws of war was in *Instructions for the Government of the Armies of the United States in the Field*, prepared by Francis Lieber and issued in 1863 during the Civil War by the US Army Adjutant General as *General Orders 100*. Laws of war are generally designed to prevent loss and suffering to neutrals and noncombatants, and to prevent unnecessary suffering to combatants and prisoners—that is, to prevent suffering that does not directly further the military objectives of the damage-inflicting side.

lay—n. The setting of a weapon for a given direction or range or both. v. 1) To direct or adjust the aim of a weapon. 2) To drop, of bombs or mines from an aircraft or mines in a minefield. 3) To spread, of a smokescreen.

leadership—The art of influencing and directing people to an assigned goal in such a manner as to command their obedience, confidence, respect, and loyalty.

leapfrog—Form of movement in which like supporting elements are moved successively through or past one another along the axis of movement of supported forces. Leapfrog movements can be made either forward or to the rear.

leather armor—See **armor**.

leather gun—A cannon developed by a Swedish colonel named Wurmbrandt for Gustavus Adolphus in 1625 and first used by the Swedes in Poland in 1627. Designed to be lighter than the cast iron or bronze guns of the time, the leather gun, of which 14 are known to have been made, consisted of a copper-lined iron barrel wrapped in layers of canvas held by three wooden rings, and with an external covering of leather nailed on. The guns weighed about 90 pounds and could be transported more easily than the conventional guns, which were much heavier and hard to move. However, leather guns became so hot after repeated discharges that a new charge would sometimes ignite prematurely in the barrel before the cannoneer had ignited the primer.

leatherneck—Term used for a US Marine. The term derives from the leather neckband on the early marine uniforms.

leave—See **furlough**.

lee gage—The naval battle position downwind from an opponent's fleet. See also **weather gage**.

left face—n. 1) A 90-degree change of stationary position to the left while at attention. 2) The command to make this change. v. 1) To execute such a change in position. 2) In the imperative, the command to do it. See also **face**.

legion—1) A formation of 4,000 to 6,000 men in the Roman Army. The size of the legion varied, but it was always the basic Roman military organization. In early Republican Rome there were normally four legions, formed in two armies, each under the command of one of the two consuls who were the chief officials of Rome. The men who composed the legions, the legionaries (sometimes called legionnaires in English) were selected at the beginning of each year by the tribunes (q.v.). They served only when needed and were discharged when an emergency was over. They were divided according to age among hastati, principes, triarii, and velites (qq.v.). Each legion was composed of 60 centuries (q.v.). In the second century B.C. the consul Marius opened the ranks to any citizen who would volunteer and this laid the foundation for a permanent professional army. Pay was raised, the lightly armed velites were discontinued, and other organizational changes were made as well. By that time many more legions had been organized, and in the Empire even more were established. Roman citizenship remained a requirement of a legionary, but the extension of citizenship to conquered territories resulted in legions with large numbers of non-Italian legionaries. 2) Designation of mixed units of infantry and cavalry (and sometimes artillery) formed in several European and the US armies during the 18th and early 19th centuries. Typical was the United States Legion, with which General Anthony Wayne defeated the Maumee Indians in the Battle of Fallen Timbers (20 August 1794). See also **Foreign Legion, French** and **Foreign Legion, Spanish**.

legionary or **legionnaire**—A soldier belonging to a legion (q.v.).

letter of marque—See **letter of marque and reprisal**.

letter of marque and reprisal—A commission issued to the owner of a private vessel, authorizing its captain to operate against enemy ships as a privateer (q.v.).

levée en masse or **levy en masse**—Universal conscription. In August 1793, during the French Revolution, the Committee of Public Safety decreed conscription of the entire male population of France, thus enabling the nation to put fourteen armies into the field very quickly. Previously, most armies had had to depend on the hiring of mercenaries (q.v.) when military expansion was required.

level—1) To aim a weapon horizontally. 2) To maneuver an aircraft into flight that is parallel to the surface of the earth after gaining or losing altitude.

level, gunner's—A spirit level used to lay a gun for elevation, or range. Usually called a gunner's quadrant (q.v.).

levy, feudal—A mustering or calling of troops into service during the medieval era. In feudal times it was the responsibility of vassal lords or landowners to raise a force of agreed size and composition upon the call to arms of the suzerain.

levy, military—n. 1) The calling of militia into service. 2) The calling of a segment of a nation's population to service. 3) The troops so levied. 4) The selective calling by an army of troops from one division or other unit to provide cadre for a new organization or to fill out one in existence. v. 1) To call troops into service. 2) To call on one organization to supply troops for another.

Lewis gun—A gas-operated, air-cooled machine gun with a horizontal drum magazine invented by US Colonel Isaac M. Lewis about 1911. It was used as an infantry weapon from the beginning of World War I and widely adopted thereafter in various modifications by ground, naval, and air forces.

Lewisite—A volatile liquid developed during World War I as a lethal, persistent, blistering poison gas, recognizable by its faint, geraniumlike odor. Similar to mustard gas (q.v.).

liaison—That contact or intercommunication maintained between parts of the armed forces to ensure mutual understanding and unity of purpose and action. It is often aided by exchange of personnel in order to facilitate an exchange of information. Liaison is lateral when maintained between two adjacent units, whether or not they are under the same commander. Command liaison is that established between echelons in the chain of command.

Liberator—In different configurations, the four-engine US Navy patrol bomber and transport, PB4Y, and the US Army Air Force's B-24, a long range bomber. The Liberator was one of the most important bombers of World War II.

liberty—In the US Navy, authorized absence from duty for less than 48 hours. See **pass**.

liburnian—A Roman naval vessel.

lieutenant—In French, literally, "holding the place." 1) An assistant or deputy. (This was the original significance of lieutenant commander, lieutenant colonel, lieutenant general.) 2) In most armies and some other services, the lowest officer rank or ranks. In the US Army, Air Force, and Marine Corps, the lowest rank is second lieutenant, the rank next above is first lieutenant. 3) In the US Navy and Coast Guard, the two ranks above ensign. The lower of these is lieutenant, junior grade; senior to this is lieutenant, sometimes called lieutenant, senior grade. 4) In other navies a junior rank at various levels.

lieutenant colonel—In most armies, including the US Army, Air Force, and Marine Corps, a rank above major and below colonel.

lieutenant commander—In most navies, including the US Navy and Coast Guard, a rank above lieutenant and below commander.

lieutenant, flight—See **flight lieutenant**.

lieutenant general—In most armies including the US Army, Air Force, and Marine Corps, a rank above major general and below general. See also **general**.

lift—See **amphibious lift**.

light artillery—Artillery, other than antiaircraft artillery, consisting in most armies of howitzers and longer-barreled cannon of 105mm caliber or smaller. See also **artillery**.

light cavalry—See **cavalry**.

light attack fighter—See **fighter aircraft**.

light cruiser—See **cruiser**.

light division—A combined arms unit of division (q.v.) type, usually smaller and more lightly and austerely equipped than standard divisions. The principal reason for establishing "light" divisions is to achieve greater transportability or tactical mobility without greatly sacrificing the striking and holding power of a standard division. An airborne division can be considered a kind of light division.

light force—A military organization which is equipped with light weapons and matériel, to facilitate mobility and transportability, but which is more limited in combat power and sustainability than more heavily equipped forces.

light horse—Light cavalry; lightly armed or armored soldiers who fight on horseback.

light infantry—See **infantry**.

lighter than air (LTA)—Designation of an airship—blimp, balloon, or dirigible—for which lift is provided by a gas-filled bag, rather than by speed and the aerodynamic effects of wings and body surface. The gases used most commonly are helium (used exclusively by the United States) and hydrogen (not used by the United States because of its extreme flammability). Military uses in the United States of aircraft lighter than air are concentrated in a branch of the Naval Air Force.

Lima—Letter of NATO phonetic alphabet. See also **alphabet, phonetic**.

limber—n. The forward, detachable portion of a horse-drawn gun carriage or caisson, comprising an axle, two wheels, and a pole for harnessing the horses. It usually carries several rounds of ammunition in a box mounted on the axle; the box also serves as a seat for one or two cannoneers. v. To attach the horses and forward section (the limber) to a gun carriage, or to a caisson.

limited duty—The duty status of a person who is physically disqualified from performing certain types of military functions.

limited objective—A political or military objective for which a nation or commander commits, or is willing to commit, only part of available resources or effort.

limited war—1) A war looked upon by one or more of its contestants as not involving its own sovereignty or most vital interests, and as being limited in at least one respect, as, for example, to a particular geographic area, to the employment of certain resources, or to the number of contestants. 2) A war considered by a detached observer as relatively limited in some key respect, especially with regard to political objectives.

line—1) A deployment of troops, tanks, ships, forts, or other elements one beside the other. In the US Navy, generally called **line abreast**. 2) The front of the combat area, usually "the line," or "front line." 3) The troops in the line. Thus, combat troops as opposed to auxiliaries. 4) In most military services, officers who are in the chain of command, as opposed to staff (supply engineers, etc.). 5) Naval term for rope, particularly when used as part of a ship's working gear, like a halyard or a hawser.

line ahead—A formation of ships one behind the other.

line company—In the British Army of the 18th and 19th centuries each infantry regiment included several (usually three or four) standard infantry companies and one company each of grenadiers (q.v.) and light infantry (q.v.). The standard companies were called "line" companies, because they usually formed the regiment's line, with the other two companies either on the flanks or, more often, detached.

line division officer—A US Navy term for an

officer in charge of a ship's deck or engineer division. He is responsible for the care and upkeep of that part of the ship assigned to his division and for training the men.

line of authority—See **chain of command**.

line of battle—The deployment of forces in a generally linear formation for battle. In land warfare troops along the two opposing parallel lines face each other. In naval warfare warships usually form a single-file line of battle so as to bring their broadside fire to bear on the enemy line of battle.

line of communications (LOC)—All the routes—land, air, and water—that connect an operating military force with a base of operations, and along which supplies and reinforcements move.

line of departure—1) A line designated to coordinate the departure of attack or scouting elements; a jump-off line. 2) A suitably marked offshore coordinating line to assist the on-schedule landing of assault craft on designated beaches in an amphibious operation. 3) A line tangent to the trajectory of a projectile at the instant of its being fired.

line of direction—A line indicated on early guns by a short point on the muzzle (the front sight) and a slight incision in the base ring (near sight) to direct the eye in pointing the gun.

line of duty—1) The performance of required military duty. 2) Any and all activity legitimately associated with military duty or service.

line officer—1) An officer in the line of command of combat units or vessels, with command precedence and seniority usually determined by date of rank. 2) An officer assigned to troop duty, as opposed to staff duty; a distinction often recognized in the expression "line and staff." 3) An officer belonging to one of the US Army's combat branches, such as the Infantry or Artillery.

line of fire—The flight path or paths followed by projectiles or bullets fired from one or more guns.

line of march—1) Direction in which a military force proceeds. 2) The roads taken in a march.

line of sight—The straight line between two points. This line is in the plane of the great circle (that is, a circle on the surface of a sphere whose plane passes through the center of a sphere, such as the terrestrial sphere), but does not follow the curvature of the earth.

line regiments—Standard regiments as opposed to special, elite, or guards regiments.

linear defense—The deployment of defensive forces evenly distributed in a line along the entire front, with all or most of the units of the defending force on the line. Such a distribution, sometimes called **cordon defense**, has often provided attacking forces with opportunities to make decisive breakthroughs. The preferred defensive deployment, because of its greater chance for success, is defense in depth (q.v.).

Linear law—See **Lanchester equations**.

linear tactics—A tactical system introduced to ground combat in western Europe by Swedish King Gustavus Adolphus in 1630. Thanks to increased firepower and rate of fire, he was able to form his musketeers and pikemen in lines six deep—half the prevailing depth. Later armies formed as few as three or two lines deep. Units were deployed abreast, along the front, with intervals between units to permit tactical flexibility. Gustavus and his successors usually used two such lines, with a third in reserve. Although much modified in detail over the centuries, linear tactics are essentially the basis of modern ground combat doctrine.

lines of operation—See **interior lines** and **exterior lines**.

linked ammunition—Cartridges that are fastened side by side by means of metal links to form a belt for continuous feeding into a machine gun.

Link trainer—A trainer for airplane pilots, named for its developer, Edwin A. Link. The Link trainer was a simulation or mock-up of the cockpit of an actual airplane, including the controls and instrument dials. The person using the trainer could replicate the effects, other than forward motion, of his manipulation of the controls. It was widely used for pilot training during World War II and was the forerunner of modern, more complicated flight simulators.

linstock—A long forked stick used before 1800 so that the cannoneer could hold the slow match close enough to a cannon's touchhole to ignite the charge while he stood far enough away to avoid the cannon's recoil.

lion—A packaged, or preformed, unit consisting of all personnel and matériel necessary for the establishment of a major, all-purpose naval base. Used in World War II, a lion included the functional components for setting up an advanced base that could repair minor battle damage to a major portion of a fleet. Such bases were delivered in units that could be assembled and functioning in a very short time.

liquid fire—See **Greek fire**.

live ammunition—Ammunition that is ready to use for lethal effect, as distinguished from inert or drill or blank ammunition. (q.v.)

Livens projector—A mortarlike device invented by US Captain William H. Livens in World War I for delivering a toxic agent. Detonated electrically, a boxed propellant charge fired 30 pounds of the agent nearly a mile in a bomb almost eight inches in diameter and 20 inches long.

LOC—See **line of communications**.

lochaber axe—A pole with an axe at the end, a weapon formerly used by Scottish highlanders. See also **axe**.

lock—The device in a firearm that causes the charge to explode. See also **flintlock, matchlock, wheellock, snaphance lock**.

lodgment—An area gained and held in territory otherwise held by the enemy or a neutral nation. See also **beachhead, bridgehead**.

loft bombing—See **toss bombing**.

logistics—1) That aspect of military activity providing for the buildup and support of a military force so that it will be efficient and effective in both combat and noncombat operations. Supplies, equipment, transportation, maintenance, construction, and operation of facilities, provision, movement, and evacuation of personnel, and other like services, are included in logistics. 2) The furnishing of supplies and equipment. See also **administration**.

logistics support—The support given by a command or other organization to a person, activity, unit, or force, by means of which all or any part of its supplies, equipment, combat matériel, maintenance, transportation, administration, or any other like service is furnished. Thus, the person, activity, unit, or force can carry out its own operations more expeditiously.

long range bomber—See **bomber**.

longbow—A bow approximately six feet long, developed in Britain (probably in Wales) before the 13th century. Made of elm, hazel, basil, and finally of yew imported from Italy or Spain, the bow, which was curved in the front and flat on the back, tapered from an inch and a half wide in the middle, where the hand grasped it, to its horn-capped ends. A skilled English archer could fire ten to 12 of its three-foot arrows a minute a maximum of 400 yards, and with considerable accuracy for almost 250 yards. Its one drawback was that the strength and skill necessary for its successful use could be acquired only after years of training. Because it was more accurate than the crossbow, and lighter, more easily handled, and adaptable either for skirmishing or for volley fire, it ultimately caused the abandonment of the crossbow and hastened the downfall of cavalry as the dominant arm of medieval warfare. See also **bow** and **crossbow**.

long-range penetration—A concept developed by British Brigadier Orde Wingate and demonstrated in two long-range penetrations behind Japanese lines in Burma in World War II. The principle was that small British ground forces, with logistic support by air drop, could operate for extended periods deep behind enemy lines, cutting communications, destroying supplies, creating general confusion, and thereby beating the Japanese at their own game of infiltration and encirclement. Wingate's two expeditions were reported to the Anglo-American public as great successes, but postwar analysis demonstrated that they were military failures.

lookout posts—Positions occupied by a small number of troops, so placed that the position and movement of an enemy may be observed.

loophole—A slit or small opening of some other shape in a wall, through which small arms may be fired with minimum exposure to the persons firing them.

loot—Any items seized from their owners in a war zone.

loran—A long-range electronic navigation system which uses the time divergence of pulse-type transmission from two or more fixed stations. (This term is derived from the words LOng-RAnge electronic Navigation.)

loss rate—The rate at which weapons or equipment are destroyed in combat. Although the rate may be expressed in absolute numbers, it is normally expressed as a daily percentage of the original strength. See also **attrition**.

Love—Letter of old phonetic alphabet. See also **alphabet, phonetic**.

low altitude bombing—Horizontal bombing with the height of release between 900 and 8,000 feet.

low angle fire—Gunfire delivered at angles of elevation below the elevation that corresponds to the maximum range of the gun. See also **high angle fire**.

low-intensity conflict—See **conflict levels of intensity**.

lozenge—See **mascle**.

lucky bag—A US Navy term for a place aboard ship or in a naval installation in which articles that are found adrift, that is, without any obvious owner, are placed, to be claimed when the lucky bag is opened.

lunette—1) A type of fieldwork open in the rear, in which the face is in the form of a half-moon, or, more usually, in which parallel flanks are connected by two outward projecting faces meeting at the front in an angle. 2) A ring set in a plate at the back end of a gun carriage, with which the pintle of limber or truck is engaged.

M

MAAG—See **military assistance advisory group**.

macana—A war club used by certain South American Indians.

mace—1) A heavy club whose large head with protruding spikes or flanges could crush or penetrate armor. Warrior clergymen usually carried maces because they were forbidden by canon law to shed blood. In the late 16th century its use as a weapon was abandoned, and it became a ceremonial instrument or sign of rank. See also **holy water sprinkler** and **morning star**. 2) A chemical anti-riot weapon that causes acute irritation of the eyes and respiratory tract.

machete—A long, heavy, broad-bladed swordlike knife, used in the West Indies and Latin America not only for cutting sugarcane and heavy vegetation but also as a weapon.

machicolation—1) A projecting parapet or gallery at the top of a castle wall with holes or openings in the floor through which defenders could drop hot liquids or rocks or fire arrows on attackers at the base of the wall. 2) One of the holes in the gallery.

machine gun—A small, fully automatic weapon of various designs. It fires small-caliber ammunition in rapid successive shots using any of a variety of mechanisms. The first such weapons, invented in the mid-19th century, were the Gatling and Lowell guns, which were multi-barreled weapons fired by crank-operated mechanisms. Their first important use was in the American Civil War. The first fully automatic machine gun was a recoil-operated gun built by Hiram Maxim in 1884. The modern machine gun is belt-fed and driven by recoil or ejected gunpowder gas. Machine guns are usually classified as light, medium, or heavy. A light machine gun

is usually air-cooled, bipod mounted, portable, and 5 to 7mm in caliber. A medium machine gun is usually tripod mounted, water-cooled, and of about 7mm caliber. A heavy machine gun is usually air-cooled or water-cooled and 7 to 13mm in caliber.

machine pistol—1) A pistol capable of fully automatic fire. 2) A submachine gun. The application of the term in the second sense has been introduced since World War II from the German and Russian words for the submachine gun, *Maschinenpistole* (machine-pistol) and *pistolyet-pulemyot* (pistol-machine gun), so-called because the submachine gun ordinarily uses pistol-type ammunition. It is called "machine carbine" by the British.

machinist—US Navy warrant officer.

machinist's mate—US Navy specialty rating, chiefs, and first, second, and third class.

magazine—1) A structure or compartment for storing supplies, particularly ammunition or explosives. 2) A sometimes detachable part of a gun or firearm that holds ammunition ready for insertion into the chamber.

magazine gun—A breech-loading weapon with a magazine tube or other container in the stock or in front of the trigger for carrying cartridges. When a round has been fired, the empty cartridge or shell case is ejected and a full cartridge is moved mechanically to take its place.

Maginot Line—The line of fortifications built by France along its frontier with Germany in 1930–34 as a defense against possible German attack. It consisted of a line of casemates, with great underground forts called *ouvrages* at five- to eight-kilometer intervals, a tank barrier constructed from pieces of rail set on end in concrete, and steel posts and barbed wire between tank barrier and casemates. The forts could house 100 to 500 men, the larger forts having both infantry and artillery capability, the smaller infantry alone. Because it was so formidable, in World War II the Maginot Line was avoided by Germany, which attacked France through Belgium in May 1940, inasmuch as French fortifications along the Franco-Belgian frontier were minimal. The defeat of France was in large part due to the positioning of most of its army behind the Maginot Line, and these troops were thus not available to block the German thrust around the line. The Germans took advantage of the Maginot Line in 1944, using its fortifications to slow the Allied drive across northeastern France after the July breakout from the Normandy beachhead.

mail—Body armor made of interlinked metal rings, sometimes known as **chain mail**. In medieval Europe, mail was usually made of alternate solid punched rings and open rings with their ends riveted together. In Asian mail armor the ends of the rings were usually butted together on heavy cloth and thus less strong.

mail coif—See **coif, mail**.

main attack—See **attack**.

main battery—On a naval vessel the guns of the largest caliber.

main battle tank—See **tank, main battle**.

main body—1) The principal portion of a military force during movement, after detachment of advance guard (q.v.), rear guard (q.v.), and flank guards (q.v.). 2) The principal portion of a military force which consists of several groups or contingents, or from which detachments have been made. See also **body**.

main command post—See **command post**.

main line of resistance (MLR)—A line close behind the forward edge of a battle position, chosen and prepared for optimum defensive capability and so designated for the purpose of coordinating the fire of all units and supporting weapons, including air and naval gunfire. It defines the forward limits of a series of mutually supporting defensive areas, but does not include the areas farther forward (toward the enemy) that are occupied or used by covering or screening forces.

main road—A road capable of serving as the principal ground line of communication to an area or locality. Usually, or if possible, it is suitable for two-way traffic in any weather at high speeds.

maintenance—1) All action taken to retain matériel in a serviceable condition or to restore it to serviceability. It includes inspection, testing, servicing, and classification as to serviceability, repair, rebuilding, and reclamation. 2) All supply and repair activity to keep a force in condition to carry out its mission. 3) The routine work required to keep a facility—plant, building, structure, ground facility, utility system, or other real property—in such condition that it may be continuously used at its original capacity and efficiency.

major—A rank in most armies and in the US Army, Air Force, and Marine Corps, above captain and below lieutenant colonel. See also **commandant**.

major attack—See **attack**.

major contingency—See **conflict levels of intensity**.

major general—A rank in the US Army, Air Force, and Marine Corps, and in many other nations, between brigadier general and lieutenant general. Some nations, like the USSR and Great Britain, do not have the rank of brigadier general; in such armies major general is the lowest general officer rank. See also **general**.

major war—See **conflict levels of intensity**.

malleolus—An incendiary weapon of classical times. A malleolus could be a burning arrow or flaming tow that had been attached to a javelin and was fired or thrown at an attacker's siegeworks.

mameluke—One of a band of Moslem cavalry soldiers from the 12th to 16th century who originally were chiefly Turkish and Circassian slaves. The mamelukes were developed into very effective soldiers, particularly horse archers, by the caliphs of Bagdad to fight the Crusaders. In time the mamelukes became so powerful that they took control of Egypt, forming a ruling military caste lasting from the 13th century until the end of the 18th century.

man—n. Enlisted man. v. To supply with enough men to operate effectively, as to man a ship, a gun, a platoon.

man-at-arms—An armored horseman of the Middle Ages.

maneuver—1) A movement to place ships, troops, matériel, or fire in an optimum location with respect to the enemy, especially in the battlefield movement of forces. 2) A tactical exercise carried out at sea, in the air, on the ground, or on a map in imitation of war. 3) The operation of a ship, aircraft, or vehicle to cause it to perform desired movements. 4) A principle of war emphasizing the importance of locating forces for optimum strategic and tactical effect in combat.

mangonel—A medieval siege engine that used the torsion principle to catapult large rocks and other missiles. Mangonels were used primarily to batter enemy stone fortifications.

Manhattan Project—To develop the atomic bomb in World War II, a special Engineer District was established in August 1942 under Major General Leslie Groves, with the deliberately misleading title of Manhattan District. This command, consolidating work begun in 1939–1940 under joint auspices of the War and Navy departments, was responsible for all aspects of the development of atomic bombs until August 1946, when the Atomic Energy Act was passed, creating the Atomic Energy Commission.

maniple—A subdivision of a Roman legion, consisting of two centuries (q.v.). The maniple and one of its centuries were commanded by a centurion elected by the legion. He appointed the centurion who commanded the other century.

Mannerheim Line—A Finnish defense line extending across the Karelian Isthmus north of Leningrad. It was an important position because it covered the only area of the Russo-Finnish frontier in which the Russians could concentrate large numbers of troops for offensive operations against Finland. The fortifications—mostly pillboxes—were well integrated with the rugged, wooded terrain of

the area and were supplemented by wire, mine, and water obstacles.

The line was attacked twice by the Soviets. The first attack, in December 1939, was easily repulsed. The second attack, during mid-February 1940, led to a breakthrough, but only after the Soviets had suffered tremendous casualties—losses so heavy that Soviet assault forces had to be regrouped and heavily reinforced before the offensive could be resumed. The fall of the line led to the defeat of Finland, and a Soviet-Finnish peace treaty was signed on 12 March. There is no doubt, however, that, had the line not existed, the Soviets would have defeated Finland in December 1939.

Mannlicher rifle—A bolt action, 11mm rifle designed by Ferdinand Ritter von Mannlicher in 1884 and adopted by the Austrian Army in 1885. Subsequently, Mannlicher designed other successful automatic and repeating weapons.

man-of-war—A warship.

manpower—The human resources available for participation in a nation's war effort. Military manpower is the proportion of a nation's manpower that is in, or available to, the armed forces.

mantelet/mantlet—A great wicker or wooden shield, sometimes mounted on wheels, used to shelter outpost guards and operators of siege engines that were within the range of weapons on the walls of the fortification.

manuballista—A small crossbow used by the Romans in the first or second century A.D.

MAP—See **Military Assistance Program**.

map, controlled—A map with precise horizontal and vertical surveyed ground control as a basis. Scale, azimuth, and elevations are accurate and can be relied upon for the calculation of firing date.

map exercise—An exercise in which a series of military situations or problems is stated and solved on a map.

map reference code—A code used primarily for encoding grid coordinates and other information pertaining to maps.

maquis/maquisards—From the French

word for "bush," the name given to the French underground organization of World War II and to its members.

marauder—1) One who raids in search of plunder. 2) The name "Merrill's Marauders" was given by newspapermen to the 5307th US Infantry Regiment, a long-range penetration unit commanded by Brigadier General Frank D. Merrill, which served in Burma in 1944.

march—n. 1) An organized movement of troops in column. It may be on foot, on horses, or by motor vehicle. A **forced march** is a march carried out at a rapid pace under duress. 2) A musical composition in regularly accented, usually duple, meter (two beats per measure), to accompany marching troops. v. 1) To walk in rhythm with a measured cadence or music. 2) To cause someone to move toward an objective.

march order—An order to a military unit, usually artillery, to take all actions necessary to be ready to move as soon as possible. See also **close station**.

marine—One of a body of troops trained for service at sea and on land, particularly in amphibious operations. Such troops have been used since at least the time of ancient Greece for landing parties and boarding parties. In the naval wars of the 17th century both the British and the Dutch organized bodies of marines for landing and boarding and for fighting from the rigging of their warships. Traditionally, marines aboard warships have been used by naval commanders to maintain order aboard ship. This was particularly important in the 18th century when crews were largely composed of sailors who had been involuntarily impressed into service and were not readily amenable to discipline.

marine amphibious unit (MAU)—A US Marine air-ground task force built around a battalion landing team and a provisional marine aircraft group usually composed of an attack squadron and a helicopter squadron. The marine amphibious unit normally employs about one-ninth of the combat resources of one marine division/wing team. It is normally commanded by a colonel and is capable of performing combat operations of relatively limited scope. The MAU provides an immediate reaction capability to crisis situations and, when committed, is normally supported from its seabase.

maritime—1) Of or pertaining to naval affairs. 2) Of or pertaining to transoceanic commerce and naval lines of communication.

maritime strategy—1) The national strategy of a nation whose security is—or is perceived to be—primarily dependent upon control of at least some portion of the seas of the world. 2) That aspect of a nation's strategy focusing on naval and overseas affairs.

mark—A symbol used to distinguish different types of weapons or other equipment. A newly developed type is designated Mark I (or MK I), and when the original design has been sufficiently modified it is designated Mark II. In the US Army designations the numeral is Arabic (Mark 1, 2, etc.).

marksman—1) One who is particularly skilled at firing a gun or other weapon at a target. 2) In the US Army and Marine Corps a rating of rifle proficiency awarded for specified achievement with a weapon on a range. There are three levels in ascending order of competence—marksman, sharpshooter, and expert.

marksmanship—The skill of firing a weapon so that the projectile hits the target at which the weapon is aimed.

marshal—n. The highest military rank in some armies and air forces (sometimes called field marshal). In the 13th century the title was given to the commander of an army in England. It apparently derives from the title of *marescalci,* held by the master of the horse of early Frankish kings. In armies with marshals there is usually only one grade or degree. In the Royal Air Force there are four levels: air vice marshal (equivalent in the US Army to major general), air marshal (equivalent to lieutenant general), air chief marshal (equivalent to general), and marshal of the Royal Air Force (equivalent to field marshal). v. To undertake the process by which forces and equipment that are to participate in an operation are put in order, so that each force may be ready for action in accordance with an overall plan. In airborne assault operations, marshaling is the process of moving ground units to their billets at or near departure airfields, preparing them for combat, and loading them into aircraft. In an amphibious or airborne operation, marshaling includes grouping together or assembling participating units when feasible or moving them to temporary camps in the vicinity of embarkation points, and completing preparations for combat and for loading.

Martello tower—A fortification of masonry with a circular ground plan. These were used for coastal defenses in many countries in the late 18th and early 19th centuries.

martial law—1) An arbitrary law proceeding directly from the military power, having no immediate constitutional or legislative sanction. 2) A condition in which military control and military law are imposed on a civilian population. Martial law is usually imposed in a war situation or when civil authority has ceased to function or is unable to function effectively. Although such a state is usually temporary, martial law may be continued indefinitely in a military dictatorship or when unrest continues for an extended period. Sometimes called "state of siege."

martinet—A strict disciplinarian. In the 17th century Jean Martinet, inspector general of the French infantry and lieutenant colonel of the king's regiment of foot, created a highly trained army for Louis XIV through intensive drilling and training, and his name has been given to strict military taskmasters ever since.

Martini-Henry rifle—A breech-loading, rifled infantry weapon adopted by the British Army in 1871 and having a caliber of .45 (11.4mm). The breech mechanism was developed by Friedrich von Martini, an Austrian, from the American Peabody action. The barrel was designed by Alexander Henry, an Edinburgh gunsmith. The act of lowering the trigger guard opened the breech, ejected the fired cartridge case, and recocked the firing pin.

mascle—1) A quadrangular device between the chevrons and the arcs of certain sergeants' sleeve insignia. (First sergeants in the US Army.) Also called a "lozenge" because of its shape. 2) In the 13th century lamellar armor (q.v.), one of the overlapping steel plates.

mask—n. A hill or other natural feature of the terrain, or an artificial structure, that conceals and so protects an individual or a force from its enemy, either visually or by screen-ing the force from enemy fire or both. v. To conceal one's position or protect it from enemy fire.

mass—n. 1) The military formation in which units are spaced at less than normal distances and intervals. 2) The concentration of means at the critical time and place to the maximum degree permitted by the situation. 3) A principle of war in which proper application of mass, in conjunction with the other principles of war, permits application of optimum strength of a force on the decisive point or points on the battlefield. Mass is essentially a combination of manpower and firepower and is not dependent upon numbers alone; the effectiveness of mass may be increased by superior weapons, tactics, and morale. v. To concentrate or bring together, as to mass the fire of all batteries.

mass bombing—Bombing with a large number of aircraft.

mass destruction, weapons of—See **weapons of mass destruction**.

massed fire—See **fire, massed**.

massing of fire—1) The fire of the batteries of two or more ships directed against a single target. 2) Fire from a number of weapons, directed against a single target. See also **fire, massed**.

massive retaliation—1) A large-scale response to an attack, possibly involving nuclear weapons. 2) A short-lived US nuclear deterrent policy, announced by Secretary of State John Foster Dulles in 1954, that called for US response to Communist-nation provocations with whatever actions the United States deemed appropriate to its interests, whether or not those responses matched the provocations in geographical location or scale of weapons. By implication, US nuclear superiority was to be used to deter all levels of aggression.

master—1) A rank in the 19th-century US Navy above that of ensign and below that of lieutenant, subsequently changed to lieutenant, junior grade. 2) An officer holding such a rank.

master-at-arms—A petty officer, usually a chief, who is responsible for the maintenance of order and discipline in a ship or naval shore station, including the custody of prisoners. Originally the master-at-arms was charged with the care and instruction in the use of small arms.

master gunner—1) The head of a gun crew on a naval vessel. 2) Formerly, the senior gunnery specialist on a warship.

master key fire—In naval gunnery, the firing of guns by a common firing circuit controlled by a firing key at a distant station.

master sergeant—The highest non-commissioned rank in the US Army, Air Force, and Marine Corps.

mast for requests—A period set aside by the executive officer of a naval ship or station during which members of the crew may present requests.

match, quick—A cord impregnated with a highly flammable mixture, such as alcohol mixed with black powder. Used as a fuze for flares.

match, slow—The means most commonly used for igniting the charge in small arms and artillery pieces before 1700. Although the match might be made of other combustible material, it was usually a braided cotton cord of three strands, which had been soaked in a solution of saltpeter and treated with lye and lead acetate to make it burn only four or five inches an hour. To ignite an artillery piece the match was held in a linstock (q.v.).

matchlock—1) A device using a match to ignite the priming powder in a weapon. Invented in the late 14th or early 15th century, and used in Europe as late as the mid-19th century, the matchlock was a pivoted lever called a serpentine because of its shape. To the free end of the serpentine the soldier clipped the glowing end of the match, which by a triggering arrangement was brought in touch with the priming powder. It was a great advance over earlier devices and many variations appeared. It was most commonly used on muskets. 2) A musket or arquebus equipped with such a lock.

matchlock arquebus—An arquebus (q.v.) with a matchlock (q.v.), used in combination with the pike in the 15th century in the West and in combination with the sword by the Turks.

matchlock musket—A weapon developed in Spain and first used as a defensive position weapon in the Italian Wars (1530–1540). It had a longer barrel and fired a heavier ball than the matchlock arquebus (q.v.) and could pierce armor and—in sufficient numbers or mass—stop a cavalry charge. A very heavy weapon at first, it was gradually modified until it replaced the arquebus. Loading the musket required some 90 different steps. Gustavus Adolphus simplified this in a lighter musket by issuing measured powder charges wrapped in paper with the ball attached. See also **matchlock**.

mate—1) In earlier times in the British Royal Navy and until the late 19th century in the US Navy, a rank between midshipman and lieutenant. 2) In naval ratings, distinguishing the assistant to a warrant rating, as boatswain's mate, carpenter's mate, aerographer's mate. In most cases there are two or three classes of mate.

matériel—1) Things of all kinds required for the equipment, maintenance, operation, and support of military activities, both combatant and noncombatant. 2) In a restricted sense, those things used in combat or logistic support operations, such as weapons, motor vehicles, airplanes, component parts of machines, and special-purpose clothing, as distinguished from items of ordinary use, such as uniforms, bedding, and housekeeping equipment.

MAU—See **marine amphibious unit**.

Mauser rifle—A rifle invented in 1871 by Peter Paul Mauser, a German weapons manufacturer. A single-shot clip-fed magazine rifle with a sliding bolt, the Mauser was the regular German Army rifle in World War I and was widely used in other armies as well. It fired projectiles from smokeless powder cartridges, approximately .30 caliber, at a muzzle velocity up to 2,800 feet per second.

Maxim machine gun—A belt-fed, recoil-operated weapon invented by Hiram Maxim in 1884; within thirty years it was used by every major power. Whereas earlier automatic guns had several barrels, the Maxim had one. Cartridges were loaded in a fabric belt, and by using the power of the barrel's recoil the gun loaded itself, fired, and ejected the spent shell. A metal jacket around the barrel containing water prevented overheating from the gun's high rate of fire. British colonial forces first used Maxim guns in the Matabele War of 1893-1894. In World War I it was one of the three determinants of the battlefield stalemate (the other two were barbed wire and high explosive artillery fire).

maximum range—See **range**.

maximum effective range—See **range**.

MBT—See **tank, main battle**.

M-day—The term used to designate the day on which mobilization is to begin.

mechanize—To equip a ground unit with trucks, tanks, armored personnel vehicles, or other means of transport. It differs from motorize (q.v.) by the inclusion of armored vehicles.

mechanized cavalry—See **cavalry**.

mechanized division—A combined arms ground force of division (q.v.) type and size, the principal combat components of which are approximately equally divided between tank-equipped armored units and motorized infantry units that can be mounted in trucks, armored personnel carriers, or a combination of both.

mechanized infantry—See **infantry**.

Medaille Militaire—A French military decoration established by Napoleon III in 1852 honoring distinguished conduct in combat.

medal—Any of many metal devices with a distinctive design, inscription, or both, and usually hung from a distinctively colored ribbon, commemorating an action, an achievement, or an event, frequently awarded with some recognition. Medals are ranked in the United States and other nations according to the reasons for which they are given (the highest US award being the Medal of Honor q.v.) and worn with the highest at the right of the top row. US medals are worn only with dress uniforms. They are represented on work uniforms by bands of the ribbons accompanying the medals.

Medal of Honor—A decoration awarded in the name of Congress to a person who, while in the armed services of the US and engaged

in conflict with an enemy of the US, has distinguished himself conspicuously by gallantry and bravery at the risk of life above and beyond the call of duty. This is the highest US decoration. There are two slightly different versions of the decoration, one for the Army and Air Force, and one for the Navy and Marine Corps. But in both the principal feature of the medal proper is a five-pointed bronze star featuring different likenesses of Minerva. The ribbon for both is pale blue with a pattern of 13 white stars just above the medal. The ribbon worn on the dress uniform is the same pale blue color with five white stars.

medium artillery—See **artillery**.

medium-range bomber—See **bomber**.

meeting engagement—A combat action that occurs when two opposing forces, advancing toward each other, usually not fully deployed for battle, engage each other at an unexpected time or place.

melee—1) In ancient and medieval battles, fighting in which the soldiers, having first thrown their javelins or fired their arrows, fought hand-to-hand with their opponents, using swords, clubs, spears or other weapons. 2) Any hand-to-hand fighting characterized by great confusion.

mention in dispatches—To recognize bravery or outstanding performance in the British services. The origin of the term was the British custom for a commander in chief to include in his dispatches to London the names of officers who had distinguished themselves in battle or campaign.

mercenary—A soldier who hires himself out for pay to fight in the army of another nation. There probably have been mercenaries as long as there have been wars, but they were particularly prevalent in western Europe in the 13th and 14th centuries after the end of the Hundred Years' War. Mercenaries and mercenary units were important elements in most European armies until the end of the 18th century.

merlon—See **battlements**.

mess—n. 1) A place where food is prepared and served, also called, on a base, a mess hall. 2) The occasion of a meal served at such a place, or the eating of such a meal, as in "he showed up at mess." 3) A group of persons who eat together regularly. 4) Particularly in the navy, a special food-related organization, or installation, such as officers' mess (q.v.), coffee mess and wine mess. v. To eat a meal.

message center—A communication office within a military organization charged with the responsibility for acceptance, and preparation for transmission, receipt, and delivery of messages.

message precedence—Designations indicating the relative order or priority that is assigned to a message of one precedence designation in respect to all other precedence designators. It establishes the relative importance or urgency of messages.

mess jacket—A short semiformal uniform jacket worn with evening dress.

mess kit—1) Equipment for eating in the field, either for an individual soldier or for an officers' mess (q.v.). 2) In British usage, uniform evening dress including the mess jacket.

mess sergeant—A noncommissioned officer responsible for the operation of a mess.

MIA—See **missing in action**.

midshipman—1) Originally a young naval cadet in the British and later the US Navy. The name derived from his battle station, which was amidships or abreast of the mainmast. The midshipmen served a sort of apprenticeship aboard sailing vessels before receiving a commission. 2) In the US Navy and Coast Guard a student at a service academy (Annapolis or New London), training to be an officer. 3) In the British Royal Navy and others that follow the Royal Navy pattern, a noncommissioned officer below the rank of sublieutenant.

midwatch—1) The period of time or watch commencing at midnight and extending for four hours or more during which one portion of the crew of a ship or naval station is on duty. 2) The members of the crew that is on duty during the midwatch.

Mike—Letter of NATO phonetic alphabet. See also **alphabet, phonetic**.

mil—A unit of angular measurement used in gunnery and the launching of bombs and guided missiles, equal to one sixty-four hundredth of a circle.

military—Pertaining to armed services of all branches—army, navy, marine corps, air force. Often used to refer only to the army, particularly in contradistinction to "naval."

Military Assistance Advisory Group (MAAG)—A joint service group, normally under the military command of the commander of a unified command and representing the Secretary of Defense, which administers the US military assistance planning and programming in the host country.

Military Assistance Program (MAP)—The US program for providing military assistance under the Foreign Assistance Act of 1961. It includes the furnishing of defense matériel services through grant aid or military sales to eligible allies, as specified by Congress.

military attaché—See **attaché, military**.

military combat—See **combat, military**.

military decoration—See **decoration, military**.

military government—The form of administration by which an occupying power exercises executive, legislative, and judicial authority over occupied territory.

military justice—See **justice, military**.

military operation—1) An action or series of actions, either combat or noncombat, conducted by a military organization. 2) In a restricted sense, an operation involving, or primarily involving, combat personnel against an enemy, as distinguished from a logistical operation.

military police—That organization within an army that carries out police functions. It corresponds to the navy's Shore Patrol. Individuals are frequently referred to as MPs.

military use of heraldry—See **heraldry, military use of**.

militia—1) Part-time military or paramilitary formations that are organized and trained to serve in defense of a nation in time of emergency. 2) Members of such formations. **Naval militia** are trained for naval duties. See also **reserve components**.

Mills bomb—A pineapple-shaped hand grenade developed by Sir William Mills and used by the British Army in World War I. Although it could be fired by a mortarlike launcher, it was usually thrown by a soldier, who first removed a pin, which released a lever, which activated a four- to five-second fuze.

mine—n. 1) A tunnel dug under an enemy position in order to gain entrance to a fortification or the area behind the front lines, or in order to lay explosives. The most extensive use of such mines was at Messines Ridge in World War I, where the Allies dug a series of mines and destroyed sections of the ridge, destroying and exposing German positions. 2) An explosive or other material, normally encased, designed to destroy or damage vehicles, boats, or aircraft, or designed to kill or otherwise incapacitate personnel. It may be detonated by the action of its victim, by the passage of time, or by controlled means. **Land mines** are used by ground forces primarily to delay an attacker or defend a position by destroying or damaging tanks (**antitank mines**) or people (**antipersonnel mines**). **Underwater** or **naval mines** are laid at sea to protect harbors, destroy enemy shipping, or blockade an enemy's ports. See also **countermine**. v. To build or lay a mine. Land mines played an important role in some World War I engagements, but in World War II both sides used them extensively, notably in the desert in North Africa, where they were easily buried. There are many types, differing in design, sensitivity, and method of detonation. Scattered or in patterns, mines are usually laid in minefields. A minefield must be at least partially cleared before an enemy can safely cross it. Clearing is generally accomplished with magnetic mine detectors (ineffective, of course, if the mine is encased in wood, cardboard, or plastic) or with tanks or other vehicles fitted with flails far enough in advance to detonate the mine without destroying the vehicle. An important development in mine warfare in the late 20th century was the introduction of scatterable mines (q.v.) Underwater mines are either floating, unanchored, below the surface, or anchored. A convention that was approved at the Hague Peace Conference of 1907 forbade the use of unanchored mines unless they become harmless within an hour after those who lay them cease to control them, forbade the laying of anchored mines that do not become harmless within an hour of breaking loose from their moorings, and defined other precautions for the protection of commercial shipping. Like land mines, naval mines differ in design, sensitivity, and method of detonation. To counteract those that are detonated by a ship's magnetic field, in World War II almost all vessels were degaussed (q.v.).

mine warfare—The strategic and tactical use of mines and their countermeasures.

minefield lane—A marked lane, unmined, or cleared of mines, leading through a minefield.

minelayer—1) A naval vessel which lays underwater mines. 2) A mechanical device which lays land mines.

minesweeper—A naval vessel equipped to search for mines and destroy them or render them harmless. Minesweepers are used to sweep up mines laid by friendly powers, as well as enemy mines, to protect commercial shipping.

minié ball—See **ball, minié**.

minimum range—See **range**.

minion—A 16th-century artillery piece, with a 3.3-inch bore, which fired a projectile weighing six pounds. Six and a half feet in length, the minion, also known as a **demisaker**, weighed a thousand pounds. Its maximum range was 2,500 yards; point-blank range was 450 yards.

minor contingency—See **conflict levels of intensity**.

minor hostilities—See **conflict levels of intensity**.

minuteman—1) An armed civilian or militiaman in the American colonies, originally in Massachusetts, before the American Revolution, who was pledged to join his unit and fight when called to service. 2) The intercontinental ballistic missile that has been the core of the US nuclear arsenal for much of the latter half of the 20th century. Designated as LGM-30, the Minuteman is a three-stage, solid-propellant ICBM. Minuteman II is equipped with a single nuclear warhead. Minuteman III is MIRVed, with three nuclear warheads. The Minuteman can be launched from silos or from mobile launchers.

miquelet—A 16th-century device to fire a weapon that used a spring mechanism to hold a match.

MIRV—Multiple independently targetable reentry vehicle, which is a single missile, or booster, equipped with several nuclear warheads that can be armed separately and sent against different targets.

misfire—n. Failure of a round of ammunition, the propelling charge of a rocket, or other explosive to fire on ignition. v. To fail to fire, or to fire improperly. See also **hang fire**.

missile—Any object that is propelled at a target. The term includes rocks thrown by hand, arrows fired by a bow, bullets from a gun, depth charges dropped from a ship, electronically complex rockets, and all other objects launched by whatever means toward a physical objective. Modern arsenals, in addition to artillery weapons, include ballistic missiles and rockets of various types.

missile, air-launched cruise—A cruise missile (q.v.) launched from an aircraft in flight.

missile, air-to-air—A missile launched from an aircraft against an enemy aircraft or missile.

missile, air-to-surface—A missile launched from an aircraft against a surface target.

missile, antiballistic (ABM)—A defensive missile designed to intercept and destroy incoming nuclear missiles.

missile, ballistic—1) Any missile or projectile that does not rely upon aerodynamic surfaces to produce lift and consequently follows a ballistic trajectory when thrust is terminated. 2) In modern usage a ballistic missile is one (other than a rocket, which provides its own thrust during all or most of its flight) that is not fired from a gun tube. Ballistic missiles in this sense do not include small arms or cannon projectiles. They are usually free-standing and get their initial thrust from a very brief episode of rocket fire. Ballistic missiles are variously classified by range, in-

cluding: **medium range**, a range of from about 600 to 1,500 nautical miles; **intermediate range**, or **mid-range**, a range of from 1,500 to 3,000 nautical miles; and **intercontinental**, a range of 3,000 or more nautical miles.

missile, cruise—A low-altitude, pilotless jet aircraft that can carry nuclear warheads and use terrain-following radar, which makes it highly accurate over long ranges. The major portion of its flight path to its target is conducted at approximately constant velocity. It is hard to shoot down because its small size and low altitude make it difficult for defensive radar to follow its flight. Types of cruise missiles include: **Air-Launched Cruise Missile (ALCM)**, which can be launched from an aircraft, like the B-52 bomber; **Ground-Launched Cruise Missile (GLCM)**, which can be launched from a land-based mobile launcher; and **Sea-Launched Cruise Missile (SLCM)**, which can be launched from a surface vessel or from a submerged or surfaced submarine.

missile, guided—See **guided missile**.

missile launcher—A weapon or structure that launches ballistic, cruise, or guided missiles.

missile, surface-to-air (SAM)—A missile launched from the ground against an aircraft or missile.

missile, surface-to-surface—A surface-launched missile designed to attack a target on the surface. In land warfare such a missile is usually considered to be, and is used as, an artillery weapon.

missing in action (MIA)—A category of personnel casualty whose whereabouts following a combat action is unknown, whether through becoming separated from his unit, being captured by the enemy, being wounded or killed and unidentified, or through desertion.

mission—1) The objective; the task, together with the purpose, which clearly indicates the action to be taken. 2) In common usage, especially when applied to lower military units, a duty assigned to an individual or unit; a task. 3) The dispatching of one or more aircraft to accomplish one particular task. 4) A group or detachment of officers, or officers and enliste men, serving in a friendly country for the purpose of providing training or support, or to perform some other function of mutual interest to both nations and their armed forces. Often called a **military mission**.

mission effectiveness—The ability of a unit or formation to accomplish an assigned mission.

mission-type order—1) An order issued to a lower unit that includes explanation of accomplishment of the total mission assigned to the higher headquarters. A concept pioneered by the German Army, particularly since the time of the elder von Moltke (c. 1860), and called *auftragstaktik*. 2) An order to a unit to perform a mission without specifying how it is to be accomplished.

mitrailleuse—An early type of machine gun, used by the French Army during the Franco-Prussian War (1870–71). The mitrailleuse was invented in 1851 and adopted by the French Army in 1867. It had 31 rifled barrels in a single jacket, which were loaded from a plate inserted at the breech and could be fired simultaneously or in sequence. A well-trained crew could fire 12 rounds per minute. It was not a success, however, and was soon abandoned. The name means, literally, "grapeshot shooter."

MLR—See **main line of resistance**.

moat—A ditch, usually filled with water, surrounding a fortification and providing extra protection against attack.

mobile—*adj.* Readily movable from one place to another, whether by its own power or aboard a vehicle.

mobile artillery—See **artillery**.

mobile defense—See **defense**.

mobility—The ability of military units or items of equipment to move as units from place to place, including the ability to support them logistically during and after movement. Strategic mobility (q.v.) is the capability of large military units for movement in or between theaters of operation. Tactical mobility (q.v.) is the capability of any military unit for movement or maneuver on the battlefield or in the presence of the enemy.

mobilization—1) The act of preparing for war or other emergencies through assembling and organizing national resources. 2) The process by which the armed forces or part of them are brought to a state of readiness for war or other national emergency. This includes assembling and organizing personnel, supplies, and matériel for active military service. See also **industrial mobilization**.

mobilization base—The total of all resources available, or which can be made available, to meet anticipated wartime needs. Such resources include the manpower and material resources and services required for the support of essential military, civilian, and survival activities.

model—1) A small, realistic representation of a real item of equipment, or system, or situation, usually to scale. 2) A variation of a weapon, piece of equipment, or other item, distinguished by later changes in the characteristics or appearance of the same basic item. See also **mark**. 3) An abstract representation of a real entity or situation by something else which has the relevant features or properties of the original. There are five basic types of models: a) verbal, b) analytic or mathematical, c) diagramatic representation, d) analog simulations, e) digital simulations. In practice the words model and simulation (q.v.) are used interchangeably. Also called **simulation model**. See also **war game, simulation**.

model, aggregated—A simulation model in which many detailed, discrete elements of a process are combined and examined as an entity. Thus, a model that treats a division as an entity in theater-level combat has aggregated platoons, companies, battalions, brigades, and supporting weapons into the division.

model, analytical—A simulation model comprised of sets of mathematical equations as models of all the basic events and activities in the process being described, and an overall assumed mathematical structure of the process into which the event or activity descriptions are integrated.

model, combat—A simulation model used to describe the basic combat processes of firepower, mobility, intelligence, logistics, and command-control in order to estimate the results—numbers of casualties and survivors, resource expenditures, terrain controlled, etc.—of battles and wars. Most modern combat models are operated by means of a computer program; thus they are often called computer models (q.v.).

model, computer (computerized)—A computer program or series of programs designed to simulate the logic of actions or interactions within an environment or a context and to provide the results for subsequent analysis. A computer model is usually designed to run without human involvement for a predetermined period of time, or until a set of predetermined conditions are satisfied. A **computer-assisted model** performs the same kind of logical calculations as a computer model, but is designed to be responsive to periodic human intervention and decision-making.

model, deterministic—A simulation model that excludes chance variables so that the output of the model is entirely determined by the input. Thus, unlike in a stochastic model (q.v.), any given set of inputs should always give exactly the same answer.

model input—The mathematical data entered in a model in order to generate model outputs.

model output—The numerical values or results generated by a simulation model.

model, probabilistic—A simulation model that contains at least one chance variable so that the output of the model is not uniquely determined by the input.

model, research—An analytical model, usually in the form of complex and comprehensive equations or algorithms, used for investigating combat or decision processes or new concepts.

model, stochastic—A computer model designed so that random variables will be introduced during its operation, to simulate the effects of chance upon the simulated operation. Thus repetitions of the operation of the model with the same inputs will almost always give different answers. If run sufficiently often with the same inputs, however, averaged results should approximate those of a deterministic model (q.v.).

Molotov cocktail—An improvised hand grenade consisting of a bottle filled with alcohol, gasoline, or other flammable liquid and stoppered with a combustible material such as a piece of cloth, which is lighted when the bottle is thrown. Named for Vyacheslav M. Molotov, the Soviet foreign minister during World War II, such grenades were widely used by partisans and other irregulars in World War II and continue to be a common weapon for use by those without access to regular weapons.

momentum—The property of continuing forward movement of a military force, as a result of previous success in combat, which is presumed to provide an additive or multiplicative bonus to its combat capability.

monitor—A type of ironclad (q.v.) warship developed by the Union in the Civil War for use in the harbors and rivers of the South. Like the original *Monitor*, which fought the *Merrimack* in 1862, monitors had a low, flat deck and one or more gun turrets that revolved so that the guns could be trained on a target regardless of the bearing of the ship.

Mongol bow—See **bow**.

mop up—To free an area of the last remnants of an enemy. To capture or destroy the remnants of an enemy in an area that has been surrounded or isolated, or through which other units have passed without eliminating all active resistance.

moral—*adj.* The behavioral, or nonmaterial, or nonphysical considerations affecting military forces and combat. In the military sense, the word is unrelated to morality and has only incidental relationship to the concept of morale (q.v.) "The moral is to the physical as three is to one," an approximate rendering of a Napoleonic concept, indicated his belief that the capabilities, performance, and motivation of soldiers and commanders were more important than weapons or other physical entities. See also **square law**.

morale—1) The state of an individual's mental attitude or feeling in respect to what he is, does, or is going to do, as determined by the presence or absence of such psychological factors as confidence in his leaders, recognition of a spirit of sacrifice in others, feeling of comradeship for associates, physical comfort, and hope. 2) A state of mind or feeling attributed to a group of individuals or to an organization.

morion—A helmet without a faceplate, worn in the 16th and 17th centuries, particularly by archers and musketeers because of the broader vision it afforded. The morion has a crest and a wide brim that comes to a high point in front and back. It is still worn by the Swiss guard at the Vatican.

morning report—The daily record or log of certain military units, usually of company size, giving daily strength and status of personnel assigned.

morning star—A type of mace (q.v.) or war club common in medieval European armies of the 14th century and the first half of the 15th century, after which it fell into disuse. The haft of the infantry version was about five feet long and was surmounted by a sphere studded with iron spikes, from which the weapon derived its name. Morning stars used by horsemen had shorter hafts, usually about two feet long. A variant type of the weapon had the spiked ball attached to the haft by a chain. Another, not uncommon, name for this weapon was "holy water sprinkler" (q.v.).

mortar—A muzzle-loading weapon that fires shells at relatively short range with a high trajectory. Named for their resemblance to a pharmacist's mortar, such weapons have been in use since at least the 15th century. Mortars are normally infantry weapons that are light, easy to disassemble and carry from one area to another, and particularly effective for lobbing shells over a hill or other obstruction. Rifled or smoothbore, they have been made in many sizes, including very large siege mortars. The largest mortar in general use today is 160mm.

mortar vessel—A successor to the bomb ketch in the early days of steam, the mortar vessel was a gunboat mounting mortars, usually of large caliber. The vessel's great extra beam enabled it to resist the nearly perpendicular force of the recoil of the huge weapon, which was mounted amidships on a massive bed.

mosaic (aerial photo)—An assembly of overlapping vertical photographs that have been matched to form a continuous photographic representation of the earth's surface. It is a useful supplement to a standard cartographer's map.

mosaic, controlled—An air-photo mosaic which—because key terrain features are related to each other by survey or by measurements from a standard cartographic map—can be used as a substitute for a map. It often has superimposed grid lines, with the same designations as standard map grid lines. A controlled mosaic can be used effectively as an artillery firing chart.

mothballs, in—Descriptive of a ship that has been taken out of active service and had its guns and other gear protected in various ways from deterioration in order to preserve it for possible recommissioning. Such ships, said to be "mothballed" or "in mothballs," are moored together in various ports, generally in or near a navy yard.

motorize—To equip a unit or formation with complete motor transportation that enables all of its personnel, weapons, and equipment to be moved at the same time without assistance from other sources. Motorization does not involve the incorporation of armored vehicles, but a motorized unit may include some armored equipment or armored components. Motorization is distinct from mechanization (see mechanize, q.v.).

motorized division—Also often called a motorized infantry division, or motorized rifle division, it is a combined arms ground force of division (q.v.) type and size. Its principal combat components are infantry units that are usually mounted in trucks or sometimes in armored personnel carriers.

motor torpedo boat—Commonly called PT boat, (for patrol torpedo) a very fast boat built in large numbers in World War II and used effectively for both combat and noncombat operations. The boats were from 60 to 100 feet long and equipped with from two to four torpedo tubes, machine guns, antiaircraft guns, depth charges, and smokemaking apparatus. One of the most famous was PT-109, on which President John F. Kennedy served. They were also called **mosquito boats** and, by the Germans, **E-boats**.

mound—1) Any pile of earth or other material built up for protection or to conceal troops or equipment. 2) A small hill or other natural elevation. 3) The basis for a siege technique whereby weapons could be elevated to over-look the defender's wall. Extension of the mound to the edge of a besieged city or forti-fication permitted an attack without need for scaling the defensive works. This technique was a favorite of Alexander the Great.

mount—v. 1) To make all preparations for launching an operation, including assem-bling troops and equipment, moving and em-barking them in ships, aircraft, or vehicles. 2) Of guns, to be furnished with, especially ships and aircraft. 3) Of guns, to put in posi-tion. 4) Of a guard, to assume responsibility for a guard system. n. A carriage or stand upon which a weapon is placed.

mountain troops—Troops trained to fight in mountainous areas, particularly in snow con-ditions. For example, US mountain troops were used in Italy in World War II and as many as possible were recruited from among men who were experienced skiers.

mounted—1) On horseback. 2) On vehicles. 3) Emplaced on (as guns on a tank).

mounted troops—1) Troops that are mount-ed, as opposed to dismounted. 2) Prior to World War II, troop units or branches of the U.S. Army normally equipped with horses, as cavalry and field artillery.

movement—A change in the location of troops, ships, or aircraft for tactical or strate-gic purposes.

movement control—1) The planning, rout-ing, scheduling, and control of personnel and supply movements over lines of communica-tion. 2) An organization responsible for these functions.

movement order—An order issued by a commander covering the details for a move of his command.

MRBM—Medium range ballistic missile. See **missile, ballistic**.

mufti—Civilian dress worn by military per-sonnel when off duty.

Mulberry—Term for an artifical port built off the Normandy beachhead in 1944 with floating caissons and a line of old vessels sunk to form a breakwater.

muleteer—A muledriver, also known as a mule "skinner."

multiple independently targetable reentry vehicle—See **MIRV**.

multiplier, force—See **force multiplier**.

munition—Usually used in the plural. In a broad sense, any and all supplies and equip-ment required to conduct offensive and de-fensive war, including guns, machines, ammunition, transport, and fuel, but exclud-ing personnel and any supplies and equip-ment used for purposes other than direct military operations.

Munroe effect—The focusing of a detona-tion into a powerful, high-velocity, narrow cone or stream of fire, resulting from a con-cavity in the face of the explosive. The direc-tion of the stream of fire is the central axis of the concavity, perpendicular to the face of the explosive charge. See also **hollow charge, shaped charge**. The effect was dis-covered by the American chemist Charles E. Munroe.

mural crown—A crown or circlet of gold awarded in Roman times to the soldier who first climbed a wall of a place under siege and planted a standard there. The mural crown was often crenelated to resemble a battlement.

mural hook—A long, heavy bar or lever sus-pended from a high vertical frame to knock down the upper parapets of a wall. The de-vice was invented Alexander's engineer Diades.

musette bag—1) A small bag with a shoulder strap in which personal items can be carried. 2) An officer's field bag.

musket—Any of the smoothbore shoulder guns that evolved from the arquebus (q.v.) in the late 16th century and were replaced by the rifle in the mid-19th century. The original musket was larger and more awkward to handle than the arquebus, and like its pre-decessor it was muzzle-loaded. Its advantage over the arquebus was its increased range and accuracy and the greater penetrating power of its bullet. The size and weight of the early musket meant that it had to be sup-ported by a forked rest, which also helped in-crease accuracy. The earliest musket was six or seven feet long and weighed up to 25 pounds. The Spanish Army standardized its bullets to weigh ten to the pound in 1565. About 1620 Gustavus Adolphus lightened the musket to about 11 pounds, and reduced the weight of the ball to 14 to the pound. He also increased the rate of fire by issuing measured powder charges wrapped in paper with the ball attached. Its awkwardness decreased with substitution of the flintlock (q.v.) for the matchlock.

musketeer—A soldier armed with a musket.

musketoon—1) A short musket with a very wide bore, sometimes bell-mouthed. 2) A soldier armed with such a weapon.

musketry—1) The firing of infantry small arms, particularly muskets or in modern times rifles, including modern automatic weapons. 2) The capability of an infantry unit to fire its small arms.

mustang—Slang expression for a US Navy officer who had begun service in the Navy as an enlisted man.

mustard gas—Dichloroethyl sulfide, an oily liquid that causes irritation and blistering and may cause death. Because of its high volatility and lethal and noxious properties, it was the most effective chemical weapon used in combat in World War I.

muster—v. 1) To assemble the members of a unit for roll call, inspection, review, or to pass information or orders. 2) To enroll. n. 1) The assembly of members of a military unit for such purposes. 2) The roll of members, also known as the **muster roll.**

mutiny—A refusal by a serviceman to obey orders or otherwise to do his duty, or the creation of violence or disturbance, with intent to override lawful authority.

mutual support—That support which units render each other against an enemy, because of the synergism of their assigned tasks, their position relative to each other and to the enemy, and their inherent capabilities.

muzzle—The end of the barrel of a gun from which the bullet or projectile emerges. Early weapons were all **muzzle-loaders,** that is, the powder and bullet were loaded from that end. Of modern weapons, mortars are the principal type of muzzle-loader.

muzzle brake—A device attached to the muzzle of a weapon that uses escaping gas to reduce recoil.

muzzle-loader—See **muzzle.**

muzzle velocity—The speed at which a projectile is traveling at the instant the projectile leaves the weapon. Muzzle velocity depends on the weight of the charge, the weight of the projectile, the amount and quality of the explosive, and other factors. Muzzle velocity is one of the fundamental ballistic properties determining a projectile's range and power.

N

Nan—Letter of old phonetic alphabet. See also **alphabet, phonetic**.

napalm—A powder (the aluminum soap of naphthenic and palmitic acids) that is mixed with gasoline to thicken it, thereby slowing the rate at which it burns and increasing the ways it can be used. Napalm was developed in 1942 for use in flamethrowers, for which pure gasoline had proved problematic because it burned too fast and could not be thrown very far. Napalm is also used in bombs and other devices.

Napoleon gun—A smoothbore bronze gun-howitzer of the early 19th century. A 12-pounder, it was served by five or six men and fired a ball of about four and one-half inches in diameter as far as a mile, accurately within a half mile. Two rounds of shell or solid shot, or four rounds of canister per minute could be delivered by an efficient crew. In the American Civil War it was the favorite piece for Federals and Confederates alike but was already obsolete in Europe.

nasal—The portion of a helmet that protects the nose and upper part of the face.

national command authorities—In the United States, the President, the Secretary of Defense, the Joint Chiefs of Staff, or their authorized successors and alternates.

national guard—1) The designation of militia (q.v.) units or military reserves of certain nations. 2) In the US armed forces, the militia units of the several states of the United States, controlled by them under federal direction and subject to call in federal as well as local emergencies. The first American use of the name "National Guard" occurred on August 16, 1824, when the 7th Regiment New York Militia took the name in honor of the Marquis de Lafayette, who was visiting in New York at the time, and the French Garde Nationale, which he had commanded in Paris in 1789. The name was gradually adopted by all the states. National Guard units have served in all the major wars in which the United States has been involved.

national objectives—The fundamental aims, goals, or purposes of a nation toward which policies are directed and the efforts and resources of a nation are applied.

national policy—A broad course of action or a statement of guidance adopted by the government at the national level in pursuit of national objectives.

national security—1) The freedom—relative and absolute—of a nation from possible military attack or effective political or economic sabotage. 2) The military protection or defense of a state.

national strategy—See **strategy, national**.

"nation at war"—See **"nation in arms."**

"nation in arms"—The concept of complete mobilization of a nation's resources, including manpower, for prosecution of a war. First introduced by Revolutionary France in 1792.

NATO—North Atlantic Treaty Organization, an alliance of several nations for collective defense and the preservation of peace and security against aggression, established in 1949. The signatory nations to the NATO treaty are the United States, Great Britain, France, Canada, Italy, Belgium, Luxemburg, the Netherlands, Norway, Denmark, Portugal, and Iceland. Subsequently, Greece, Turkey, the Federal Republic of Germany, and Spain were admitted as member nations after the treaty's original ratification.

nautical chart—See **chart**.

naval base—1) A base primarily for support of forces afloat, contiguous to a port or anchorage and consisting of activities or facilities for which the navy has operating responsibilities, together with internal lines of communication and the minimum surrounding area necessary for local security. 2) An installation that supports land-based units of the naval air force, more accurately called a naval air base or naval air station.

naval campaign—An operation or a connected series of operations conducted by naval forces, including surface, subsurface, air, and amphibious forces, for the purpose of gaining, extending, or maintaining control of the sea.

naval district—See **district, naval**.

naval gunfire operations center—The agency established in a fleet or task force to control the execution of plans for the employment of naval gunfire, to process requests for naval gunfire support, and to allot ships to forward observers. The center's ideal location is in the same ship as the Arms Coordination Center of a fleet or task force.

naval militia—See **militia**.

naval mine—See **mine**.

naval stores—Any articles or commodities used by a naval ship or station, such as equipment, consumable supplies, clothing, petroleum, oils and lubricants, medical supplies, and ammunition.

navy—All of a nation's warships and naval aircraft and their administrative and support structure, including personnel and equipment, whose purpose is maintenance of control of the sea in a nation's areas of dominance, and defense of a nation's territory from attacks by sea and by air.

Navy Regs—Slang for US Navy Regulations; they are compiled for the organization and functioning of the navy.

navy yard—A naval shore establishment which provides a variety of important services to fleets. The largest navy yards have complete facilities for building, refitting, modernizing, repairing, docking, storing, and providing logistic support for ships. Only a few yards in any naval power are so equipped, however. Most yards provide some of these services—generally refitting, modernizing, and repairing—and most specialize in providing limited services to a particular class of warships.

NBC—1) Nonbattle casualty (q.v.). 2) Collective abbreviation for nuclear, biological, and chemical, as in NBC warfare. See also **CBR**.

nebelwerfer—A six-tube rocket launcher used by the German Army in World War II. The tubes were fired separately in a total of ten seconds and could be reloaded in 90. The nebelwerfer was used for short-range bombardments and for laying smoke screens—the range was 7,300 yards with high explosive and 7,750 with smoke shells.

needle gun—A bolt-action gun invented by the Prussian gunmaker Johann Nikolaus Dreyse in 1829 and used by the Prussian Army from 1848 to 1871. In the needle gun a long, sharp firing pin pushed through the powder at the back of the cartridge to strike the fulminate primer at the base of the bullet.

need to know—Reason to see classified information. A criterion used in security procedures requiring the custodians of classified information to establish strictly, prior to disclosure, that the intended recipient must have access to the information in order to perform his official duties.

neutrality—1) The status of a country that refrains from taking part in a war between other powers. 2) The condition of immunity from attack, invasion, or use by belligerent powers given to an area, a country, or a waterway.

neutralization fire—Fire delivered to prevent or interrupt effective hostile activity, particularly movement or the firing of weapons.

neutralize—As pertains to military operations, to render ineffective or unusable for a critical period of time.

neutron bomb (neutron warhead)—A tactical nuclear weapon (q.v.), more often a missile warhead than a bomb, designed to achieve its lethal effects primarily by radiation rather than by blast. Proponents of the weapon believe that by thus reducing collateral damage a neutron weapon is both more humane and more effective than is a standard tactical nuclear weapon.

New Model Army—The army organized by Oliver Cromwell in 1645. Having long felt the need for major military reform, Cromwell had urged Parliament to adopt a "frame or model of the whole militia," a proposal accepted by the Commons and the Lords which provided for a standing army of 22,000. This force was to be raised by impressment and supported by regularized taxation. The 12 regiments of infantry totaled about 14,000 men, and the cavalry's 11 regiments totaled about 6,600 men plus 1,000 dragoons. Artillery, a stepchild of the British weapons family, underwent some reorganization and expansion, largely on the basis of continental experience. For the first time red garb became the general uniform for a British army. Most significant, the traditional aversion of the local militia to campaigns beyond their home territories was bypassed through the creation of this mobile professional force.

nickeling mission—Slang in World War II for an air mission for the purpose of dropping propaganda leaflets which were called "nickels."

night operations—Operations taking place at night.

no fire line—A line short of which artillery or ships do not fire on targets except on request of the supported commander, but beyond which they may fire at any time without danger to friendly troops.

no man's land—A term that originated in World War I for the land between the two opposing entrenched armies.

nonbattle casualty (NBC)—A person who has not become a casualty in combat, but who is lost to his organization by reason of disease or injury, or by reason of being missing where the absence does not appear to be voluntary or due to enemy action or to being interned. See also **casualty**.

noncombatant—Not normally engaged in or assigned to combat duties.

noncommissioned officer (NCO)—An enlisted member of the armed forces who holds a rank or rating in which he exercises authority by appointment rather than by commission or warrant. In the US Army, for example, the noncommissioned officers are corporal, sergeant, sergeant first class, technical sergeant, first sergeant, and master sergeant. In the Navy, noncommissioned officers are called petty officers. See also **officer** and **petty officer**.

nonrated—1) In the US Air Force, not having a currently effective aeronautical rating. 2) In the US Navy, referring to enlisted men who have not yet attained a specialty rating.

Norden bombsight—A device invented by C. L. Norden in 1920 and used in a 1931 version by American bombers in World War II. See also **bombsight**.

Nordenfelt gun—A gun with from one to 12 barrels, developed from an 1873 design by Helge Palmerantz and financed by Thorsten Nordenfelt, both of Sweden. The barrels were in a horizontal row, with individual locks, and cartridges were fed by gravity from a magazine that sat above the carrier block.

normal barrage—See **barrage**.

normal zone of fire—The area within the zone of fire (q.v.) for which an artillery unit is normally responsible and within which its fire is normally directed.

November—Letter of NATO phonetic alphabet. See also **alphabet, phonetic**.

nuclear safety line—A geographical line selected to follow well-defined topographical features if possible and used to delineate the level of protective measures, degree of damage or risk to friendly troops, and/or possible limits to which the effects of friendly tactical nuclear weapons may be permitted to extend.

nuclear threshold—Level of violence beneath which military operations can be conducted without causing nuclear war.

nuclear warfare—Warfare involving the employment of nuclear weapons.

nuclear weapon—A device in which the explosion results from the energy released by reactions involving atomic nuclei, either fission or fusion or both.

nuclear yield—The energy released in the detonation of a nuclear weapon, measured in terms of the kilotons or megatons of trinitrotoluene (TNT) required to produce the same energy release. Yields are categorized as: very low, less than 1 kiloton; low, 1 kiloton to 10 kilotons; medium, 10 kilotons to 50 kilotons; high, 50 kilotons to 500 kilotons; and very high, over 500 kilotons.

numerus—See **bandon**.

O

objective—1) A goal to be obtained that, when defined and understood, gives direction to an operation or a series of operations. Sometimes used in such phrases as "national objectives" or "strategic objectives." 2) A place to be reached, especially a target or target area, or destination. 3) A principle stating that success in war or battle requires that the goal to be achieved should guide and determine every action taken in the conduct of the war or battle.

oblique order—1) A formation of troops in which all are facing 45 degrees from the direction in which they would face in normal marching order. Such a formation is ordinarily part of a drill routine. 2) A tactical procedure for bringing a part of an army into contact with a portion of an enemy army while the remainder of the attacker, being echeloned to the rear, threatens to engage the remainder of the defender's force. First used by Epaminonidas of Thebes at the Battle of Leuctra, 371 B.C.; most skillfully employed by Frederick the Great at the Battle of Leuthen, 1757.

Oboe—Letter of old phonetic alphabet. See also **alphabet, phonetic**.

observation—To watch an action or activity, usually of the enemy. The two principal forms of observation in combat are ground observation and aerial observation. See also **observe**.

observation post—A position selected or located primarily for the purpose of observation. Often abbreviated as O.P. An aircraft intended primarily for observation is called an air observation post.

observe—To employ visual means (eyes or vision enhancement equipment) to gain maximum information about a military situation and, particularly, to ascertain the movements and positions of a hostile military force.

observed fire—Fire for which the points of impact or burst can be seen by an observer. The fire can thus be controlled and adjusted on the basis of observation. Observed fire can be both area fire and precision fire (*qq.v.*). See also **fire**.

observer—1) An individual whose mission is to observe general activity. Neutral nations often send observers to learn about military developments in belligerent armies in wartime. 2) An individual whose mission is to observe fire. See also **forward observer** and **observation post**.

obstacle—Any obstruction that stops, delays, or diverts movement. Obstacles may be natural—deserts, rivers, swamps, or mountains—or artificial—barbed wire entanglements, pits, concrete, or metal antimechanized devices. Obstacles may also be fixed or portable, and may be issued ready-made or be constructed in the field.

obturate—To close up the breech of a gun so that the gas generated in firing cannot escape.

occupied territory—Territory of a nation held by foreign armed forces.

OD—See **olive drab** and **officer of the day**.

Oerlikon—Any of various Swiss-developed 20mm automatic aircraft or antiaircraft guns, firing greased ammunition and manufactured—or licensed—by the Oerlikon arms manufactory near Zürich.

offensive—1) The state or condition of a person or military force which is attacking; an attitude of attack. 2) An attack, especially a large-scale or sustained attack. 3) A principle of war embodying the thesis that success in war requires offensive action to achieve objectives and to exploit the advantages of initiative, maneuver, and surprise.

offensive à outrance—Attack at all times, at all costs, and under any circumstances. This doctrine of warfare was developed in the French Army before World War I by Colonel Louis L. de Grandmaison, a disciple of General Ferdinand Foch, who also stressed the importance of the offensive, but with less fanaticism.

offensive grenade—See **grenade**.

officer—A person who by virtue of rank exerts authority over other members of the armed services. There are three principal classes of military officers—**commissioned officers**, **warrant officers**, and **noncommissioned officers**. A commissioned officer holds a document (commission) from the duly constituted government (in the United States, from the President), appointing him an officer and awarding him a rank. A warrant officer holds a warrant of rank, issued by proper authority, and typically wears a uniform identical, or similar, to that of a commisioned officer; his duties are usually administrative, rather than operational. A noncommissioned officer (q.v.) is appointed to his rank by an officer of the unit or formation in which he serves. See also **company grade officer, field grade officer, flag officer, general officer, line officer, staff officer**, and **warrant officer**.

officer-in-charge—1) The officer temporarily commanding a unit or facility in the absence of a commanding officer. 2) The officer responsible for an office or an activity.

officer of the day (OD)—The officer in a military command who has initial responsibility for handling all internal security problems that may arise during his 24-hour tour of duty as well as for overseeing the performance of the interior guard during the period.

officer of the deck—The naval officer on a warship who, during a prescribed period (generally 24 hours), has initial responsibility for the operation of the ship and the carrying on of ship routine. This duty is performed by a line officer, and during his watch everyone

on board ship except the captain and the executive officer is subject to his orders.

officers' country—Slang for that portion of a ship or naval shore station that is designated for the use of commissioned officers only, as living or working quarters or for recreation.

off limits—Not to be entered, patronized, or used by military personnel.

offshore patrol—A naval defense force operating in the outer areas of navigable coastal waters. It is a part of the naval local defense forces, and consists of naval ships and aircraft operating outside those areas assigned to the inshore patrol (q.v.).

Oka—A Japanese rocket-propelled one-man guided missile of World War II. It was also called "Cherry Blossom."

OKH—*Oberkommando des Heeres*. The German Army High Command in World War II. Included the German Army General Staff.

OKL—*Oberkommando der Luftwaffe*. German High Command of the Air Force in World War II.

OKW—*Oberkommando der Wehrmacht*. High Command of the German Armed Forces in World War II. The staff reported directly to Hitler. Most of its officers were from the Army General Staff.

"Old Contemptibles"—Nickname of the first contingent of the British Expeditionary Force (BEF) in 1914. Originally coined by the troops in response to the supposed reference by Kaiser Wilhelm II to them as that "contemptible little army."

"Old Guard"—The elite veterans of Napoleon's Imperial Guard.

"Old Ironsides"—The frigate USS *Constitution*. A 44-gun American frigate active in the Barbary Wars and the War of 1812.

olive drab (OD)—1) A standard olive-green color used for US Army uniforms and equipment since the late 19th century. 2) Material of this color.

onager—Developed about the third century A.D., and popularly known as the Roman onager, it was a torsion-driven catapult, consisting of a stout, heavy rectangular frame, with an upright throwing bar and two solidly braced uprights with a heavy crossbeam on top. The base of the throwing arm was thrust through the twists of a horizontal skein of heavy cord stretched between the two sides of the frame just behind the uprights. At rest, this cord held the throwing arm tightly against the crossbeam. The top, or end, of the throwing arm was usually in the form of a spoon-shaped ladle, or had a leather pouch attached to it. In firing, the arm was cranked back to a near-horizontal position by a windlass, then a rock or other missile was inserted into the receptacle at the end of the arm and a trigger mechanism was released. The throwing arm returned with great force to a vertical position and hit the crossbeam. The force of inertia could hurl the 40- to 60-pound rock about 450 yards. Roman soldiers called this engine "onager" (wild ass) because of the tendency of the rear end of the frame to lift or buck when the throwing arm hit the crossbeam.

onset—A violent attack or assault, especially the assault of an army or body of troops upon an enemy or a fort.

on the double—An expression, generally attached to an order or a command, indicating that the mission should be accomplished rapidly, in double time, that is, at the run.

"on the way"—A message to an air artillery observer alerting him that a round has been fired and will soon be observable to him.

O.P.—Abbreviation for observation post (q.v.).

open flank—See **flank**.

open order—A formation of troops deployed as for combat, in contradistinction to close order formations for drill and marching.

operation—A military action or the carrying out of a strategic, tactical, service, training, or administrative military mission.

operational—1) Ready and fit for action, used of troops, installations, naval vessels, weapons, and equipment. 2) Having to do with operations, as a directive or an order.

operational art—See **operations**.

operational chain of command—See **chain of command**.

operational maneuver group—A large Warsaw Pact formation designed to carry out a Soviet doctrinal concept of deep penetration (q.v.). The force—probably a tank army with two or three tank divisions and one or two motorized rifle divisions, supported by substantial tactical air units—is expected to be prepared to exploit a breakthrough by front line, probably first echelon (q.v.), forces. The anticipated mission is a rapid and deep strike into the NATO rear area, to create maximum destruction and disruption of NATO defensive operations. It represents an updating of the Soviet mobile group of World War II. Although it does not fit the definition of a second echelon (q.v.) force, it is considered by NATO to fall into the category of a follow-on force (q.v.).

operational priority—A category of precedence reserved for messages requiring expeditious action by the addressee. See also **precedence**.

operational research—See **operations research**.

operation map—A map showing the location and strength of friendly forces involved in an operation. It may also indicate predicted movements and location of enemy forces.

operation order—A directive, usually formal, issued by a commander to subordinates for the purpose of effecting the coordinated execution of an operation. In the US Army it is usually in five paragraphs which set forth (1) the situation, (2) the mission, (3) the commander's decision, (4) his plan of action, and (5) details of the method of execution. Thus coordinated action by the whole command is ensured. Also called **operations order**.

operation plan—A plan for operations extending over a considerable space and time and usually based on stated assumptions. It may cover a single operation or a series of connected operations to be carried out simultaneously or in succession. It is the form of directive employed by higher echelons of command in order to permit subordinate commanders to prepare their supporting plans or orders. The designation "plan" is often used instead of "order" in preparing for operations well in advance. An operation plan may be put into effect at a prescribed time, or on signal, and then becomes the operation order. See also **operation order**.

operations—1) The process of carrying on combat, including movement, supply, attack, defense, and maneuvers needed to gain the objectives of any battle or campaign. 2) A conceptual division of the process of waging war, involving combat activities between those of tactics (q.v.) and those of strategy (q.v.). Originally devised by the Germans during World War I to distinguish between the performance of battlefield activities (tactics) and multifront or multitheater aspects of waging war (strategy). As visualized by the Germans, operations was just above the lower boundary of strategy, and pertained to control of the activities of large ground forces—usually armies or army groups—in a discrete theater of combat. The Soviet armed forces adopted the concept of operations during World War II, but have treated it much more rigidly than the Germans, viewing it as restricted to activities of armies and army groups, and as a distinct form of activity lying between tactics and strategy. They call it **operational art**. This concept, closer to the Soviet than the German pattern, was adopted by the US Army in 1982.

operations analysis—See **operations research**.

operations research—The analytical study of military problems, undertaken to provide responsible commanders and staff agencies with a scientific basis for improving military operations. Also known as **operational research** and **operations analysis**.

orb—Formerly in military tactics, the forming of a number of soldiers in a circular defense, generally six deep. This defensive configuration was considered by the French Marshal de Puysegur, in his *Art of War* (1749), to be particularly strong because the enemy could expect no part of the circular formation to be weaker than any other.

order—1) A communication, written, oral, or by signal, that conveys instructions from a superior to a subordinate. In a broad sense, the terms "order" and "command" are synonymous. 2) With "arms," as in "order arms," the drill command to rest the rifle butt on the ground against the right foot with the barrel perpendicular to the ground, or leaning slightly to the rear, while standing at attention.

orderly—1) An enlisted man in personal attendance on one or more officers. 2) A person assigned to a specific task to keep things in order, as a latrine orderly or a medical orderly.

orderly book—1) The record of activities of a unit or a guard. Similar to the log of a naval vessel. 2) An official record of attendance or use.

orderly room—A room in a barracks used as the office of a company or other unit of similar size.

order of battle—A statement of identification, strength, command structure, and disposition of the personnel, units, and equipment of any military force.

order of march—The sequence in which the elements of a military force move from one location to another.

ordinary, in—Used formerly of naval vessels not in actual use, which must be roofed over, with their masts and running gear stowed ashore. The ships were formed in divisions, with a lieutenant in command, and a line-of-battle ship, called a "guardship-of-ordinary," responsible for the several divisions in a port. Ships similarly stored in modern times are said to be "in mothballs" (q.v.) or "mothballed."

ordnance—Collective term for all military weapons, ammunition, explosives, combat vehicles, and battle matériel, together with the necessary maintenance tools and equipment.

organic—Assigned to and forming an essential part of a military organization. Organic parts of a unit are those listed in its table of organization for the army, air force, and marines, and those assigned to the administrative organizations of the operating forces of the navy.

organization—A group of individuals with responsibilities, authorities, and relationships defined for the purpose of effectively accomplishing a mission.

organization and equipment, tables of

(T/O&E)—A listing of the numbers of units, troops, weapons, vehicles, and other major items of matériel with which any army organization is normally equipped.

orientation—A training program for recruits or troops reporting to a new command to introduce them to the organization, functions, procedures, and other aspects of the command.

oriflamme—The red or orange-red silk banner of the Abbey of St. Denis in France. Early kings of France used this banner as their standard (q.v.) in combat.

Oscar—Letter of NATO phonetic alphabet. See also **alphabet, phonetic**.

other ranks—British term for soldiers. See also **enlisted personnel**.

Out—A signal communications transmission indicator signifying the end of a message, with no response expected. See also **Over**.

outflank—In ground combat, to maneuver an attacking force around and behind the flank of the opposing force. Thus, a tactical advantage might be gained because the opponent will be forced to divert strength from the front line to defend the flank and rear.

outline plan—A preliminary plan that outlines the salient features or principles of a course of action prior to the initiation of detailed planning.

outpost—1) A military post, base, station, or other installation located far from its national homeland, having the primary responsibility of guarding the nation against surprise attack or of protecting a territory or possession. 2) A detachment of guards or sentries posted outside a stationary military force to protect it against surprise by the enemy.

outpost line of resistance—The outer line of defense of a defensive position or zone. Usually held by a covering force (q.v.).

outrank—To be senior because of higher rank or more time in grade than others. Sometimes shortened to "rank."

outwing—To extend the flanks of a combat force in action in order to gain an advantageous position against the right or left flank of an enemy.

outworks—A smaller fortification built beyond (that is, closer to a potential attacker than) the main defense structure.

Over—A signal communications transmission indicator signifying the end of an element of a message and requesting a response from the party at the other end. Its use with the word "out" (q.v.)—as in "Over and Out"—is improper procedure.

over the top—An expression used in World War I for the movement out of the trenches toward the enemy lines as an attack began.

overhead fire—Fire over the heads of friendly troops.

overkill—The provision of more resources or fire power than reasonably required to achieve an objective.

overlay—A drawing or printing on a sheet of translucent material, keyed to a map or chart so that it may be placed accurately above it and which shows deployments of forces or equipment.

overseas bar—A stripe worn on the sleeve for each six months of overseas duty that had been served between December 7, 1941 and September 2, 1946. Popularly called a "Hershey bar," for the former head of the Selective Service, General Lewis B. Hershey.

overseas cap—A visorless uniform cap of the material of which the uniform suit is made, which when not on the head folds flat from side to side so that it can fit in a pocket or under a belt. Such caps were generally optional for wear by all services in World War II.

P

pace—1) The regulated marching speed of a military column. For quick time (q.v.) this is 120 steps per minute; for double time (q.v.) this is 180 steps per minute. 2) A step in marching; for quick time in the US Army, the length of stride is 30 inches. For double time it is 36 inches.

pacification—A mission assigned to a military force to establish or maintain governmental authority and order in a geographical region which is or is potentially hostile.

pack—1) A container designed to carry items of equipment on the back of a soldier or animal. Sometimes called a knapsack (q.v.). 2) The packaged equipment carried by a soldier on his back; the infantryman's pack generally contains shelter-half, blanket, underwear, socks, toilet articles, etc. 3) Submarines working together on an operation, for example, the German "wolf packs" of World War II.

pack transportation—The personnel, animals, and equipment used for the transportation of matériel and supplies loaded on pack animals (generally horses and mules).

palintonon—A type of ballista (q.v.) with the guide beams on an inclined plane. It is estimated that some versions were as large as 10 yards long, 5 yards high, and 4 yards wide, and could fire stones weighing one to eight pounds as far as 300 yards.

palisade—n. A fence made of stakes placed close together to form a defensive barrier. v. To construct such a fence.

palisade gun—See **tarasnice**.

pan—A shallow receptacle that holds the powder in a matchlock, flintlock, or other gunpowder weapon.

pandour—One of the lightly armed border troops used in the early 18th century for defense of the Austrian frontier against the Turks and by Maria Theresa in the War of the Austrian Succession.

pannel—A carriage on which mortars and their beds used to be conveyed on a march.

panzer—In German, armor or armored. The word was originally used for body armor, but since the Second World War, in English and German, panzer has been used primarily to refer to German tanks.

panzer grenadier—n. An infantryman in a German unit composed of both infantry and armored elements (that is, a mechanized unit). adj. Descriptive term for such a unit.

Papa—Letter of NATO phonetic alphabet. See also **alphabet, phonetic**.

parachute troops—See **paratroops**.

parade—n. 1) A ceremonial procession of marching personnel, sometimes interspersed with vehicles. 2) The area, more formally called the parade ground, where troops march in review. v. To cause to march, as to "parade the troops" or "parade the colors."

parallel—A trench; usually one of a series dug approximately paralled to the fortification being attacked, so that each new trench line would bring the attackers closer to their objective. This aspect of siege warfare was developed in the 17th century by the Marquis de Vauban.

parallel sheaf of fire—See **sheaf of fire**.

paramilitary force—A force or group that is distinct from a country's standing professional forces, but resembling them in such respects as organization, equipment, training, or mission.

paramilitary operation—An operation undertaken by a paramilitary force.

parapet—A defensive wall of earth or stone.

paratroops—Troops carried into battle by aircraft and dropped by parachute, as distinguished from airborne infantry, who are flown to battle, but may be landed from aircraft or helicopter. Paratroops are a special category of troops within airborne forces. Also called **paratroopers**.

paravane—A naval mine defense device. It is a type of water kite that is towed from the forefoot of a vessel, riding out to one side at the end of a wire rope usually 56 yards long, to deflect mines that may be moored in the path of the vessel and cut the mooring lines so that the mine rises safely to the surface.

Paris Gun—The designation of any of three long-range 210mm German guns used to bombard Paris from March 23 to August 7, 1918. Most of the several emplacements were in the St. Gobain Wood near Laon, 70 miles from Paris. The Paris Gun, *Pariskanonen* in German, designed by a team of Krupp engineers, had a barrel almost 120 feet long and could fire a 275-pound projectile to a range of approximately 80 miles. When the guns' first shells landed on Paris it was thought that they had been dropped from German bombers. In all, the Paris Guns fired 203 shells against the city but did not produce the panic and population exodus the Germans hoped to achieve. The advance of the Allied armies in August 1918 ended the shelling.

park, artillery—1) The area where artillery pieces and vehicles are stored and serviced. 2) The artillery so stored.

parley—A meeting of representatives of opposing forces to discuss terms for ending or suspending combat. Before the 20th century, the side wishing to have a parley announced it by bugle or drum, and fighting was usually halted during the parley.

parole—n. 1) A password. 2) Status of an individual released from confinement or arrest on his acceptance of terms for such release. 3) Status of a prisoner of war released upon agreement of the opposing side and his ac-

ceptance of the terms. v. The act of releasing a person from arrest or confinement.

parry—To prevent a weapon or a projectile from hitting its target, used especially of thrusting weapons, like swords, spears, and lances.

partisan—n. 1) A spear with a two-edged blade and a lug curving out from the base of the blade at either side. It was developed in the 15th century and is still carried ceremonially by the British Yeomen of the Guard at the Tower of London. The French name for this weapon is *langue de boeuf*. In 15th century England this became *langedebeve*. 2) A member of an irregular or guerrilla group operating within occupied territory to harass and inflict damage on the occupying forces. *adj.* The form of activity conducted by partisans. See also **guerrilla** and **guerrilla warfare**.

partridge—A very large bombard, formerly used in sieges and in defensive works.

partridge mortar—A weapon developed near the beginning of the 18th century. It had a central bore for firing a standard large 110mm mortar shell, surrounded by 13 smaller bores that, arranged in a circular pattern, fired smaller grenades or bombs almost simultaneously. It was used by the French in 1702 at Bouchain and in 1708 at Lille.

party—A small group performing a specific task, as an inspection party, a boarding party, or a shore party (q.v.).

pass—1) A short tactical movement, usually a dive, by an aircraft over a target, either air or ground. 2) A card or certificate that permits one to enter a gate, door, or other area requiring control. 3) Temporary release from duty, certified by written permission or pass.

pass in review—To march in formation before a commander or reviewing dignitaries.

passed midshipman—In the 19th century in the US and Royal Navies, a midshipman who had passed his examination and was a candidate for the rank of lieutenant.

passive—In defensive activities, passivity describes any or all noncombatant measures taken to defend an area or installation against a ground or air attack or airborne assault, including the use of cover, concealment, dispersion, etc.

passive defensive—See **defense**.

passport—A document giving a ship permission to leave port in time of war or to pass through restricted waters. It is issued officially, especially to commercial vessels belonging to neutral nations.

password—A secret word or distinctive sound used for friendly identification in reply to a challenge.

patch—1) An embroidered cloth badge worn on a uniform sleeve indicating the wearer's rank or unit. 2) Before the invention of the cartridge, a piece of cloth or leather that was greased and then wrapped around a ball or a bullet so that the projectile fit more tightly inside the barrel of a gun. The infantryman carried a number of patches in a **patch box**, which was cut into the butt of his weapon or carried on his belt.

patch box—See **patch**.

paterero—See **pedrero**.

pathfinder—1) A specialized military team, dropped or airlanded in advance of a main force, whose mission is to mark a drop zone or landing zone and to operate navigational aids to guide incoming aircraft. 2) The aircraft carrying such a team. 3) A radar device used for navigating or homing to an objective when visibility precludes accurate visual navigation. 4) An aircraft—usually one of several—leading a World War II bombing mission, particularly for the Royal Air Force Bomber Command. A pathfinder's mission was to designate the target for the main body of following aircraft on the raid or mission by dropping bombs or flares or a combination of both.

patrol—1) A detachment of ground, sea, or air forces dispatched by a larger unit to gather information or carry out a destructive, harassing, mopping up, surveillance, or security mission. 2) The checking of an area, usually on a periodic basis, by moving from one point to another to see if a normal condition has been disrupted. 3) The person, vehicle, or vessel that performs such a check. See also **inshore patrol** and **offshore patrol**.

patrol, fighting—A tactical unit sent out from the main body to engage in independent combat activity; a detachment assigned to protect the front, flank, or rear of the main body.

patrol, offensive—A patrol with an offensive mission.

patrol wing—A US Navy organization, later changed to fleet air wing.

pattern bombing—The systematic covering of a target area with aerial bombs uniformly distributed according to plan.

pavise—A large shield carried by medieval infantrymen, and particularly by crossbowmen. The frame was usually covered by hide or painted canvas. Some pavises were large enough to cover the whole body and could be set upright to protect the bearer as he fired or loaded his weapon. Crossbowmen sometimes wore pavises on their backs and turned around for protection.

pay clerk—US Navy warrant officer.

peacekeeping forces—A military force established by international agreement to maintain peace in a contested area. Contemporary forces of this type are usually set up by the United Nations and have been used several times in the Middle East and Africa.

pedrero—A 16th-century mortar-type weapon that fired a stone projectile—pedrero means "stone mortar" in Spanish—rather than a heavier iron projectile. Despite its thin walls, the pedrero could fire a stone cannonball almost as far as a cannon, with a maximum range of 2,500 yards. Pedreros varied greatly, but a typical one weighed 3,000 pounds, was nine feet long with a ten-inch bore, and fired a 30-pound projectile. Also known as a **perrier**, derived from the French word **pierrier**, and as **paterero**, derived from the Latin for rock or stone.

peep—See **jeep**.

pelta—See **peltast**.

peltast—A lightly armed, javelin-throwing infantryman of ancient times who carried a **pelta**, a light skin-covered wicker shield, crescent-shaped, circular, or oval. The first peltasts were Thracians, who proved so successful that similar troops were introduced into Greek armies. They were cheaper to arm than hoplites (q.v.) and, because lightly armed, they could move more quickly.

penetrate—1) To break through a defense. 2) To force a way into a target or other substance. Said of a shell, bomb, or other projectile, and also of rays and radioactive dust.

penetration—1) A form of offensive maneuver that seeks to break through an enemy's defensive position, widen the gap created, and destroy the continuity of his positions. 2) The depth to which a projectile or a force has advanced after entering its target.

pennant—1) A flag. Also known as **pennon**, the term is most frequently used for small flags. 2) A small, long, narrow flag, flown on a ship for use as a signal or identification. 3) Such a flag carried on a lance.

Pennsylvania rifle—An 18th- and 19th-century muzzle-loading weapon that was first made by German settlers in Pennsylvania. Later, the weapon came to be known as a Kentucky rifle (q.v.).

Pentagon—A large, low, five-sided, five-floor building on the Virginia side of the Potomac River just outside Washington, D.C., in which are housed the principal executive offices of the Department of Defense, including those for the military departments of the Army, the Navy, and the Air Force. Also called "the National Defense Building," the Pentagon was completed on January 15, 1943. It contains over 17 miles of corridors and has provided space for as many as 32,000 workers. It is often personified, as in "the Pentagon says" or "the Pentagon ordered," to refer to defense policy or the US defense community.

penteconter—A 50-oared galley.

penthouse—1) A shedlike structure for the protection of a gun and its carriage from the weather. 2) A shelter used to protect weapons and troops from hostile fire in ancient and medieval siege operations.

pentomic division—The name given to the standard US infantry division during the late 1950s and early 1960s, replacing the triangular division (q.v.) of World War II and the Korean War. Its principal maneuver elements

were five battle groups (q.v.), larger than the standard battalion and smaller than the standard brigade or regiment. Experimental in nature, the pentomic division was soon replaced by an updated version of the older triangular division. See also **division**.

percussion cap—A device containing a material that explodes on being struck and is used to ignite powder or other explosive material. The first caps were developed about 1814 after the Scottish Reverend Alexander Forsyth's discovery in 1807 of this property in a mercuric fulminate. They were small cylindrical cups resembling a tall hat and were made of iron, pewter, and copper variously. The cap was partially filled with the fulminate, which was, in turn, held in place by a covering of tin foil and shellac. The cap fitted over a tube protruding perpendicularly from the rear end of the barrel of a gun. When struck with a hammer, the fulminate exploded, producing a spark that fired the powder in the gun. Adoption of this device greatly increased the effectiveness of handguns, permitting them to be fired in wet weather, thus altering the tactics of warfare. Percussion caps of various types were used until early in the 20th century.

perimeter defense—An all-around defensive system of field fortifications and weapons to protect an area or locality.

permanent fortifications—Fortifications of a lasting, elaborate nature, usually erected in time of peace and constructed of hard materials like stone, masonry, and concrete. Modern examples include the French Maginot Line, the German West Wall, and the Finnish Mannerheim Line (qq.v.), all erected in the 1930s for the defense of frontiers. See also **fortifications**.

permanent rank—A rank that does not cease with a change of duty or a change of circumstances, as opposed to temporary rank.

perrier—See **trebuchet** and **pedrero**.

personnel—1) Individuals, usually members of the armed services. 2) Collectively, the people of the armed forces. 3) A staff function, activity, or division concerned with military individuals and policies relating to them.

personnel carriers—Armored vehicles,

tracked or wheeled, designed primarily to transport infantry into combat, but usually having no provisions permitting a majority of the transported personnel to engage the enemy while fully enclosed.

petard—An explosive device, often in the shape of a small bell, designed to be detonated against a gate or a wall in order to force a passage through.

Peter—Letter of old phonetic alphabet. See also **alphabet, phonetic.**

petraria—See **trebuchet.**

petrary—See **trebuchet.**

petronel—A 15th-century firearm designed for use by a mounted man, steadying it against the breast (whence the French name **poitrinal**) and a rest attached to the bow of the saddle.

petty officer—A naval noncommissioned officer (q.v.).

phalanx—An ancient formation of infantry in which soldiers stood so closely together that their shields overlapped, with spears projecting forward, the spears of soldiers in the rear ranks sometimes resting on the shoulders of the men in front. Such a formation was used in Sumeria as early as the third millenium B.C. The Greeks used it from about the 7th century B.C. The phalanx, with a depth of 8 to 24 ranks, was greatly improved by Philip II of Macedon and Alexander the Great. In battle the normal deployment was a long, solid line with narrow intervals through which the psiloi, (q.v.) could pass. Battle was waged on the flattest ground available. However, the Macedonian phalanxes of Philip and Alexander, 16 men deep, were so well trained that they could perform complex maneuvers even in rugged terrain.

pharmacist—US Navy warrant officer.

pharmacist's mate—US Navy enlisted specialty, chief, first, second, and third class. See **hospital corpsman.**

phase line—A line used for control and coordination of military operations, often a terrain feature extending across the zone of action, or an imaginary line connecting two prominent terrain features.

phosgene—A poisonous gas (carbonyl chloride) harmful to the lungs. It is colorless, extremely heavy, and smells like new cut hay or corn. It was introduced in World War I, and generally supplanted chlorine as a gas.

photographer—US Navy warrant officer.

photographer's mate—US Navy enlisted specialty rating, chief, first, second, and third class.

PIAT (Projector, Infantry, Antitank)—A British World War II-vintage individual antitank weapon that was fired from the shoulder. Spring operated, it fired a three-pound projectile that could penetrate four-inch armor at a distance of 50 yards.

pickelhaube—German term used for the ceremonial spiked helmet of World War I vintage.

picker—A small pointed brass wire, formerly supplied to every infantry soldier to use for cleaning the vent hole of his musket.

picket—1) A pointed stake that, generally used with others, provides a defensive barrier. 2) A group of persons detached from a main party and serving as a line of outposts to guard the others from surprise attack. 3) A radar beacon set up in a peripheral area. 4) A ship that patrols an outer defense point to obtain early warning. 5) Aircraft posted around a sensitive area to provide early warning of the approach of enemy aircraft, to track them, and to control friendly aircraft sent to intercept.

picklebarrel bombing—1) Slang expression indicating the bombing of a very small target. 2) Highly accurate bombing.

piece—A general term applied to any weapon, individual or crew-served.

piecemeal attack—A maneuver in which units of a command enter into battle in small increments, as they become available on the battlefield. See also **attack.**

pierrier—See **pedrero.**

pig's head—A Roman tactic in which the ten maniples (q.v.) of a legion were formed into a wedge—four, three, two, and one maniple from back to front—as a device to break through an enemy's ranks.

pike—A long spear, used by infantry for offensive and defensive purposes, from ancient times until the late 17th century.

pikeman—A soldier, usually an infantryman, armed with a pike (q.v.).

pillage—To plunder, used especially of the seizing of goods by an occupying armed force. See also **loot**.

pillbox—A small, low fortification that houses machine guns, antitank weapons, and other weapons together with the crew to fire them. A pillbox is usually made of concrete, steel, or filled sandbags.

pilot officer—A commissioned rank in the Royal Air Force and other air forces, corresponding to second lieutenant in the US Air Force.

pilum—A Roman javelin, developed in the third century B.C. There were many types, partly because javelins were made by hand, and partly because of technical innovations. In the second century the Romans seem to have standardized a light pilum that was easily handled and capable of great penetration of hard-surfaced objects. It was basically a seven-foot long weapon made half of metal and half of wood—a four-and-one-half-foot wood shaft into which was inlaid (with a two-foot overlap) a four-and-one-half-foot iron rod. Marius made one of the two pins that bound the wooden shaft and metal rod together of wood, so that the shaft would break off in an enemy's shield. Caesar and others made the iron rod of soft metal and the tip hard: thus the tip would penetrate and the shaft would bend so that the javelin could not be easily withdrawn. By the 1st century B.C. the pilum was as important a legionary weapon as the sword.

pincer movement—A type of double envelopment in which two columns drive on each side of an enemy position and converge at its rear to isolate and destroy it.

pinfire cartridge—A cartridge developed by the French gunsmith B. Houllier in 1846. It consisted of a metal case in which a small percussion cap was mounted sideways in the base. A pin protruded through the case, and

when struck by a hammer, it drove into the cap, exploding and detonating the charge, thereby propelling the bullet. This was the first completely self-contained ammunition for small arms, and was the prototype of the center-fire and rim-fire cartridges.

pintle—An upright pin on the rear of a towing vehicle that serves as a coupling device for a trailer or other vehicle being towed. For instance, the trail of a gun carriage attaches to the pintle of the limber or prime mover (qq.v.).

pioneer—Formerly, pioneers were soldiers in European armies detailed from the different companies of a regiment and furnished with saws, felling axes, spades, mattocks, and billhooks for construction tasks, clearing of obstacles, and the like. The term is still used in the British Army for a member of units corresponding to engineer units in the US Army.

pipe—To deliver an order or other message by the musical notes of a boatswain's pipe (q.v.). Piping aboard, for example, is one of the honors rendered when such important persons as the ship's captain come aboard ship.

pipe clay—A clay used for making tobacco pipes. Because of its white color, pipe clay, in its powdery form, was frequently used to brighten the white portions of uniforms, such as the belting.

pistol—A small firearm that is held and fired with one hand. The origin of the name is uncertain, but it is thought to have derived from the city of Pistoia, Italy, where it is claimed that the weapon was developed. However, it is also claimed that the first handguns had a bore the diameter of the ancient coin called a *pistole*. In early times those armed with pistols were called pistoleers.

pitched battle—1) A battle or, loosely, any form of combat, that is highly intense. 2) A battle of forces in close contact, using carefully planned tactics.

pivot gun—A gun, usually of long range and high muzzle velocity, mounted on the bow or poop deck of a warship or armed merchantman, capable of 360-degree traverse. See also **swivel gun**.

plain langauge—Unencoded, used for messages not converted to coded form.

plan—The scheme or proposed method for accomplishing a mission or attaining an objective. A plan for a military operation is usually formulated on the basis of an estimate of the situation (q.v.), and in turn is the basis for an operations order (q.v.).

plan, contingency—A plan for a major contingency which can be reasonably anticipated by a national or regional command. A contingency plan is usually related to a geographic region or to one of the principal geographic subareas of a command. There are usually a number of such plans, for different possible situations and different areas.

plashing—An obstacle similar to abatis (q.v.), in which were trees and limbs of trees were intertwined to form a thick hedge. It differed from abatis in that it was more hastily prepared and the tree trunks were not embedded in the ground. See also **zareba**.

plate armor—Armor in the form of solid sheets of metal, normally shaped to the contour of that which it protects. Plate armor of bronze was used for body armor at least as early as the 15th century B.C., but it was not until the 13th century A.D., long after bronze armor had been abandoned, that medieval European armorers began producing plate armor of iron. At first, these iron plates, covering vital and vulnerable areas such as the shoulders and thighs, were worn under the chain mail (q.v.) that had been the armor commonly used since Byzantine times. By the middle of the 13th century, however, plate armor was replacing chain mail, and in the next century whole suits of plate armor had been developed. The armor built into or added to ships, aircraft, tanks, and other items of military equipment is also known as plate armor. See also **armor**.

platform—The flat base upon which a gun is mounted.

platform wagon—A wagon used for transporting heavy ordnance.

platoon—The next to the smallest unit in a modern military organization. Composed of 25 to 50 men and usually divided into squads or sections, platoons are the smallest unit commanded by a commissioned officer (sec-ond lieutenant or ensign), although they are often commanded by noncommissioned officers. From two to four platoons form a company. The term "platoon" was used in the 17th century to refer to a group of musketeers who fired their weapons together.

plunging fire—Fire that strikes a target at an angle greater than 45° from the horizontal. This can be direct fire from a superior elevation, but is usually one of the following: high angle fire from a gun or howitzer at an elevation greater than 45°, or fire from a mortar under most circumstances. Plunging fire is more likely to be indirect than direct fire (qq.v.).

pocket battleship—A lightly armored, diesel-powered warship of which three were built for the German Navy prior to World War II. They were designed to comply with in the limits of 10,000 tons displacement set for the German navy by the Versailles Treaty, but actual displacement was about 12,000 tons. They carried six 11-inch guns and had a cruising range of over 20,000 miles without refueling.

poilu—A World War I slang term for a French soldier, particularly those in the front line. The origin is not clear, although the word means "hairy" in French, and may have been applied because of the unshaven appearance of many of these soldiers in prolonged combat.

point-blank—Directly in front of the target. Point blank implies that a target is so close that it is not necessary to allow for the drop in the trajectory of a projectile and that it is almost impossible to miss it.

point-blank range—The distance to that point where an artillery projectile, fired from a level or horizontal bore, first strikes the ground in front of the cannon.

point d'appui—A position, usually fortified and often containing supplies, from which a military force can operate as a base for maneuver in order to apply pressure against an enemy.

point defense—Defense of positions or areas that are critical or important to the security or viability of a force, such as a dominating ridge, an artillery emplacement, or the force commander's headquarters.

point fuze—A fuze at the forward end or point of a shell. See also **fuze**.

point target—A target so small that it requires a highly accurate placement of ordnance, usually by precision fire (q.v.), in order to neutralize or destroy it.

poitrinal—See **petronel**.

POL—Abbreviation for a class or category of automotive-related supplies: petroleum, oil, and lubricants.

pole-axe—A medieval weapon consisting of a long shaft with a head in the form of an ax or a combined ax, pick, and hammer.

pom-pom—1) A rack of light antiaircraft cannon, usually mounted in fours, as on a ship. 2) An automatic cannon. The term was first applied to an automatic cannon used by the Boers in the Boer War, because it sounded like the beating of a drum.

poniard—A dagger, particularly one with a blade that has a triangular or square cross section.

ponton—A floating structure, either a boat or similar to a boat, that serves to support something, such as a dock or a bridge. Can be spelled **pontoon**, but pon-ton is the correct pronunciation.

ponton bridge—A bridge built over pontons placed at intervals, parallel to the current of a stream, and connected by planks or other material, to afford a temporary means of crossing. Pontons serve both to provide buoyancy and to connect the horizontal members of the bridge. The parts of such bridges are carried by forces anticipating a river crossing and can be assembled quickly.

ponton carriage—The truck or trailer on which a ponton is carried.

pontonnier—French term for the engineer responsible for pontons.

pontoon—See **ponton**.

port arms—Drill command for an infantryman to hold his rifle in a diagonal position in front of his chest.

portcullis—In a fort, a grille constructed of iron or wood that is suspended above the gateway and, in case of attack, can be lowered more quickly than the gate can be closed.

portfire—Another name for the artilleryman's linstock (q.v.), a long, forked stick, which before about 1800 held the slow match the gunner used to ignite the charge in an artillery piece.

port of embarkation—The port, sea or air, where military personnel board ships or aircraft for transport to a combat area or to an overseas assignment.

position defense—The type of defense in which the bulk of a defending force is disposed in selected tactical localities where the decisive battle is to be fought. Principal reliance is placed on the ability of the forces in the defended localities to maintain their positions and to control the terrain between them. Can be successful only if a mobile reserve is available to add depth and to restore the battle position by counterattack.

post—n. 1) A military base, including the grounds and buildings. 2) An assigned or designated position or station. v. 1) To assign to a military organization. 2) To assign an individual (usually a sentinel) to his designated position or station.

post exchange—A retail store at a US post or installation, operated for military personnel and their dependents under the supervision of the commander. Sometimes called **PX**, also **base exchange**.

postern—In a fort or castle, a back or side entrance door or gate, where entrance or exit could be inconspicuous, including that of defenders mounting a surprise attack on a fortification's besiegers.

posture, military—1) The disposition, strength, and condition of readiness of military forces as these factors affect capabilities. 2) The particular position of, or the attitude assumed by, an organized force or nation from which it can act or react immediately, as in **posture of offense**, or from which it may be unable to react as in **posture of unawareness**. 3) The state of preparedness of a defending force, as in hasty defense, prepared defense, fortified defense (qq.v.).

posture of offense—See **posture, military**.

posture of unawareness—See **posture, military**.

pot de fer—From the French meaning "iron jug," it was a name given to early cannon because of their bottlelike shape. The earliest pot de fer is shown in an illuminated manuscript of 1326–27.

POW—Prisoner of war. A military person taken captive during a period of belligerence or threatened belligerence.

powder—A dry explosive or explosive propellant, whose form may be very fine particles or larger pieces. See also **black powder**, **gunpowder**, **smokeless powder**.

powder monkey—In sailing vessels, the name given to a young boy who carried cartridges from the magazine (q.v.) to the guns.

power, military—The capability to use military force effectively in support of national policy.

Praetorian Guard—Elite personal corps of guards of the Roman emperors, abolished by the Emperor Constantine in 312 A.D. The Praetorian Guard was in many instances the military force employed by rebellious leaders to overthrow the reigning emperor.

prearranged fire—See **fire, prearranged**.

precedence—1) Priority in rank and preference in privileges, or in position in ceremonies, based on rank and length of service. In the US armed forces the order of precedence is: Army, Navy, Air Force, Marines, Coast Guard. 2) An indication in the heading of a military message or dispatch which indicates its degree of priority or the speed with which the dispatch is to be handled. See also **message precedence**.

precision bombing—1) In a restricted sense, horizontal bombing done with the appropriate precision instruments and equipment so as to strike a target of comparatively small bulk or area. 2) In a general sense, any type of bombing against a small or restricted target. In sense one, precision bombing is usually a strategic operation and is executed either to destroy a target with a minimum expenditure of force or to hit a target near other areas

or installations not considered desirable to hit. When the term is used in sense two, it is usually applied to tactical attacks by dive bombers or fighter-bombers against tanks or other identifiable targets.

precision fire—Artillery fire so conducted that the center of impact (q.v.) is placed precisely on a target. This method is used for precise registration of a gun or battery on a base point or checkpoint (qq.v.), as well as for the physical destruction of individual targets, such as immobilized tanks, pill-boxes, or other structures to be destroyed by long-range fire. Precision fire *must* be observed.

preemptive attack or **preemptive strike**—An attack initiated on the basis of evidence that an enemy attack is imminent. Thus, through surprise, the initiative may be seized from the enemy.

preparation fire—Fire delivered on a target preparatory to an assault.

prepared defense—See **defense**.

present arms—1) A form of salute by a soldier armed with a rifle; the rifle is brought to a vertical position in front of the body, muzzle up, trigger side forward, and butt off the ground. 2) The command for a salute by a unit whether or not the troops are armed with rifles. See also **salute** and **honors**.

presidio—Spanish term for a garrisoned post. In the United States, specifically, Army posts in San Francisco and Monterey, California.

press gang—A military detail whose mission was to force men into military service, particularly but not exclusively into naval service. Press gangs are now obsolete, but until well into the 19th century they were a major source of manpower for armies and navies. See also **impressment**.

preventive war—A war initiated by a nation in the belief that military conflict, though possibly not imminent, is inevitable, and that to delay action would involve greater risk to the national interest. See also **preemptive attack**.

prime—To insert a small amount of explosive, or **primer**, in a weapon as a detonator for the main explosive charge. The primer is the highly explosive material, or a device containing a highly explosive material, that causes the detonation of a weapon's main explosive charge. An important early type was the Maynard tape primer, invented in 1845 by Dr. Edward Maynard, an American dentist. Consisting of small pellets of fulminate (usually of mercury) sealed between two narrow strips of paper, it greatly assisted in the conversion of flintlocks (q.v.) to percussion, and was used in the US government's 1855-model pistol-carbines, rifle-muskets, and rifles.

prime mover—A vehicle, including heavy construction equipment, possessing military characteristics, designed primarily for towing heavy, wheeled weapons and frequently providing facilities for the transportation of the crew and ammunition.

priming pan—A shallow pan in the (qq.v.) lock mechanism of a matchlock or flintlock gun which holds a small quantity of priming powder.

princeps—A Roman soldier who wore full armor, carried two javelins, and fought in the second line of the Roman legion. (Plural, **principes**.)

principle of war—A principle or axiom considered fundamental for conducting war successfully. Principles have been variously stated, by Clausewitz, Jomini, Douhet, and other writers on the subject. Modern versions of the principles of war, as understood in the British and American armed forces, are based upon a formulation by British theorist J. F. C. Fuller in 1920. The simple summations of these principles usually include the following: objective, offensive, unity of command, surprise, mass or concentration, economy of force, security, maneuver, and simplicity. The principles are interdependent, although some analysts subdivide these principles, call them by different names, or conceive of them in fewer or greater numbers.

priority—A category of precedence reserved for messages that require expeditious action by the addressee or provide essential information for the conduct of operations in progress when routine precedence will not suffice. See also **precedence**.

prisoner of war—In accordance with the Geneva Convention of 1949, such a prisoner is anyone who falls into the hands of the enemy in time of war and who is a member of an armed force or of certain civilian categories as established by the Convention. Prisoners of war of nations signatory to the 1949 convention are subject to the disciplines and guaranteed the rights and privileges that are set forth in the articles of the Convention.

private—The basic, initial military rank of enlisted or conscripted soldiers in most armies. (Derived from the term, "private soldier.") See also **private first class** and **seaman**.

privateer—1) A privately owned vessel commissioned by a nation at war to attack and seize enemy vessels, as a means of destroying enemy commerce. Privateers operated from ancient times into the 19th century, when the practice was abolished by the 1856 Declaration of Paris. 2) Member of the crew of a privateer. See also **letter of marque and reprisal**.

private first class—In the US Army, an enlisted man ranking immediately above private and below corporal, corresponding to airman 2d class in the Air Force.

prize crew—A crew put aboard a captured vessel to take it into port.

prize law, naval—Law applied by a prize court, based on custom and conventional international law, to determine the ownership of ships and goods captured at sea, and related questions. Naval prize law is determined by **prize courts**.

probable error—The error in range, deflection, or radius that a weapon may be expected to exceed as often as not. In other words, half of the rounds fired from a gun at a given, fixed setting, will be between the target and a line drawn at the distance of the probable error from the target. The other half will be farther from the target than that line. Horizontal error of weapons making a nearly vertical approach to a target is usually described in terms of circular error probable (q.v.).

procurement—In a broad sense, the process of acquiring personnel, matériel, services, or property from outside a military service by means authorized in pertinent laws, regulations, or directives. Though procurement is not the same as production, see **production, military**.

production, military—The conversion of raw materials into weapons or equipment or components thereof, through a series of manufacturing processes. It includes functions of production engineering, quality assurance, and the determination of resource requirements.

projectile—An object projected by an applied exterior force and continuing in motion by virtue of its own inertia, as a rock, bullet, bomb, shell, or grenade. See also **ballistics**.

promote—1) To elevate a person to a higher rank. 2) To appoint a person to a more important position, especially one leading to or entailing an elevation in rank.

propaganda—Any information, ideas, doctrines, or special appeals in support of national objectives, designed to influence the opinions, emotions, attitudes, or behavior of any specified group in order to benefit the sponsor, either directly or indirectly. Types of propaganda include: **black propaganda**—propaganda that purports to emanate from a source other than the true one; and **white propaganda**—propaganda disseminated and acknowledged by the sponsor or by an accredited agency thereof. See also **psychological warfare**.

propaganda bomb—See **bomb**.

propellant—An explosive charge for propelling a bullet, shell, or other projectile; also, a fuel, either solid or liquid, for propelling a rocket or missile.

protected flank—See **flank**.

proving ground—An area where military weapons and equipment are tested to determine their capabilities for combat performance.

provost guards—A detail of soldiers assigned to guard duty, usually over prisoners, under a provost marshal (q.v.).

provost marshal—1) A staff officer at command or subordinate level who advises on, and exercises supervision and inspection over, the maintenance of discipline, enforcement of security, and the confinement of prisoners. 2) In the US Navy, an officer who has responsibility for prisoners who are to go before a court martial.

proximity fuze—A fuze for a projectile designed to activate the explosive charge when close to or at the target. Also called an **influence fuze**, proximity fuzes may incorporate photoelectric cells, radar devices, or other devices as activating elements. See also **variable time fuze** and **fuze**.

psiloi—Lightly armed Greek foot soldiers. Psiloi generally came from the lower classes of society and many were mercenaries.

psychological warfare—The planned use of propaganda and other psychological actions having the primary purpose of influencing the opinions, emotions, attitudes, and behavior of hostile foreign groups in such a way as to support the achievement of national objectives. Psychological warfare may complement armed warfare and be pursued concurrently, or it may be used to exploit armed events before or after they occur.

PT boat—See **motor torpedo boat**.

pursuit—An offensive action designed to maintain pressure on, catch, cut off, or annihilate a withdrawing hostile force attempting to escape, usually after having suffered a defeat in battle.

puttee—A leather, canvas, or other cloth gaiter covering the lower leg. 2) A strip of cloth wound spirally around the leg from ankle to just below the knee, usually called "leggings" in the US Army.

Pyrrhic victory—A victory in which the side that attains its immediate objective in a battle suffers such heavy losses that the victory is virtually worthless. The name derives from the battles at Heraclea (280 B.C.) and Asculum (279 B.C.), in which Pyrrhus, the King of Epirus, defeated the Romans but suffered extremely heavy losses.

PX—See **post exchange**.

Q

Q-boat—An armed British vessel of World War I disguised as an unarmed merchant vessel to deceive and attack surfaced German U-boats. Also called **decoy ship**.

Q clearance—A clearance granted by the US Atomic Energy Commission (later the Department of Energy) to persons needing access to specially restricted data related to nuclear energy and particularly to nuclear weapons.

quadrant—An area, usually within an arc of about 90 degrees, assigned to a unit or individual for surveillance, defense, or control.

quadrant, gunner's—1) A device to indicate the elevation of the barrel or tube of a cannon (the angle with respect to horizontal). See **quadrant elevation**.

quadrant elevation—The angle between the level (horizontal) base of the trajectory of a cannon and the axis of the bore when laid. It is the algebraic sum of the elevation, angle of site and complementary angle of site. See also **elevation** and **site, angle of**.

quadrireme—A galley of ancient times that had four banks of oars on each side. See also **bireme** and **trireme**.

quaker gun—A wooden log cut and finished to resemble an artillery piece in order to decieve an enemy. Quaker guns were mounted on ships and in forts. For example, in 1861, during the US Civil War, the Confederates mounted logs painted black in their fortifications at Centreville, Virginia. To Union observers the quaker guns resembled cannon, and the Confederate fortifications were judged much stronger than they actually were.

quarrel—A short, heavy arrow or bolt fired from a crossbow. See also **crossbow**.

quarter—n. 1) The sparing of an enemy's life. 2) An assigned or proper station; specifically, the position at which officers and men aboard ship or in shore stations are posted during battle. Usually used in the plural. 3) Housing for military personnel, such as barracks. v. To provide shelter.

quarter bill—A list assigning officers and crew to stations during naval action.

quarter or **quarto cannon**—See **cannon**.

quarter deck—The name for that part of a ship's upper deck—generally the after part—reserved for the use of officers. From ancient times the quarter deck was the station of the ship's commanding officer, from which he conducted official and ceremonial functions. On Roman galleys, for example, an altar was located between the sternpost and the tiller; on this altar was placed a helmet symbolizing the emperor's authority, and sacrifices were offered before a battle. Roman seamen coming aboard a vessel faced aft and saluted the authority represented by altar and helmet. On a modern warship the national ensign is flown at the stern when a ship is at anchor, and the custom of saluting the quarterdeck persists as a gesture of respect.

quarter guard—A British term for the contingent that guards a force or military post.

quartermaster—1) In the US and other armies and the US Marine Corps, an officer with responsibility for supplying food, clothing, and equipment. The Quartermaster Corps which serves this function is the equivalent of the Supply Corps in the Navy. 2) In the US and other navies, a petty officer with responsibility for navigation of a naval vessel. Quartermaster derives from the Latin *quartarius magister*, master of the deck from which the Roman trireme (q.v.) was steered.

Quartermaster Corps—See **quartermaster**.

quartermaster general—1) In the Prussian Army, a senior officer of the General Staff. 2) In pre–20th century British and American armies, the officer responsible for the supply of the army, under the army commander or commander-in-chief.

quarters—1) Lodgings or accommodations for military personnel. 2) The structure or shelter affording such lodgings or accommodations, or the place where they are located. 3) An assigned station on a warship. 4) The mustering of all sailors in a division aboard ship for inspection, usually held in the morning and followed by drill. See also **general quarters**.

quarters, winter—A camp set up by a campaigning army for housing during the months when snow and cold made fighting difficult or impossible. With some notable exceptions this practice was followed from ancient times until the invention of the railroad and, especially, of all-weather roads for automobiles; these facilitated winter travel, making possible the provision of supplies to a mobile army.

quarter staff—A long wooden staff used in hand-to-hand combat before the introduction of gunpowder.

Quebec—Letter of NATO phonetic alphabet. See also **alphabet, phonetic**.

Queen—Letter of old phonetic alphabet. See also **alphabet, phonetic**.

quick-match—See **match, quick**.

quick time—The standard marching cadence of foot soldiers on the march or at drill or in parade. In the US Army this is 120 steps, 30 inches in length per minute.

quillon—The bar at the base of the handle of a sword or dagger that prevents the hand from sliding onto the blade.

quincunx—See **checkerboard formation**.

quinquireme—An ancient galley with five banks of oars. See also **bireme** and **trireme**.

quintain—In medieval times, a target, often a half-length figure designed to look like a Moor, with a shield in one hand and a lance or sword in the other. The quintain was set vertically on a post past which a horseman would ride, attempting to spear the quintain with his lance as he passed. Unless he hit it squarely the figure would rotate and hit him.

quoin (gunnery)—A block in the shape of a wedge placed under the breech of a gun to hold the barrel at a selected elevation, set by a quadrant (q.v.).

R

rabinet—A kind of early cannon. Also given as **robinet**. See also **esmeril**.

radar—An acronym formed from RAdio Detection And Ranging. Radar was developed just before World War II by scientists on both sides of that conflict, although much credit belongs to the British scientist Robert Watson-Watt. In 1940 the detection of approaching German bombers by a series of radar stations on the coast of England, enabled British fighters and other defenses to meet the attacks and was a significant factor in the British victory in the Battle of Britain.

radar countermeasures—See **electronic countermeasures**.

radar picket—A ship or aircraft equipped with early warning radar, stationed on patrol at a distance from the force protected so as to extend the range of radar detection.

radarman—US Navy specialty rating, chief first, second, and third class.

radeau—A large, unwieldy, bargelike gunboat, driven by sail, such as those constructed during the American Revolution for service on inland waters. One radeau, the British *Thunderer*, was part of Sir Guy Carleton's squadron that defeated the American squadron of General Benedict Arnold at the Battle of Valcour Island on Lake Champlain in 1776. She mounted 14 guns and had a crew of 300 men.

radio deception—The employment of radio to deceive the enemy. Radio deception includes sending false dispatches, using deceptive headings, employing enemy call signs, and so forth. See also **electronic deception**.

radio electrician—US Navy warrant officer.

radioman—US Navy specialty rating, chief first, second, and third class.

radio recognition—The determination by radio of the identity or the friendly or enemy character of an otherwise unidentified individual or entity.

radio silence—A period during which all or certain radio equipment capable of transmission is kept inoperative. Radio silence may be imposed for all equipment for all frequencies or may apply only to certain types and designated frequency bands.

radius of action—The maximum distance a ship, aircraft, or vehicle can travel away from its base with normal combat load and return without refueling, allowing for all safety and operating factors.

raid—An operation, usually small scale, involving a swift penetration of hostile territory to secure information, seize hostages, confuse the enemy, destroy his installations, or carry out some other specific task. It ends with a planned withdrawal upon completion of the assigned mission.

railhead—The section of a railroad where supplies are unloaded. Also, the position on a railroad line where military supplies, usually for front-line forces, are unloaded for delivery by other means.

raise a blockade—To cause a blockade to be broken up, by dispersing the ships, one's own, voluntarily, or the enemy's, by force.

raise a siege—To abandon the attempt to take a place by besieging it, or to cause the attempt to take a place by siege to be abandoned.

rake—To sweep a target, especially the

length of a ship or a column of troops, with gun or cannon fire; thus **raking fire**. See also **enfilade**.

raker—A gun so placed on a ship as to rake (q.v.) the enemy's ship.

rally—1) To achieve order and to halt or slow a retreat, particularly in a unit that has crumbled in the face of enemy attack. 2) To assemble a group or force for some purpose.

rally point—A point at which, or over which, aircraft reassemble and reform after making an attack.

ram—A device protruding from the bow of a warship with which to damage an enemy vessel. Rams were used on ancient and medieval vessels, particularly on rowed galleys, which came in direct contact with each other in battle. Rams were again introduced into the design of some ironclad warships in the later 19th century.

rammer—A device to push home shot or powder in a gunpowder weapon. Originally, it was used with muzzle-loading firearms and cannon to push the powder charge, the wad, and the shot to the base of the gun, at which point the weapon could be ignited and fired. The rammer was a wooden cylinder the size of the bore on the end of a stick or staff, which often was marked to indicate when the elements were in place. It is used with modern breech-loading cannon using separate-loading ammunition (q.v.) to set both the projectile and powder charge in place before firing. See also **ramrod**.

rammerman—The member of a gun crew who handles the rammer; the rammerman also handles the sponge used to assure that all sparks in a tube are extinguished after firing.

ramp—See **apron**.

rampart—A wall or earthen embankment comprising the main defensive work of a permanent or field fortification. A rampart is usually surmounted by a parapet (q.v.).

ramrod—1) A rod, originally of wood and later of metal, that drove home the charge in a muzzle-loading firearm. Because they were long and thin, wooden ramrods tended to break frequently. 2) A metal rod used to clean the barrel of a firearm. In the early part of the 18th century, according to some historians, Frederick William I of Prussia introduced an iron ramrod to replace the fragile wooden ramrod. The new implement, when exploited by training, permitted an increase in the average rate of fire of the Prussian infantry from two to at least three shots a minute. See also **rammer**.

randing (webbing) of gabions—See **gabion**.

random shot—A shot fired at no particular object.

range—1) The distance between any given point and an object or a target. 2) The maximum distance to which a ship, aircraft, or land vehicle can travel before requiring refueling. 3) The maximum distance (**maximum range**) a weapon may fire its projectile. The **maximum effective range** is the maximum distance at which a weapon may be expected to deliver its destructive charge with the accuracy needed to inflict prescribed damage. The **minimum range** is the shortest distance to which a gun can fire effectively from a given position. 4) An area equipped with targets for practice firing. Also called a **firing range**.

range deviation—The distance that a projectile's point of impact is short of, or beyond, the target, measured along the gun–target line or along a line parallel to the gun–target line.

range table—A table showing the relationship between the range of a cannon and the elevation settings. See also **firing tables**.

ranger—A soldier belonging to a specially selected group of soldiers in the US Army, trained especially in raiding tactics. The British equivalent of the ranger is the commando. (q.v.). The first such unit was Rogers's Rangers, a bush-fighting unit formed by the American colonist Robert Rogers to fight against the French and the Indians in North America during the French and Indian War. Rogers's Rangers was revived by Rogers during the American Revolution and fought on the British side as a Loyalist unit. Another such unit was the Queen's Rangers. Irregular light troops in 18th-century European armies, like the pandours (q.v.) of Austria-Hungary, were equivalent to the North American rangers. See also **commando**.

ranging—1) The process of establishing the distance of a target. The term usually describes the process of adjusting weapon fire. Types of ranging can also include echo, intermittent, manual, navigational, explosive echo, optical, and radar. 2) Old term for disposing troops in proper order for an engagement, a march, or maneuver.

rank—n. 1) The status of a person as determined by the grade to which he belongs. 2) The relative status of a person within a grade. 3) The grade to which a person belongs. 4) A line of persons, side by side. 5) *pl.* As in "in the ranks": in military service, especially, as an enlisted person. v. To be senior in rank to.

rank and file—A collective term for an army's soldiers, derived from military formations, which consist of ranks (lines of individuals) and files (columns of individuals).

rapid fire—Shots fired in rapid succession.

rapier—A long, slender sword. In the 16th and 17th centuries a rapier had two cutting edges as well as a sharp point. Its utility for cutting was recognized as less important than its importance as a thrusting weapon in the 18th century, however, and rapiers no longer had cutting edges.

rappel—The beat of a drum to call soldiers to arms.

rated—1) In the US Navy, the term refers to an enlisted person above the grade of seaman first class. 2) In the US Air Force, and air contingents of the Navy and Marine Corps, it refers to a person, officer or enlisted, with a currently effective aeronautical rating, usually as a pilot.

rate of advance—The average distance in miles or kilometers that a force or unit engaged in combat moves forward in a given period of time, usually days or hours. The advance may be opposed or unopposed.

rate of attrition—See **attrition**.

rate of fire—The number of rounds fired per weapon per unit of time (usually a minute).

rate of march—The average distance in miles or kilometers that a unit travels in an administrative (as opposed to combat) movement in a given period of time, including all scheduled or unscheduled halts. It is usually expressed in miles or kilometers per hour.

rating—1) British term for a navy enlisted man. 2) In the US Navy, rating is the equivalent for enlisted personnel to rank for officers.

rating badge—The device worn by a naval enlisted man on his sleeve indicating his specialty and his rating.

ratio—A mathematical or numerical comparison of two entities, such as force ratio, strength ratio, power ratio.

ration—1) An allowance of provisions for one person for one day. 2) A monetary equivalent for such an allowance. 3) The provisions, especially food, in the amount of this allowance. 4) *pl.* Any provision issued in whatever amount, as in a "can of rations." v. 1) To provide food. 2) To limit the amount of provision of any kind of supply.

ravelin—An outwork on the far side of the ditch or moat of a fortification and located between two bastions, built in the shape of an arrowhead, facing away from the fortification, with an opening toward the wall. Ravelins served many defensive purposes, including the protection of the ground in front of the main fortification.

raven—See **corvus**.

raw troops—Inexperienced or untrained soldiers.

RAZON—A World War II bomb with movable control surfaces in the tail adjusted by radio signals to control the missile's range and azimuth.

readiness, operational—The capability of a unit, ship, weapon system, or equipment to perform the missions or functions for which it is organized or designed.

ready reserve—That portion of the US military reserve component that may be called to active duty immediately in an emergency declared by the Congress, proclaimed by the president, or otherwise authorized by law.

rear admiral—1) A naval rank between captain (or commodore when such rank is in use) and vice admiral. 2) An officer of that rank. See also **flag rank**.

rear area—An operational area immediately behind the forward area (q.v.), containing administrative and service establishments, communications, and other support elements. The communications zone lies within this area. See also **area**.

rear echelon—That echelon (q.v.) of a headquarters or of a unit that is located further from the front than the forward, or main, echelon. A rear echelon is not necessarily in an operational area.

rear guard—A security detachment that protects the rear of a column from hostile forces. During a withdrawal, it delays the enemy by armed resistance, destroying bridges, and blocking roads.

recce—(pronounced reck ee) Short for reconnaissance (q.v.).

recognition signal—A signal sent out over airwaves or visibly displayed to enable a person, body of troops, ship, or aircraft, to be recognized by those acquainted with the signal.

recoil—n. The backward movement of a gun when it is fired. Recoil is caused by the backward pressure of the propellant gases. 2) The distance that a gun or part travels in this backward movement. 3) The force of this movement as exerted against something else, as in, "the recoil of the gun was felt throughout the plane." v. 1) To move rapidly backward, as a gun upon being fired. 2) To move backwards rapidly as a result of some other action, usually hostile.

recoilless rifle—A light crew-served infantry cannon first introduced in World War II, with holes in the breechblock that permit the propellant gases to escape. Recoilless rifles are light in weight (because they have no recoil mechanism) with lower muzzle velocity and shorter range than other guns or howitzers of the same caliber (because part of the propellant gas pressure is used to prevent recoil). They are often used as antitank weapons, fired either from a ground mount or a vehicle.

reconnaissance—A mission undertaken to obtain, by visual observation or other detection methods, information about the activities and resources of an enemy or potential enemy, or to secure data concerning meteorological, hydrographic, or geographic characteristics of a particular area.

reconnaissance, aerial—Reconnaissance undertaken with an airplane or other aircraft to secure information about the enemy, or about the terrain or weather, by visual observation, aerial photography, or electronic methods.

reconnaissance, contact—Aerial reconnaissance used in a fast-changing tactical situation, such as an amphibious invasion or where a unit is isolated, in order to keep commanders informed, provide assistance where needed, and minimize the danger of shelling or bombing friendly forces.

reconnaissance in force—An offensive operation, larger and more deliberate than a raid (q.v.), designed to discover or test the enemy's strength or to obtain other information.

reconnaissance patrol—A small patrol used to gain information about the enemy, preferably without his knowledge.

reconnaissance vehicles—Armored vehicles, tracked or wheeled, designed primarily to scout or conduct reconnaissance.

reconnoiter—To make a reconnaissance of an area, place, airspace, or the like.

recruit—n. An enlisted person newly entered in military or naval service, especially such a person who has voluntarily entered. v 1) To strengthen or fill up a military or naval organization or reserve with new personnel, especially through voluntary enlistment. 2) To induce a person to enter voluntarily into active military or naval service, or into a reserve component.

redan—A simple type of field fortification consisting of two parapets joined to form a salient angle facing the enemy. Redans are open at the rear. Another name for the redan is "flèche."

redeployment—The transfer of a unit, an individual, or supplies deployed in one area to another area, or to another location within the area, or to the zone of interior (q.v.) for the purpose of further employment.

red-hot shot—A cannon ball heated to redness to be fired by a cannon at a ship, a wooden structure, or other combustible target.

redoubt—A small earthwork, generally outside a main fortification and usually covering the approaches to the areas most likely to be attacked.

reduced charge—A charge of powder for a cannon that is less than the maximum charge for which the weapon is designed. See also **semifixed ammunition**.

reentrant—An angular feature, either manmade, as a fortification, or natural, as a terrain feature (usually a defile [*q.f*] or draw) pointing inward or refused from the front.

reflex bow—See **bow**.

refuse a flank—The act of swinging back, away from the enemy, usually in successive echelons (each facing front, toward the enemy), the elements of a wing or flank of an army or other force, which refrains from fighting when the remainder of the battle line is engaged. The purpose is to freeze in place the opposing portion of the enemy force through the obvious threat of sending the refused flank or wing into action, while the other flank or wing is carrying out a decisive attack. First used by Epaminondas of Thebes against the Spartans in the Battle of Leuctra (371 B.C.). Another famous use of a refused flank was by Frederick the Great at the Battle of Leuthen (1757). See also **echelon**.

regiment—A military unit, particularly of infantry, cavalry, or artillery, that can be either administrative or operational. The origin of the modern regiment may be found in the permanent (usually regional) infantry formations established by French monarchs in the later Middle Ages. Since the late 18th century, a regiment, commanded by a colonel, has usually been an operational unit of two or more battalions, and has functioned most often as a component of a division. In the modern US Army, until the mid-1960s, infantry regiments consisted of three battalions, in strengths varying from about 1,500 to 3,000 men. Artillery regiments (until World War II) consisted of two or three battalions of three gun batteries each, about 1,000 to 1,500 men, and 24 to 36 guns. In World War I the standard US division consisted of two infantry brigades of two regiments each, and an artillery brigade of one medium and two light regiments. In World War II the standard US division consisted of three infantry regiments and four artillery battalions (three light and one medium). In the early 20th century, the British Army adopted a regimental system, with the regiment only an administrative unit, located regionally, with separate battalions being combined into brigades, which in turn were combined into divisions. In the 1960s the US Army began to move toward the same relationship of individual battalions to administrative regiments, with the battalions combined into brigades of size and function comparable to the old operational regiment. See also **battalion**, **brigade**, **division**.

regimental combat team—See **combat team, regimental**.

regimental landing team—A US task organization for an amphibious landing, comprising a Marine or Army infantry regiment reinforced by those elements required by its combat function ashore, such as engineers, communications personnel, etc.

regimentals—1) The military uniform and insignia worn by a certain regiment. 2) In general, dress or semidress military uniform.

registration—Also known as **registration fire**, precision artillery fire at a registration point or checkpoint to determine the exact settings for range and deflection in order to place the center of impact of fire on the target. From this data can be calculated the settings for other guns so that the centers of impact of their fire can be trained on these points, and, by transfer and calculations from a map or chart, on nearby targets.

regular Army (Navy, Air Force, Marine Corps, Coast Guard)—In the United States, that portion of the armed services that is permanently on active duty in the armed forces, not including the reserve elements.

reinforce—To strengthen by the addition of units, personnel, ships, aircraft, or military equipment. Also spelled **reenforce**, particularly by the US Navy.

reiter—A German mercenary heavy cavalryman of the 16th and 17th centuries, who usually wore black armor and accoutrements and was called **schwarzreiter**—black rider—or reiter for short. Reiters were usually armed with a brace of wheellock pistols (q.v.), although originally they were armed with a boar spear and a sword. The early reiters wore chain mail (q.v.); this was replaced by a breastplate, an open helmet, and heavy thigh-length leather boots. In combat the reiters advanced at a slow trot in a line of small, dense columns, each several ranks deep, and fired their pistols one rank at a time, in a maneuver called a caracole (q.v.). The reiters were the cavalry equivalent of the landsknecht (q.v.).

relief—n. 1) Replacement in a duty station or post. 2) The person or unit relieving.

relief in place—A combat operation in which all or part of a unit is replaced in a combat area by an incoming unit, to which the combat responsibilities and assigned zone of operations of the replaced elements are transferred. The replaced elements are withdrawn prior to the resumption of operations by the incoming unit.

relieve—1) To rescue or reestablish contact with a force that is besieged or cut off by the enemy. 2) To take the place of a person or unit at a duty station or post.

remote control—Control from a distance over the operation of an aircraft, gun, or other piece of equipment, especially using electrical or electronic apparatus.

remount—A horse that replaces another in a mounted (cavalry or horse-drawn) unit or that is available as such replacement. Generally applied to all new horses delivered to a unit; mounted units have throughout history carried extra horses with them when moving from one area to another.

rendezvous—n. 1) A prearranged meeting at a given time and place, from which to begin an action or phase of operation, or to which to return after an operation. 2) A place, point, or spot on the ground, on the sea, or in the air where such a meeting occurs, or is planned to occur; a rendezvous point. v. To meet at an appointed time and place. See also **rally point**.

repeating firearm—A firearm that may be fired several times without having to be reloaded.

replacement—An individual, or unit, or item of equipment, or group of individuals, units, or items of equipment, assigned to a combat unit or force to bring it to or toward its authorized strength after it has incurred losses of personnel and/or equipment.

report—Any transmission or presentation of data or information, whether in oral, tabular, graphic, narrative, questionnaire, punchcard, or other form, regardless of the method of transmission. There can be many kinds of reports: periodic, after-action, morning (daily, by company-sized units), etc.

reprisal—1) In international law, resorting to force short of war to redress a grievance. 2) In war, a retaliatory act, usually of the same kind or degree as that committed by the enemy.

requirement, operational—That which is needed in order for a unit or force to be able to carry out the mission assigned to it as part of a tactical concept.

requisition—n. An authoritative demand or request, especially for personnel, supplies, or services authorized but not made available without specific request. v. 1) To make such a demand or request. 2) To demand or require services from an invaded or conquered nation or region.

reserve—1) That portion of a force that is held out of combat in anticipation of its being used later to influence the outcome of a battle, engagement, or operation. Such a reserve held out by an army is called the **army reserve** (or the army's reserve); that held out by a corps is the **corps reserve**, and so forth. 2) A military organization of people not on active duty but holding ratings or commissions and available to be called to active duty when needed. Members of a reserve may or may not attend regular training meetings and serve periodically for short periods of active duty. 3) A member of such an organization.

reserve army—1) A military force organized or designated as an army not committed to battle. 2) An army held in reserve that is a component of a theatre or army group. 3) A force organized and active in the interior of a nation from which replacements or reserves can be sent to the support of forces engaged in combat, as in Germany in World War II.

reserve components—Nonregular military forces of the nature of a militia (q.v.), available to augment the regular forces in time of war or emergency. In the United States the reserve components include: Army National Guard, Army Reserve (formerly the Army Reserve Corps), Naval Reserve, Air National Guard and Marine Corps Reserve.

Reserve Corps—Former designation of one of the principal reserve components (q.v.) of the US Army.

reserve, general—See **general reserve.**

reserves, floating—See **floating reserves.**

reservist—A member of a military reserve component, in contradistinction to a "regular."

resistance movement—An organized effort by some portion of the civil population of a country or region to resist the legally established government or an occupying power, and, at a minimum, to disrupt civil order and stability or, as a maximum, overthrow or displace the legal authority.

rest—A device upon which to place a hand weapon for carrying or for aiming. On medieval armor a rest for bearing the weight of the lance was found on the side of a breastplate. The early matchlock musket (q.v.), because of its weight and awkwardness, required a rest in the form of a stick with a fork or curved top to steady it on the ground when being fired.

restricted—1) Limited in access or use to a specified class of people. 2) A former security classification for the protection of information, documents, or material. In US military security classifications, the term was used for material not sensitive enough to require a classification of "confidential" (which is now the lowest grade of classification) or "secret."

restricted data—All data concerning: a. design, manufacture, or use of atomic weapons; b. the production of special nuclear material; or c. the use of special nuclear material in the production of energy, but not including data declassified or removed from the restricted data category.

resupply—1) To furnish a force, organization or the like with supplies, or to replenish supplies or equipment that have been consumed, lost, destroyed, or damaged. 2) To supply by air a force, unit, or the like that is cut off from its normal supply force.

retire—n. Movement to the rear by a force that is not engaged to avoid contact with a hostile force that is approaching from the front or flank. v. 1) To make such a movement. 2) To leave active duty after a certain number of years of service, or to change status from active duty to civil or reserve status.

retirement—1) A retrograde movement in which a force out of contact moves away from the enemy. 2) The status of a military man who has retired.

retreat—n. 1) A term imprecisely used for a retrograde movement (q.v.) a force in the face of an enemy attack or threatened attack. 2) The signal to a force to make such a movement. 3) A bugle call played at the lowering of the colors at sunset. v. To move to the rear in the face of an enemy attack or threatened attack.

retrograde movement—Any movement of a command to the rear, or away from the enemy. It may be forced or voluntary. Types of retrograde movement include: withdrawal, retirement, and delay, or delaying action (q.v.). The term "retreat" (q.v.) is sometimes used (but technically incorrectly) as synonymous with retrograde movement.

reveille—1) A signal, usually a bugle call or other music, for rousing personnel at a military post at the start of day. 2) The bugle call sounded for this purpose. 3) A ceremony held in early morning, usually attended by all personnel.

reverse slope—The slope of a terrain feature that falls away from the enemy, that is, the back slope of a hill from the point of view of one who faces the hill.

revetment—Reinforcement of an earthern wall with hard material (wood, sandbags, or concrete). Usually applies to the side of a trench, but can apply to any earthen wall for the purpose of protecting aircraft, gun emplacements, storage areas, or other material or places against bombing, shelling, or strafing attacks. The term is sometimes applied to a barricade against explosives. Types of revetment include: demi, fascine, gabion, leaning, rectangular, and solid.

review—1) A formal inspection of a military unit or organization, during which the unit marches in parade formation past the inspecting or reviewing official or party. 2) A ceremony to honor a person or persons or to honor an event, to include a march past the person or persons being honored.

revolt—1) The use of military force in a violent effort to overthrow a constituted government or authority. In such a conflict the opposing forces are often classified as rebels, those endeavoring to overthrow the government, and loyalists, or governmental forces. If the governing authority is indigenous, this is a common form of civil war (q.v.). 2) The acts undertaken by a rebel or a group of rebels who attempt to bring about a revolution (q.v.).

revolution—A successful revolt.

revolver—A handgun with a rotating cylinder in which several rounds of ammunition can be placed, each in its separate chamber. As each round is fired the next automatically moves into line with the barrel.

ribaud—A crossbow mounted on a cart. Some ribauds had steel arms and fired two-foot iron bolts 500 yards.

ricochet—To skip, bounce, or fly off at an angle after striking an object or surface. Usually applied to a projectile from firearms or cannon.

ricochet fire—A form of antipersonnel artillery fire, so directed that the projectiles (equipped with delay fuze) will ricochet off the ground, then explode in the air to provide a wider lethal area of shell fragmentation than would result from point detonation.

rifle—n. A gunpowder weapon with spiral grooves cut into the surface of its bore so that the projectile will leave the weapon with a spinning motion that provides greater accuracy and stability to its flight. Although modern cannon and small handguns usually have rifled barrels, the term "rifle" is most commonly used in reference to a firearm fired from the shoulder. Such rifles were originally developed for hunting in Europe and were not rapidly adopted as military firearms because of the greater time and expense required to produce such precision weapons (in comparison to the smoothbore musket) and because it took longer to load them than to load the musket. A typical late-20th-century military rifle is a magazine-fed, gas-operated, air-cooled, semiautomatic, shoulder-type weapon designed for semiautomatic fire, but convertible to automatic fire. It usually fires a round between 5.5mm and 7.62mm (about .22 to .300 caliber in inches). v. To cut these spiral grooves in a gun barrel.

rifle grenade—See **grenade**.

rifleman—A man armed with a rifle. In the American Revolution the Continental Congress called for the raising of six companies of expert riflemen from Pennsylvania and two each from Maryland and Virginia. Riflemen served as scouts, sharpshooters, and for similar special or elite duties, since the musket continued to be the weapon of the majority of troops.

rifle-musket—The standard infantry small arm weapon in the mid-19th century, after the introduction of the conoidal bullet revolutionized weapons and infantry combat.

rifle pit—A hole, or short trench with a mound of earth in front of it, to protect one or two soldiers. The connection of several rifle pits by digging between them creates a trench or trench line. See also **foxhole**.

right face—n. 1) A 90-degree change of stationary position to the right while at attention. 2) The command to make this change. v. 1) To execute such a change in position. 2) In the imperative, the command to do it. See also **face**.

rimfire—1) Of a cartridge, having the primer in the rim of the base. 2) Of a firearm, using rimfire cartridges.

riposte—A blow or attack in response to a hostile blow or attack.

riverine area—An inland or coastal area comprising both land and water, characterized by limited land lines of communication and extensive water routes for surface transportation and communications.

riverine operations—Military operations conducted by forces organized to exploit and to cope with the unique characteristics of a riverine area. Joint riverine operations combine land, naval, and air operations, as appropriate to the nature of the specific riverine area. The operations on the rivers in the west in the American Civil War were classic riverine operations.

road space—The length of roadway allocated to or actually occupied by a column of troops or vehicles on a route, expressed in miles or kilometers.

robinet—See **rabinet**.

rocket—A device that is propelled by the ejection of gases or other material formed by chemical action within the device itself. Although the origin of rockets is obscure, it seems likely that they were developed in the 13th century in China, where the first gunpowder weapons may have been rockets. Various methods of propulsion and various designs were tried in succeeding centuries, but it was not until the 18th century that rockets were used seriously as weapons. It was Hyder Ali and his son, Tipu Sahib, fighting the British in India, who first used rockets with impressive combat results. With the invention of the Congreve rocket (q.v.), at the beginning of the 19th century, rocket corps appeared in European armies. Most rockets were propelled by black powder, although other propellants were experimented with, until Robert H. Goddard successfully launched the first liquid propellant rocket in 1926. Since then many propellants, both solid and liquid, have been developed, and rockets have been developed for many purposes, from the hand-held bazooka of World War II to the interstellar vehicles of the late 20th century.

rocket launcher—A device for launching rockets. Rocket launchers use rails, posts, tubes, or other devices to carry and guide the rocket. Rocket launchers can be wheel-mounted, motorized, aircraft mounted, and in the case of the bazooka, designed to be carried and fired by an individual.

Roger—A signal communications transmission indicator, signifying understanding or agreement with the other party's previous transmission. See also **affirmative**. 2) Letter of the old phonetic alphabet. See also **alphabet, phonetic**.

rolling barrage—See **barrage**.

rolling fire—Musket fire by a line of soldiers

in quick succession from one end of the line to the other.

Romeo—Letter of NATO phonetic alphabet. See also **alphabet, phonetic**.

round—A single bullet, shell, missile, or other projectile, including its propelling charge (and its case, in fixed or semifixed ammunition). Originally, a general discharge of firearms by a body of troops, each of whom fires his weapon once. A round of cartridges was a supply giving one cartridge to each man. See also **ammunition**.

round ship—A broad-beamed cargo ship of the pregunpowder era, called on to serve as a warship during time of war.

rout—n. 1) The disorderly and hasty departure of a defeated force from the combat area. 2) A defeat so massive that the defeated force flees in disorder. v. To inflict an overwhelming defeat that causes the enemy to flee in disarray.

route—The prescribed course to be traveled from a specific point of origin to a specific destination.

route march—An organized movement of troops in column by road or cross-country. The troops move in **route order**, or **route step** that is, they are not required to march in cadence; or (generally) in silence, and are permitted to carry their weapons at will, provided safety precautions are strictly observed. Route marches are usually conducted away from the immediate presence of the enemy; their purpose generally is to assemble the elements of a command expeditiously while conserving the comfort of the troops.

ruffle—A continuous drumbeat on a military drum. Ruffles are commonly heard combined with a bugle fanfare, in a combination known as **ruffles and flourishes**, played to announce the arrival of an important person, notably the President of the United States.

ruffles and flourishes—See **ruffle**.

rules of engagement—Directives promulgated by the US Department of Defense which delineate the circumstances and limitations under which US forces will initiate and/or continue combat engagement with other forces encountered.

rules of war—See **laws of war** and **jus beli**.

ruse de guerre—Strategem of war intended to mislead or deceive an enemy.

S

saber or **sabre**—A sword with a single-edged, slightly curved blade, with a curved hand guard, designed for cutting rather than thrusting; it was the standard horse cavalry weapon. Also worn by most officers, up to the middle of the 20th century, for ceremonial purposes.

sabot—1) Artillery. A support or container around a projectile which permits it to be fired in a gun of larger caliber. 2) An aluminum case or jacket fitting around the steel core of certain armor-piercing ammunition. In sense one, the sabot is of a size to fill the larger bore and falls from the projectile in flight. The sabot may sometimes incorporate a rocket motor for giving additional velocity or range to the projectile. Originally borrowed (c. 1600) from the French term for a wooden shoe, the word "sabot" has been in military use since about 1850, as applied to various shoelike devices used in fitting projectiles to the bore of cannon or rifles, or in providing projectiles with an outer casing to take the grooves of rifling.

sabotage—An act with an intent to injure, interfere with, or obstruct the national defense of a country by willfully injuring or destroying any national defense or war material, premises, or utilities, or to disrupt normal processes or procedures.

sabretache—A pouch or case made of leather, worn by cavalrymen, suspended from the saber belt.

saddler—One who makes, repairs, or sells saddles. Every mounted company-size unit of the US Army before World War II had a saddler as a military specialist.

safe conduct—A guarantee of passage without hindrance through a military or combat zone, honored by one or both belligerents.

safeguard—A means of protection, specifically: a. a convoy or escort; b. a pass, safe conduct; c. a precautionary measure or stipulation; d. a technical contrivance to prevent accident.

sagittarius—A Roman bowman, named for his weapon, the sagitta, or arrow. See also **veles**.

sailor—An enlisted man serving in a navy. Also called a seaman (q.v.).

saker—A small cannon used in the 16th century which fired a ball weighing about three and one-half pounds. The name may derive from a species of falcon commonly called saker (Latin *falco, sacre*). See also **cannon**.

salient—1) A portion of a battle line that projects into enemy-held territory. 2) A part of a fortification that projects outward toward enemy territory.

sally—A sudden charge of forces from a defended position for the purpose of attacking the enemy, especially an attack against an investing force by besieged troops. An opening in a fortification through which troops can issue is a **sally port**. Such gates are often in the rear of the fortification.

sally port—See **sally**.

salute—1) In a general sense, a greeting or sign of courtesy passed between parties, as by gesturing with the hand, firing a cannon, dipping a flag, or dipping an aircraft wing. 2) Among military personnel, an outward sign, mark, or act of courtesy, respect, or honor given in a specific and prescribed manner, specifically the hand salute.

salvage—n. 1) Property that has some value in excess of its basic material content but is in such condition that there is no reasonable prospect of its use for any purpose, and its repair or rehabilitation is impracticable. 2) The saving or rescuing of condemned, discarded, or abandoned property, and of materials contained therein for reuse, refabrication, and scrapping. *v.* To save or rescue such material.

salvo—1) In naval gunfire, a method of fire in which a number of weapons are fired at the same target simultaneously. 2) In close air support/air interdiction operations, a method of delivery in which all ordnance of a specific type is discharged simultaneously. 3) In artillery fire, the discharge of all pieces in a battery or larger unit in quick sequence with a uniform time interval between rounds. See also **volley**.

SAM—See **missile, surface-to-air**.

Sam Browne belt—A uniform belt with a diagonal strap from the left waist across the right shoulder, worn with the dress uniform by officers of the US and British armies from World War I until World War II. The belt was used by American officers when in field or parade uniform to carry personal weapons, such as saber or pistol, cartridge box, canteen, and first aid kit. The belt originated as a sword belt, invented by British general Sir Samuel James Browne in the late 19th century, to help carry the weight of his sword, which he could not support because he had lost his left arm.

samurai—The military aristocracy of feudal Japan and its warrior members. Many of the practices and formalities of the samurai survived into World War II, including the ritual death that many Japanese officers chose over surrender.

sap—1) A trench, usually in the form of a zigzag, dug toward an enemy's defensive position. A **single sap** is a trench from which the dirt is thrown to one side for added protection. In a **double sap** the dirt is used to form parapets (q.v.) on both sides. If built under fire, two gabions (q.v.) are used side by side for protection, pushed forward as the work advances, and the sap is called a **flying sap**. If the gabion is on a roller it is called a **sap roller**. 2) A tunnel dug under an enemy position in order to place a mine and destroy the fortification. See also **approaches**.

sapper or **sapeur**—One who builds saps

(q.v.), or, more commonly, in the British Army, a military engineer.

sap roller—See **sap**.

sarissa—A spear, introduced by Philip of Macedon and used by the Macedonian heavy infantry and cavalry during the third and fourth centuries B.C. Under Philip and Alexander, the sarissa was 13 to 14 feet long; later, they were as long as 20 feet. The men in the first lines of a phalanx (q.v.)—Polybius says the first five ranks—held the sarissa pointing forward, so that an enemy faced a wall bristling with sharp-pointed weapons. Spearmen in the rear ranks held their sarissas upright, to help ward off missiles.

saturation bombing—Intense bombing intended to leave no place in a given area free from destructive effects.

saucisson or **saucisse**—A large fascine (q.v.).

scabbard—A sheath for a sword, dagger, or other cut or thrust weapon.

scalade—Climbing over a fortified wall or rampart with the help of ladders, especially during an attack.

scale—*v.* To climb a wall, cliff, or other obstruction. *n.* The ratio of distance represented on a map to actual distances.

scale armor— See **armor** and **lamellar armor**.

scarp—The side of a ditch or moat that is next to the wall.

scatter bomb—Any bomb, such as a fragmentation bomb, that scatters its effect over a wide area. However, the term is usually applied to bombs that propel submunitions, small bombs or mines that are carried inside the larger bomb.

scatterable mines—Mines constructed so that they can be laid quickly and in large numbers by artillery fire, by a mechanical minelayer, or even by air drop. Because of their numbers, and their unconcealed but unexpected appearance along enemy movements routes, they are difficult for troops to avoid, and they delay troop movements until they are cleared. In particular, they make all night movements hazardous. Scatterable mines are one of the most important of post–World War II weapons developments.

scattered drop—A bomb drop in a pattern that was not intended.

scenario—A narrative description of an event or events—actual or hypothetical—usually as an element of an analysis of the event or of matters related to the event. Often the basis for analysis by means of a combat model (q.v.).

scheduled fire—Prearranged artillery fire, particularly used during preparation fires (q.v.). Provides for a series of artillery concentrations by batteries, battalions, or several battalions on different targets at times related to the attack plans of the supported maneuver elements. See also **fire, prearranged**.

schiltron—Medieval Scottish formation of pikemen (q.v.), similar to a phalanx (q.v.).

schnorkel—1) A retractable air-intake tube on a submarine that permits the intake of air and the escape of gases so that the vessel can stay submerged for long periods. The device was introduced by the Germans in World War II. 2) A tube attached to the exhaust pipe of a tank to permit the vehicle to cross rivers even when completely submersed, a technique and device exploited by Soviet forces in the post–World War II era. The term is often spelled **snorkel** in English.

schwarzreiter—See **reiter**.

schwerpunkt—A German term (literally, "point of strength") referring to that geographical point on a battlefield at which maximum force is employed to achieve an objective; an expression of the combination of the principles of objective, offensive, and mass (or concentration).

scimitar—A Middle Eastern saber with a strongly curved blade and a single cutting edge on the outside. The scimitar was used for the most part by Moslem soldiers.

scissoring—A flight technique by escort fighters in which their patterns continually crisscross each other above or below the formation they are escorting.

scorpion—An ancient catapult similar to the onager (q.v.).

scout—n. A person, unit, ship, or aircraft sent out from an armed force, sometimes in secret, to ascertain what lies ahead in the way of terrain or enemy forces. v. To search for information about the enemy or the land ahead: to make a reconnaisance (q.v.).

scouting force—A contingent of naval vessels, usually small and fast, whose function is to search ahead of and around the main body to obtain information about the enemy.

scramasax—A broad, double-edged dagger about 18 inches long, used by medieval Franks and Saxons.

scramble—v. Of aircraft, to take off as quickly as possible, generally to intercept approaching hostile aircraft. n. The whole action involved in getting interceptor aircraft into the air in the shortest time possible.

screen—n. Anything used to protect or shield a country or friendly force from hostile attack, interference, or observation. v. To use forces to protect other forces or to prevent an enemy from gaining information about forces or activities.

screw-gun—A British artillery piece, usually a howitzer (q.v.) for mountain warfare, designed to be transported in separate pieces by pack animals. Since the barrel weight of a cannon even as small as 3-inch caliber was too great for one pack mule or horse, the howitzer was made so the barrel came in two parts that could be screwed together when the piece was assembled. The British 3.7-inch screw-gun was used in mountain warfare in World War II.

scrub—1) A slang expression meaning to cancel a plan, mission, or operation, especially after a long wait. 2) To wash a person out, dismiss him from training.

scutatus—A Byzantine heavy infantryman, known as scutatus from the shield (scutum, q.v.) that he carried. Scutati wore helmets, mail (q.v.) shirts, gauntlets, greaves (or knee-length boots), and surcoats. They carried a lance, shield, sword, and sometimes an ax. Uniform appearance was achieved by color of surcoat, helmet tuft, and shield.

scuttlebutt—Originally a cask of fresh water, later a water cooler, used by the crew of a ship for drinking purposes. Because crewmen gathered around the scuttlebutt and talked, the term came to be a slang expression meaning unofficial reports and rumors.

scutum—A large, rectangular, curved Roman shield, made of wood, leather-covered and iron-edged, and often having in the middle a knob with which the legionary (q.v.) was expert in pushing and striking his enemy.

Seabee—Member of a naval construction battalion (CB). Such battalions were established in World War II, composed of men with expertise in various construction specialties, in order to build naval bases on islands in the Pacific.

seaborne—Carried on a ship or ships. Of an air assault, launched from a ship or ships.

sea frontier—In the US Navy, the command of a coastal frontier, including the coastal zone and adjacent sea areas.

Sea Lord—One of six high-ranking British naval officers who served or were commissioners or members of the ten-man Board of Admiralty (q.v.). The First Sea Lord was the chief of the naval staff and the senior officer of the Royal Navy. The Second Sea Lord was chief of naval personnel. The Third Sea Lord, the controller, was responsible for equipment, ordnance, and research. The Fourth Sea Lord concerned himself with supplies and transport. The Fifth Sea Lord was responsible for naval aviation. The Sixth Sea Lord was the vice chief of the naval staff.

seaman—In the US Navy, a nonrated enlisted man. The lowest grade is the apprentice seaman, or seaman third class. The next two grades are seaman second class and seaman first class, above which is the lowest grade of petty officer. See also **private**.

seaman branch—In the US Navy, sailors working on and above decks, as boatswain's mates, fire controlmen, gunner's mates, quartermasters, signalmen, torpedoman's mates, turret captains, minemen, and coxswains. Men with these ratings formerly wore their rating badges on their right sleeves. All other ratings wore their badges on their left sleeves. See also "**black gang.**"

sea pay—A special pay allowance received by personnel assigned to ships.

seapower—1) That power that arises from a nation's ability to put ships to sea and to exercise control over significant sea and ocean areas. 2) An instance of this power as it exists in a particular nation, especially with reference to its adaptation to national policy and naval support thereof.

sear—That part of the lockwork of a firearm that engages the hammer or striker to hold it in a cocked position.

search and rescue—The use of aircraft, surface craft, submarines, and other special equipment in the search for or rescue of personnel.

second—1) To support a person or a unit in combat. 2) In British usage, to attach a military person temporarily to some unit other than his own or to assign some temporary duty.

second echelon—In Soviet Army doctrine, as displayed in World War II, that portion of a command that is not part of the first echelon (q.v.) and which is a follow-on force (q.v.). The second echelon is not considered to be a reserve, and is always given a specific mission to perform in following up on the anticipated success of the first echelon. In the event the first echelon fails to accomplish its mission, and thus does not create the conditions for the second echelon to perform its assigned task, the second echelon can be used as a reserve, and be committed to reinforce the first echelon and help it accomplish its mission. Like the first echelon, the second echelon at one level of command may be part of either the first or the second echelon of a higher echelon of command.

second in command—An officer whose rank and responsibilities are next in importance to those of a commanding officer, whose duties he performs in his absence. See also **deputy** and **executive officer**.

second lieutenant—1) The lowest commissioned rank in the US Army, Air Force, and Marine Corps, comparable to an ensign in the Navy. 2) An officer holding that rank.

second line—1) Of aircraft, having officially recognized limitations for combat or other military use, though employed in an emergency and in situations for which first-line aircraft are not available. 2) In ground combat, that portion of a force that supports the first line of troops and follows it into battle.

second strike—Nuclear counterattack by the victim of a first strike (q.v.).

secondary attack (holding attack)—An attack of only moderate intensity, designed to fix the enemy in position, forcing him to commit his reserves prematurely in a sector other than that in which the main attack will take place.

secondary battery—On a warship, all guns except main battery (q.v.) and antiaircraft guns.

secret—A security classification (q.v.) of material whose unauthorized disclosure would endanger security, seriously injure the interests or prestige of the United States, or be of great advantage to a foreign nation with respect to the US.

section—1) As applied to ships or naval aircraft, a tactical subdivision of a division (q.v.). It is normally one-half of a division in the case of ships, or two aircraft. 2) A subdivision of an office, installation, territory, works, or oganization; especially a major subdivision of a staff. 3) A tactical unit of the Army and Marine Corps. A section is smaller than a platoon (q.v.) and larger than a squad (q.v.), and in some organizations, the section, rather than the squad, is the basic tactical unit.

sector—1) An operational area for attack or defense, designated by boundaries within which a unit operates and for which it is responsible. 2) One of the subdivisions of a coastal frontier.

sector of fire—An area, limited by boundaries, assigned to a unit or to a weapon to cover by fire.

secure—v. To gain posession of a position or terrain feature, with or without force, and to make dispositions to prevent its destruction or loss by enemy action. *adj.* 1) An indication that everything is as it should be under pertaining circumstances. 2) Protected against access by unauthorized personnel, as a secure area for discussion or use of classified material.

security—1) Measures taken by a command to protect itself from espionage, observation, sabotage, annoyance, or surprise. 2) A principle of war emphasizing the vital importance of such measures. 3) A condition that results from the establishment of protective measures which insure a state of inviolability from hostile acts or influences. 4) With respect to classified matter, it is the condition that prevents unauthorized persons from having access to official information safeguarded in the interest of national defense. 5) Protection of supplies or supply establishments against fire, theft, sabotage, or enemy attack.

security, communications—Measures designed to deny to unauthorized persons information of value which might be derived from the possession and study of telecommunications, or to mislead unauthorized persons in their interpretations of the results of such a study.

security classification—A category or grade assigned to defense information or material to indicate both the degree of danger to national security that would result from its unauthorized disclosure and the standard of protection required to guard against unauthorized disclosure. The normal security classifications in the US armed forces, in ascending order of sensitivity, are: for official use only, confidential, secret, and top secret.

security clearance—A clearance given by an investigative authority, permitting access to classified information (q.v.) and material up to a specified level of classification, provided there is a need to know.

selective service—see **conscription**.

self bow—See **bow**.

self-propelled—1) Of a gun, mounted on a vehicle with its own motive power. 2) Of a missile, propelled by fuel carried within the missile itself. 3) Of a military unit, having self-propelled guns. 4) Of a vehicle, having its own means of mobility.

self-propelled artillery—See **artillery**.

semaphore—A system of transmitting messages by moving hand-held signal flags, or other signalling devices, into various positions.

semiautomatic—1) The capability of a weapon to be fired either in the single-shot or automatic modes. 2) Using part of the force of an exploding cartridge to extract the empty case and insert the next round in the chamber, but requiring a separate pull on the trigger to fire each round.

semifixed ammunition—Ammunition in which the cartridge case is not permanently attached to the projectile (q.v.). This permits the withdrawal of a powder packet or packets before firing, in order to fire the weapon at less than full charge. See also **ammunition, fixed**, and **separate-loading and reduced charge**.

seniority—A higher rank or position. One military individual has seniority over others by holding a higher rank or having held the same rank longer than another. He is said to be **senior** to these subordinates.

sensing—An observation of fall of shot with respect to the target.

sensor—A device that detects such things as: terrain configuration, the presence of military targets, and other natural and man-made objects and activities by means of energy emitted or reflected by such targets or objects. The energy may be nuclear, electromagnetic (including the visible and invisible portions of the spectrum), chemical, biological, thermal, or mechanical, including sound, blast, and earth vibration.

sentinel—A person posted as a guard to protect a thing or area, as to give warning of an approaching enemy. Also known as **sentry**.

sentry—See **sentinel**.

separate-loading ammunition— Ammunition in which the projectile (q.v.) and charge (q.v.) are loaded into a gun separately. See also **fixed ammunition** and **semifixed ammunition**.

separation—The process of leaving military service to return to civilian life.

sepoy—An Indian soldier serving in British forces.

sergeant—1) An enlisted rank of noncommissioned officers in various military services, including the US Army, Air Force, and Marine Corps. There are several grades of sergeant. In the US Army these are, in ascending order: sergeant, staff sergeant, technical sergeant, master sergeant. 2) A mobile, inertially guided, solid propellant, surface-to-surface missile (q.v.) in the US arsenal in the late 20th century.

sergeant, color—See **color-sergeant**.

sergeant, flight—See **flight sergeant**.

sergeant major—A designation for the chief administrative noncommissioned officer of a unit headquarters. He is also normally the senior noncommissioned officer of the unit. The sergeant major usually assists the adjutant.

sergeant-major-general—1) In the 16th century, the senior administrative officer of an army under operational command of a monarch or his lieutenant general. 2) Commander of the militia, by election, in the Massachusetts Bay Colony.

sergeant of the guard—The senior noncommissioned officer of a guard detachment.

serial number—1) The number given a member of the service upon entry to military service. The same serial number identifies a service member throughout his career. 2) Designation of a unit of march when a large military force moves by road. 3) The number assigned to a piece of military equipment for identification and control.

serpentine—1) A cannon used from the 15th through the 17th centuries. A typical English serpentine—the name was metaphorical, referring to the weapon's deadliness, not its shape—had a caliber of seven inches and fired a 42-pound shot. 2) The S-shaped arm that held the match of the matchlock (q.v.) and pivoted to touch the lighted end to the primer in the priming pan (q.v.).

service ammunition—Ammunition intended for combat rather than for training purposes.

service dress—The normal uniform worn by service personnel except for special duty requirements. Also simply called **dress**.

service medal—A medal recognizing participation in a war, campaign, or special operation. It is distinguished from a decoration (q.v.) by being automatically awarded to any-one who participated. (The term medal is sometimes used for both service medals and decorations.)

service, military—1) Collectively, the armed forces of a nation. 2) One of the separate branches of the armed forces. 3) Formal participation in military activity in the armed forces.

service practice—The training process of firing ammunition in order to increase proficiency. Generally applied to artillery.

service ribbon—A ribbon worn on the uniform to represent a service medal.

service star—1) A small metal star worn in the US services on a campaign or war ribbon, representing a clasp on the ribbon holding the medal and signifying membership in a unit that participated in a war campaign or a separate battle of the campaign. Each bronze star is a symbol of one battle or campaign, each silver one represents five. 2) When worn on a decoration ribbon, the stars represent additional awards of the same decoration.

service stripe—In the US armed forces, a diagonal stripe representing three years of active service worn by enlisted persons on the left sleeve of the uniform, except for Navy ratings, who wear them on the right. Also known as **hash marks**.

service test—A test of an item, system, matériel, or technique conducted under simulated or actual operational conditions to determine whether specified military requirements or characteristics are satisfied.

service troops—Those units designed to fulfill supply, maintenance, transportation, evacuation, hospitalization, and other services required by air and ground combat units.

service unit—A noncombat army unit. This includes supply, medical, and engineer units.

serviceman—A person in the armed services.

set-piece battle—The conduct of a battle by a military force in accordance with carefully prepared plans, which have usually been repeatedly rehearsed before the battle.

shako—A military dress hat made of leather or other stiff material, cylindrical in shape, with a short visor, a chin strip, and a plume.

shaped charge—An explosive charge one side of which is so hollowed as to cause a focusing of the blast energy into a high-velocity jet of flame that will penetrate armor. It is sometimes called a **hollow charge** or **beehive-shaped charge**. See also **Munroe effect** and **hollow charge**.

Sharps rifle—A .50 caliber breech-loading (q.v.) rifle with a vertically sliding breechblock (q.v.), invented by Christian Sharps in 1848. It was in use for about 45 years, and was a major infantry weapon of the US Civil War.

sheaf, converged—The lateral distribution of fire of two or more artillery pieces so that the planes of fire intersect at a given point.

sheaf of fire—Designates the planes of fire of the trajectories of two or more artillery pieces in order to achieve a desired pattern of bursts in the target area. Sheafs of artillery batteries are normally parallel, unless some other effect is desired. See also **sheaf, converged; sheaf, open**; and **sheaf, parallel**.

sheaf, open—1) The lateral distribution of the fire of two or more pieces so that adjoining points of impact or points of burst are separated by the maximum effective width of burst of the type of shell used. 2) Term used in a call for fire to indicate that the observer desires a wider sheaf than the one being employed.

sheaf, parallel—A sheaf in which the planes of fire of all pieces are parallel. See also **sheaf of fire**.

sheaf, width of—A field artillery term meaning the lateral angle between the shell bursts at the sides of the sheaf (called flank bursts). The comparable naval gunfire term is **deflection pattern**.

shell—n. A projectile, usually of iron or steel, containing a high explosive, chemical, or atomic charge. A high explosive shell is designed either to carry the explosive charge for destructive effect against an object or to scatter shell fragments for antipersonnel effect. *v.* To fire shells at a target.

shell case—A cylindrical case of metal or cardboard that contains charge, primer, and shot.

shell fragment—A piece of jagged metal resulting from the explosion that tears apart the case or shell of a high-explosive projectile. The explosion propels the fragments with lethal force in a roughly circular or oblong area around the point of impact. During World War I some military doctors, unaware of the difference between shrapnel (q.v.) and a high-explosive shell, referred to shell fragments, and the wounds they caused, as shrapnel and shrapnel wounds. This erroneous nomenclature has persisted stubbornly, despite the efforts of linguists to correct it.

shell shock—See **fatigue, combat**.

shield—A protective device for the body, used by warriors since antiquity, carried on the nonweapon-bearing forearm, and maneuvered to fend off blows or missiles, or to set on the ground. Shields were carried by most soldiers before the introduction of gunpowder reduced both the incidence of close encounter warfare and the effectiveness of the shield. Over the centuries shields were made of many materials, including leather, wood, and metal, and in many shapes and sizes, from the small circular target to full-length rectangular shields that covered the entire body. Still in use on artillery pieces and antitank guns is a thin, bulletproof metal sheet designed to protect gunners from small-arms fire and shell fragments, commonly known as a **gun shield**.

shield-wall—A circular or square formation, similar to a phalanx (q.v.), used by several Germanic tribes of the early Middle Ages, in which warriors maintained a defensive position, usually against hostile cavalry, by putting together the shields of the front ranks to form a kind of wall across the front or around the formation.

ship of the line—In the days of sail, a ship-rigged warship that was large enough and with enough guns to form part of the principal battle line, originally so called in the British Navy. During the 18th and early 19th centuries, the typical ship of the line carried 74 or more guns; a few of the largest carried 100 guns.

ship's armor—See **armor**.

ship's clerk—A US Navy warrant officer whose duties are administrative. Senior to him is the chief ship's clerk.

ship's company—All the personnel assigned to duty aboard a naval vessel (often also used on naval shore stations), as opposed to those others carried aboard, such as naval aviation personnel on an aircraft carrier, observers, or personnel being transported from one place to another. See also **company**.

ship's service—A service aboard ship and at US naval shore stations that includes such facilities as a store that sells items unavailable in the ship's store or canteen, a laundry, barber shop, tailor shop, cobblers, and photography shop. All items are sold at slight profit, and the proceeds put in the ship's or station's welfare fund. (Comparable to Post Exchange and Base Exchange in the US Army and Air Force.)

ship's store—A store on a ship or naval shore station that carries personal items for sale to all personnel. Also known as the **canteen**.

ship-to-shore movement—That portion of the assault phase of an amphibious operation that includes the deployment of the landing force from the assault shipping to designated landing areas.

shock action—Military action employing shock tactics (q.v.).

shock tactics—Tactics, especially of cavalry or armor, in which the physical overrunning of the opponent is contemplated or attempted. Shock implies both the physical imposition of force upon an enemy and the psychological effect upon the enemy of the threat of being overrun.

shock troops—A German term for troops selected and trained to lead attacks; elite troops. See also **storm troops**.

shock wave—The pulse formed by the blast from an explosion. Can apply to any explosion, high explosive or nuclear in origin.

shore fire-control party—A specially trained

unit for control of naval gunfire in support of troops ashore, consisting of a spotting team to adjust fire and a naval gunfire liaison team to perform liaison functions for the ground unit commander.

shore party—1) A group of people going ashore from a naval vessel for whatever reason. 2) A task organization of the landing force (q.v.), formed to facilitate the landing and movement off the beaches of troops, equipment, and supplies; to evacuate casualties and prisoners of war; and to facilitate the beaching, retraction, and salvaging of landing ships and craft. It comprises elements of both the naval and landing forces.

shore patrol—The section of the US Navy that carries out police functions on shore; it corresponds to the Army's Military Police.

short-range ballistic missile (SRBM)—See **missile, ballistic.**

shot—1) The inert material (such as a pellet or bullet) discharged from a firearm. 2) A solid projectile, as opposed to a shell (q.v.)

shot locker—A strongly built compartment in a ship's hold for containing shot (q.v.).

shoulder—n. 1) In a fortification, the angles formed by the front wall and the side walls. 2) The angle between the general front line and the side of a salient. v. To lay against the shoulder, as in "shoulder arms."

shoulder belt—A belt supported by one or more suspenders over the shoulder or shoulders, on which various items of equipment are carried (such as small arms, ammunition, canteen, and first aid kit). Can be made of webbing or leather. In the latter case, it is usually called a Sam Browne belt (q.v.).

shoulder board—An insignia of rank worn by military officers on the shoulder of their uniform blouse, coat, or shirt. The board is a stiff material covered with cloth, and is attached to the shoulder of the uniform through loops or buttons or both.

shoulder patch—A cloth insignia worn by military personnel just below the shoulder seam (either right or left) of the uniform. Shoulder patches usually indicate the unit to which a person belongs or has belonged.

show of force—The marshaling or deployment of substantial military forces to demonstrate to potential adversaries (and potential allies) a readiness or willingness to fight. A show of force is usually intended as a deterrent (q.v.).

shrapnel—1) A projectile invented by British General Sir Henry Shrapnel in the late 18th century. It had an outer case within which were metal balls as submunitions, equipped with a fuze designed to cause the projectile to explode in midair and scatter the balls with great destructive force on enemy personnel. 2) The material scattered by a shrapnel shell. 3) A designation of shell fragments (q.v.) that is technically incorrect.

shuttle bombing—Bombing of objectives using two bases. By this method, a bomber formation strikes its target, flies on to its second base, reloads, and again bombs a target on the return to its home base.

sick bay—The medical treatment center of a ship or naval shore station.

sick call—1) A call on drum, bugle, or trumpet announcing that sick personnel should go to a medical facility. 2) The period during which sick personnel may seek treatment at a medical facility.

side arm—An individual weapon that is carried at the side or the waist. Side arms include such weapons as swords, bayonets, daggers, and pistols.

side boys—Nonrated enlisted men aboard a US Navy warship whose duty is to fall in at the gangway in appropriate numbers whenever naval etiquette calls for it to be tended, as it usually is when an officer or civilian dignitary comes aboard or leaves the ship. The side is tended by from two to eight "boys," depending on the rank of the person being honored.

siege—1) The process of besieging a fortress or town. 2) The activities of a besieging force in the process of a siege.

siege artillery—Artillery whose purpose is to breach the walls of a fortress or town that is under siege or to drop projectiles on targets within them. Siege artillery includes not only guns but also, in ancient and medieval times, the whole array of catapults, arms, towers, and other devices.

siege gun—See **siege artillery**.

siege tactics—See **siegecraft**.

siege towers—Movable structures, usually built slightly higher than the walls of a besieged fortification, with which to raise archers and missile-throwing machines to a level from which they could fire at defenders on the top of the walls or fire over the walls into the enclosure. Such towers were always constructed on the spot, although prefabricated portions were sometimes carried in the siege train (q.v.). They were used from earliest times until about the mid-15th century, when gunpowder weapons made them both unnecessary and too vulnerable.

siege train—The equipment for conducting a siege, carried on wagons and carts with an attacking army.

siege works—The engineering field fortifications used in the process of conducting a siege, particularly the construction of saps (q.v.) and parallels (q.v.) used to bring the attacking forces and their siege cannon close enough to the defender's fortifications for effective fire.

siegecraft—The art of conducting sieges.

Siegfried Line—1) Allied nickname during World War II for the German Westwall fortified zone, which protected Germany's western frontier, from the Swiss border to Ostend on the English Channel. 2) German designation, properly given as Siegfried Position, for a deep fortified zone that General Hindenburg ordered constructed behind the German front lines on the Western Front in the fall of 1916. During February-April 1917 the German armies withdrew to this line, which was shorter and more defensible than the salient position they had previously occupied. The Allies called the Siegfried Position the **Hindenburg Line**.

Sierra—Letter of NATO phonetic alphabet. See also **alphabet, phonetic**.

sight—n. 1) Actual visual contact, contrasted with radar or sonar or other contact in which the object is not actually seen. 2) A device on a weapon by which the person firing it deter-

mines where to aim. v. 1) To see. 2) To look at a target in order to establish the point at which to aim.

signal—n. 1) A message transmitted from one military unit to another, by whatever means. 2) Any transmitted electrical impulse. v. To send a message.

signal communications—The transmission of messages between military headquarters and units. Usually refers to electric and electronic transmissions.

signal flag—1) Any flag that is flown or used to convey a message. 2) Usually refers specifically to the set of flags of various shapes and colors used for signaling at sea. (At night, lights are used instead.) 3) Flags for signaling on land, by Morse code or semaphore.

signalman—1) US Navy specialty rating; chief first, second, and third class. 2) In the Army, a person using signal flags.

signals—See **signal communications**.

significant local war—See **conflict levels of intensity**.

simplicity—A principle of war emphasizing that success in war is, in part, dependent upon the use of organizations, lines of communication, plans, and operations that are relatively free of complexity.

simulated rank—A status given a civilian in terms of a military rank.

simulation—1) A representation of a combat situation, usually by means of a computer program. 2) A kind of combat model. See also **model** and **simulation, computer**.

simulation, computer—An analytical technique using mathematical and logical models to represent and study the behavior of actual or hypothetical events, processes, or systems. It offers opportunities to: test theories and proposed modifications in systems and processes; study organizations and structures; examine past, present, and future events; and to represent the use of forces that are impractical to mobilize in peacetime.

single sap—See **sap**.

single fire—1) One ship firing alone at a single target. 2) Firing an automatic or semiautomatic small arm in such a way that only one round is fired for each pull of the trigger.

site, angle of—The vertical angle from a horizontal plane through an artillery gun position that is formed whenever the location of the target is higher or lower in altitude than that of the gun or guns that are firing at the target. If the target is higher than the guns, a positive angle of site is created, and a round fired at the elevation corresponding to the range to the target would strike the ground in front of the target. If the target is lower than the guns, there is a negative angle of site, and a round fired at an elevation corresponding to the range would strike the ground beyond the target. Thus, in order to have the round strike at or near the target, a positive angle of site has to be added to the range elevation, or a negative angle of site has to be subtracted. By this process the trajectory is considered to pivot around the gun, and to be tilted up or down by the magnitude of the angle of site. Since the trajectory is not really as rigid as this concept implies, when the angle of site is large, a compensating value, called **complementary angle of site** and found in firing tables, is applied. See also **quadrant elevation**.

situation map—A map showing the tactical or administrative situation of a military organization at a particular time. In combat, situation maps in unit command posts are constantly updated as new information is received.

situation report—A report giving the status of a unit or formation at a given time.

six-bell hammocks—The hammocks belonging to members of a ship's crew who had night watches and were allowed to sleep until 0700 hours.

skip bombing—A method of aerial bombing in which the bomb is released from such a low altitude that its forward motion causes it to slide or glance along the surface of the water or ground and strike the target at or above water or ground level.

skirmish—An encounter in which few soldiers are involved, usually lasting a short time. A skirmish may be isolated or occur as part of a larger engagement or battle.

skirmisher—A lightly armed or specially trained soldier whose purpose is to engage in small actions, often in front of the main body or on the edges of the battlefield. Skirmishers have been used throughout the recorded history of warfare.

skirmishing order—See **open order**.

skirmish line—A line of soldiers preceding an attacking force, or deployed in front of a defending force, prepared to engage enemy skirmishers or units and prevent them from surprising the main body.

slant range—The line of visual distance between two points that are at different elevations.

SLCM (submarine-launched cruise missile)—See **missile, cruise**.

sleigh—1) A vehicle on runners used for conveying loads, especially over snow and ice. 2) The portion of a gun carriage on which the barrel slides in recoil and returns to battery from recoil.

slice—An average logistic planning factor used to obtain estimates of requirements for personnel and matériel. A unit slice generally consists of the total strength of the stated basic combatant element, in personnel and equipment, plus its proportionate share of all supporting and higher headquarters personnel and equipment. For instance, if an overseas theater contains ten divisions of 15,000 men each, but the total manpower in the theater—including higher headquarters and all support personnel—totals 300,000, then the division slice for that theater is 30,000 men, that is, the total manpower divided by the number of divisions.

sling—A device used by ancient warriors for firing rocks or other projectiles at an enemy. A sling was usually a rectangular piece of leather or hide with a thong or cord attached to each end. The slinger (q.v.) set a rock or other missile in the sling and, holding it by the cords, whirled it rapidly around above his head. By loosing one cord at the proper moment he released the missile with considerable force. It was this kind of sling that David used to slay Goliath. Projectiles included baked clay balls, and during the Roman Republic pellets of lead in acorn shape were used.

slinger—One who uses a sling. From prehistoric times into the medieval period, slingers were an element of all combat forces. Famous in Roman times, Balearic slingers were equipped with long-, medium- and short-range slings and organized in special auxiliary army units. Slingers, often recruited from mountainous regions, formed part of the mercenaries in the service of the French crown in the Middle Ages. They are specifically mentioned in action during the Battle of Navarrete (1367).

slit trench—A narrow trench in which one or more soldiers may take refuge in combat or during an air attack. See also **foxhole** and **trench**.

sloop—A modern naval vessal smaller than a destroyer, used as a convoy escort, in antisubmarine operations, and as minesweepers.

sloop-of-war—A small warship of the 17th, 18th, and early 19th centuries. Smaller than a frigate, it was usually brig-, or square-rigged, but could also be schooner-, or fore-and-aft-rigged, with one, two, or three masts.

small arms—Individual and small-caliber (less than 12mm) crew-served weapons which are not a part of another weapon system. Examples are: pistols, rifles, shotguns, submachine guns, and light machine guns. Machine guns mounted on tanks are not considered small arms.

small stores—Personal items and clothing issued to enlisted naval personnel.

small wars—A general term used for wars of less than major importance; usually armed conflicts such as colonial wars, punitive expeditions, and the like, that are not likely to precipitate international crises. See also **conflict levels of intensity**.

smoke grenade—See **grenade**.

smoke screen—A cloud of smoke used to mask either friendly or enemy installations or maneuvers. Smoke screens may be generated on the ground by use of smoke shells, smoke grenades, or smoke pots. At sea, smoke screens are usually formed from warships' smokestacks by burning smoke-producing fuel.

smoke tank—A tank on an aircraft from which an in-flight smoke screen can be laid down.

smokeless powder—A powder containing nitrocellulose that burns explosively but produces very little smoke. Introduced in about 1885, smokeless powder permitted gunners to continue to see their targets, which formerly had been rapidly obscured by the smoke of black powder. Its widespread adoption marked a significant increase in the lethality of weapons, which also became less vulnerable to countermeasures inasmuch as their concealment was not betrayed by smoke.

smoking lamp—A lamp aboard ship which, when lit, indicates that smoking is permitted. Now only an expression, the smoking lamp is lit when there is little hazard of fire or when smoking might not be considered disrespectful. Thus, the smoking lamp is out during religious services, and when ammunition or fuel oil is being taken aboard. Although once an actual lamp, the smoking "lamp" today may simply be an announcement over a vessel's intercom system.

smoothbore—Said of guns, small arms or cannon, when the bore or interior surface of the barrel or tube is smooth, that is, not rifled. Prior to about 1850, most small arms were smoothbores; the transition from smoothbore to rifled artillery cannon was made between 1865 and 1885. With the exception of grape (q.v.) and canister (q.v.), smoothbore artillery guns usually fired only spherical projectiles with a diameter smaller than that of the bore. With the exception of shotgun-type rounds, the same was true of smoothbore small arms. Some modern high velocity tank guns are smoothbore.

snaphance lock (snaphaunce)—A type of flintlock, invented in the 16th century, in which a piece of flint was held by an S-shaped cock. The cock pivoted to strike a piece of steel (the battery or frizzen) on the other side of the pan (q.v.), producing a spark which fired the powder in the pan, which in turn ignited the charge in the gun chamber.

snaplock rifle, Swedish—A 16th-century Swedish gun lock, distinguished from the matchlock (q.v.) in that the match was touched to the powder in the pan (q.v.) by a spring mechanism rather than by manual means.

snipe—To fire a rifle from a concealed position at individual soldiers. The person detailed for this duty is a **sniper**.

sniper—See **snipe**.

sniperscope—A telescopic sight used to enable snipers to fire accurately at long range.

snorkel—See **schnorkel**.

soften—1) To weaken the resistance of an enemy by an attack or series of attacks in preparation for a major assault. 2) To bomb or strafe enemy positions in preparation for another attack or assault. 3) To shell an enemy-held position or area with artillery or naval guns in preparation for an attack or assault. Often **soften up**. See also **preparation fire**.

soixante-quinze—See **French 75**.

soldier—1) In a general sense, any person in an army or ground combat force. 2) An enlisted person in an army, as distinguished from an officer.

solid square—A body of troops in which the ranks and files are equal in number.

SONAR (SOund NAvigation and Ranging)—A method or system, analogous to radar, in which high frequency sound waves are emitted, so as to be reflected back by objects. SONAR is used especially by ships for detecting underwater objects, such as submarines or mines. See also **ASDIC**.

sonarman—US Navy enlisted specialty rating chief, first, second, and third class.

sonobuoy—A buoy containing an acoustic receiver and radio transmitters; it is dropped in the water from a ship or aircraft to detect and transmit underwater sounds, especially the noise of a submerged submarine.

SOP—See **standard operating procedure**.

sortie—n. 1) A sudden attack made from a defensive position. In this sense, it is sometimes called a sally (q.v.). 2) An aircraft airborne on a mission against the enemy or in direct support of such a mission v. To depart from a port or anchorage, with an implication of departure for operations or maneuver.

spahi—1) A member of a Turkish cavalry corps. 2) A member of an Algerian or Moroccan cavalry corps in the French Army.

Spanish square (tercio)—A 16th-century formation developed by the Spaniards, consisting of several colunelas (q.v.), eventually standardized at three, with just over 3,000 soldiers in total. It is not clear whether the name *tercio* came from the three colunelas, or from the fact that the square comprised about one-third of the infantry component of the average Spanish field army. Like the ancient Roman legion and the modern division, it became the basic combat unit of Spain's army, large and diverse enough to fight independent actions on its own. By the time this formation had become standardized, the sword and buckler troops (q.v.) and halberdiers (q.v.) had been eliminated, leaving pikemen (q.v.) and arquebusiers (q.v.) as the components of the tercio, which dominated European battlefields for the remainder of the century.

SPAR—See **Women in the Coast Guard**.

sparum—A kind of dart with a triangular point fired by a crossbow.

spatha—A long cutting sword of the Roman Empire cavalry and of Germanic tribal cavalry of the Roman era.

spear—Any stick with a pointed head used as a thrusting or throwing weapon. The man armed with such a weapon is a **spearman**.

spearhead—n. Any force that precedes others in an attack. v. To lead an attack, to provide the drive for launching an attack, to take the lead in an undertaking.

special court martial—See **court martial**.

special force—Any military force trained for unusual or unconventional operations. The US Army's Special Forces consist of personnel with cross-training in basic and specialized military skills. They are organized into small, multipurpose detachments with the mission to train, organize, supply, direct, and control indigenous forces in guerrilla warfare and counterinsurgency operations, and to conduct their own unconventional warfare operations.

special orders—Orders issued by a headquarters that apply to a particular person or situation, and do not have the general application of general orders (q.v.).

special operations—Secondary or supporting operations that may be adjuncts to other operations and for which no one service is assigned primary responsibility.

special purpose vehicle—A vehicle incorporating a special chassis and designed to meet a specialized requirement: for example, an amphibious vehicle.

special staff—All staff officers having duties at a headquarters and not included in the general, coordinating staff group or in the commander's personal staff group. The special staff includes certain technical specialists and heads of services, for example: quartermaster officer, ordnance officer, transportation officer.

specialty mark—The device on a naval enlisted man's rating badge that indicates his specialty.

specified command—An operational US force of major size and importance that has a broad continuing mission and that is established by the President through the Secretary of Defense with the advice and assistance of the Joint Chiefs of Staff. It can be composed of elements from one service or two or more services. The Strategic Air Command is an example.

spectrum of conflict—The conceptual portrayal, usually in diagrammatic form, of the various types of violent hostility, from least violent to most violent.

Spencer rifle—A repeating rifle invented by Christopher M. Spencer in 1860 and used by Union forces in the Civil War. It had a tubular magazine in the butt that held seven cartridges. By releasing the trigger guard the rifleman inserted a cartridge in the chamber and expelled the empty cartridge case from the preceding shot. The same device was also incorporated in a carbine, with which Union cavalry was armed.

spike—1) To drive a spike or similar object into the vent of a muzzle-loading (q.v.) cannon so that the gun is useless. 2) The term is often used for any action to disarm artillery

pieces, such as removing the breechblock (q.v.) or removing the firing pin.

spin—An aerial maneuver or performance, either controlled or uncontrolled, in which an airplane makes a vertical descent with its nose pointed sharply, or slightly, downward, while revolving about the line of descent in any of various **attitudes**. An attitude is a flyer's term for an aircraft's position as determined by the inclination of the axes to a specific frame of reference; unless otherwise specified, the frame of reference is fixed to the earth.

spin stabilization—The directional stability of a projectile imparted by the spinning of the body about its axis of symmetry. The spin is usually imparted by rifling in the tube of the weapon that fires the projectile.

Spitfire—A low-wing fighter of the Royal Air Force with a single in-line engine. Used throughout World War II, the Spitfire had the speed, rate of climb, maneuverability, and firepower that made it one of the most effective fighters of its day. In its long operational use it underwent numerous modifications, and some Spitfires were used by the US Army Air Forces.

spoiling attack—A tactical maneuver employed to impair or interfere with an anticipated hostile attack while it is being prepared by the enemy. Usually employed by mobile defending units which attack enemy assembly positions in front of the defender's main line of resistance or battle position.

spoils—Whatever the victor takes from the vanquished in time of war. Among ancient Greeks the general got the largest share, with the rest divided evenly among his troops. Spoils taken by Roman legions belonged to the republic.

sponge—A device for cleaning the bore of cannon, usually consisting of lambskin mounted on a wooden cylinder on a long stick, sometimes on the same stick with the rammer on the other end, which just fitted the gun bore. After the gun had been fired the wet sponge was swabbed around inside the barrel to make sure there were no sparks remaining. A sponge is used to clean the chamber of modern weapons firing separate-loading ammunition after each round is fired.

spontoon—A short pike that was used by boarding parties in the 17th century and became a badge of rank of infantry officers in England and in the US Continental Army, where it was called a half-pike, feather staff, or leading staff. Various forms appeared, but the characteristic one of the 17th to the mid-19th centuries had a broad blade with a rounded base and was heavily decorated with insignia.

spot—1) To determine by observation deviations of gunfire from the target for the purpose of supplying necessary information for the correction of fire. 2) To place in a proper location.

spotter—1) A person who observes and reports on naval gunfire and often designates targets. 2) A person in an aircraft who looks for people, for other aircraft, troops, artillery hits, or other objects.

Springfield rifle—1) A .30 caliber US Army rifle, first manufactured at the armory in Springfield, Massachusetts, in 1903. Breech-loading and magazine fed, with bolt action, it was the most accurate infantry weapon of the early 20th century. It was the American Army's chief arm from World War I until the beginning of the Second World War although it underwent some modification in the intervening decades. 2) In the Civil War the same armory had given its name to a rifle-musket that was the standard weapon on both sides of that conflict. It was a .58 caliber muzzledloader. An 1873 model was the carbine used by the US Cavalry at the time of the Battle of Little Bighorn. Its caliber was .45 inch.

spy—One who acts clandestinely to obtain information from one side in a conflict, or in anticipation of a conflict, and passes it to the other side.

squad—In a military organization, the smallest personnel unit. In the US Army, for example, a platoon generally consists of three or four squads, each with about 8 to 15 soldiers.

squadron—1) A naval organization consisting of two or more divisions of ships, generally of the same type. In use since the late 16th century, the term originally indicated a number of ships that a single flag officer could command. 2) The basic administrative aviation unit of the US Army, Navy, Marine Corps, and Air Force. 3) A basic administrative cavalry unit (horse or armored), comparable to an infantry battalion (q.v.), with two to four troops (q.v.) and administrative and supporting elements.

squadon leader—1) The pilot in command of a squadon (q.v.) of aircraft. 2) A commissioned rank in the Royal Air Force and other air forces, corresponding to major in the US Air Force.

square—A body of troops forming a square or rectangle, with several ranks on each side. A square may be hollow or solid. Hollow squares have a hollow space or gap in ranks and files in the center. Solid squares have the ranks and files equal in length, or no gap in ranks and files. Squares were formerly offensive and defensive tactical formations (see, for example, Spanish square (q.v.) but with the advent of linear tactics (q.v.) in the 17th century, they were largely abandoned as offensive formations and retained only for defense, particularly against cavalry, since they had no flanks that cavalry could attack. The improved accuracy of rifled weapons in the mid-19th century rendered square formations obsolete, because the massed troops were an easy target for such weapons.

square law—1) One of the two Lanchester Equations (q.v.). 2) A relationship between physical factors or considerations in war, and moral or behavioral factors, as shown in historical experience Because of this relationship—first hypothesized by T. N. Dupuy in 1980—the effect of human behavioral considerations is much more important (by an exponential factor of two) than physical considerations. In other words, instead of the relationship suggested by Napoleon—"The moral is to the physical as three is to one"—this new Square Law indicates that: "The moral is the equivalent of the physical squared."

squib—A small pyrotechnic device which may be used to fire an igniter, as in a rocket, or for similar purposes. Not to be confused with a detonator, which explodes.

squid—A World War II sea warfare device that fired three contact depth charges that would explode close to a submarine and, if it did not sink, would force it to the surface.

stability—1) In the context of the balance of

power, a high probability that no single nation will become dominant, that most of the nations will continue to survive, and that large-scale war will not occur. 2) In the context of strategic nuclear forces, usually a low probability of nuclear war.

staff—1) The group of officers assigned to the headquarters of a military command to assist in planning and administration but with no authority to command. Staff of large organizations are usually divided into two major elements: the general staff and the special staff (qq.v.). 2) A long pole or stick that could be used as a striking or thrusting weapon, or bear a cutting device or other weapon at the end.

staff, allied—A staff or headquarters composed of officers of two or more allied nations working together.

staff, combined—A multinational, multiservice staff.

staff sergeant—In the US Army, Air Force, and Marine Corps, a noncommissioned officer. In the Army, a staff sergeant is above a sergeant and below a sergeant first class. In the Air Force he is between an airman first class and a technical sergeant. In the Marine Corps he is between a sergeant and a gunnery sergeant.

staff sling—A simple weapon used in ancient and medieval warfare consisting of a sling (q.v.) attached to the end of a wooden staff. The use of the staff allowed the slinger to project the missile with greater force and attain greater ranges.

stage—1) To process, in a specified area, troops that are in transit from one locality to another. 2) That part of a missile or propulsion system that separates from the missile at burnout or cutoff. Stages are numbered chronologically in order of burning.

staging area—1) In amphibious or airborne expeditions, a general locality between the mounting area and the objective of the operation, through which the expedition or parts of it pass after mounting, for refueling, regrouping of ships, exercise, inspection, or redistribution of troops. 2) In other operations, a general locality, containing accommodations for troops, established for the concentration of troop units and transient personnel

between movements over the lines of communication.

stalemate—A situation in which neither of two opposing combatants holds a clear advantage and neither is capable of mounting the effort required to achieve a major victory.

stand (of arms)—A rack for storing small arms, usually rifles.

standard—A flag, banner or other device that identifies an individual, an office, a unit, a nation, or a place.

standard bearer—The individual who carries a standard (in a modern military unit it is a **color bearer**).

standard operating procedure (SOP)—A set of instructions covering those features of operations which lend themselves to a definite or standardized procedure without loss of effectiveness. This facilitates rapid, coordinated action in the event of emergency or confusion of battle. The procedure is applicable unless prescribed otherwise in a particular case. Thus, the flexibility needed in special situations is retained.

standardization—1) Use of identical weapons for the sake of economy and flexibility within any large military force, and particularly in an alliance of military forces of different nations. 2) The process by which the Department of Defense achieves the closest practicable cooperation among the services and defense agencies for the most efficient use of research, development, and production resources, and encourages on the broadest possible basis the use of: a) common or compatible operational, administrative, and logistic procedures; b) common or compatible technical procedures and criteria; c) common, compatible, or interchangeable supplies, components, or weapons and equipment; and d) common or compatible tactical doctrine with corresponding organizational compatibility.

standing army—See **army**.

standing barrage—High density artillery fire—frontal or flank—fired on call on a preselected line to form a barrier obstructing enemy movement. Usually an element of final protective fire. See also **fire, final protective** and **barrage**.

standing order—An order within a command that remains effective until it is withdrawn or countermanded.

star fort—A fort with several bastions, whose ground plan resembles a star.

star rank—In the US Army, Air Force, and Marine Corps, the rank of brigadier general and higher, comparable to flag rank in the Navy.

star shell—See **illuminating shell**.

start line—1) A line designed to coordinate the departure of attack or scouting elements; a jump-off line. 2) A marked off-shore coordinating line to assist assault craft to land on designated beaches at scheduled times. See also **line of departure**.

station—n. 1) A general term meaning any military or naval shore installation, location, activity, function or group of functions. 2) A particular kind of activity or function to which other activities or individuals may come for a special service, often of a technical nature, for example, an aid station. 3) An assigned or prescribed position in a naval formation or cruising disposition; or an assigned area in an approach, contact, or battle disposition. 4) Any place of duty or post or position in the field to which an individual, or group of individuals, or a unit may be assigned. v. 1) To assign to a billet or post. 2) To place in a specific position for a specific purpose, as a guard at the gate.

stick—1) Missiles or bombs fired quickly, one after another, or released separately at predetermined intervals from a single aircraft. 2) A number of parachutists who jump from one aperture or door of an aircraft during one run over a drop zone.

stiletto—A small, pointed dagger; the word is a diminutive of the Italian word for dagger, *stilo*. It is thought to have first been used in Italy. The stiletto blade is slender and triangular or sometimes square in section, and useful only for stabbing.

stirrup—A device attached to a saddle that holds the foot of a horseman. The origin of the stirrup is unknown, although it seems to have come from eastern Asia early in the Christian era. Its introduction enabled the mounted horseman to maintain his seat and control his horse while wielding a weapon, to an extent hitherto impossible. It significantly increased the lethality of the lance by putting the total weight of horse and rider behind its impact, and the lethality of the bow and the sword by giving the horseman a secure mounting from which to use them. It was the principal reason for the predominance of cavalry in warfare for about 1,000 years.

stock—1) The portion of a hand or shoulder-fired weapon, usually of wood, to which the barrel and the other parts of the weapon are attached. 2) The portion of a rapid-fire weapon that connects the shoulder piece and the slide.

stockade—1) A fortification, one commonly built in American colonial and pioneer days, consisting of posts or stakes driven close together upright into the ground. 2) An area so enclosed and used for a prison; the term is now often applied to any military prison within a fenced area.

Stokes mortar—A type of mortar invented by F. W. C. Stokes in 1915 and used by the Allies in World War I. Often called a trench mortar. The cartridge, with a firing pin at its tail end, was dropped into the barrel and immediately fired. It was the prototype for most modern mortars.

stone bow—A crossbow designed for throwing stones.

stone mortar—An early mortar used to throw stones about 150-250 yards, or six-pound shells 50-150 yards. The stones were put in a basket fitted to the bore and the basket placed on a wooden bottom covering the chamber. See also **pedrero**.

storekeeper—In the US Navy, a specialty with four ratings: third, second, and first class, and chief.

stores, military—Arms, ammunition, clothing, provisions, and other items pertaining to an army.

storm—n. A violent assault on a fortified place. v. To assault, particularly a fortified place.

storm troops—In the German Army, an elite body trained to be hardhitting and efficient. See also **shock troops**.

straddle—A naval gunnery term describing a situation in which several rounds of a salvo or volley have fallen on each side (over and short) of a target. See also **bracket**.

stradiots—Light cavalry, originally Albanians, introduced by the Venetians into their wars in northern Italy late in the 15th century. Similar cavalry were adopted promptly by the French. The stradiots were lightly armored and carried bows or lances.

strafe—1) To rake a body of troops or other persons with gunfire or rocket fire from a flying aircraft at close range. 2) To attack a roadway, rail yard, factory, or other installation with bullets, shells, or rockets fired from a low-flying airplane.

straggle—1) To fall behind a moving body of troops, ships, or aircraft. 2) To fall back individually to the rear in combat, without command. 3) To return late to duty. One who straggles is a straggler.

straggler line—A line designated by a force commander in a combat situation where stragglers are to be apprehended, collected, and returned to their units.

stratagem—Any action or maneuver whose purpose is to deceive or surprise an enemy.

strategic—1) Of, or pertaining to, strategy. 2) Often used to refer to long-range air attacks against the industry or population of an enemy nation. See **strategic air warfare**.

strategic air force—An air force organized and equipped to carry out long-range air operations deep into the heartland of an enemy nation.

strategic airlift—The sustained air transportation of personnel and cargo between theaters, to provide long-range logistic support for a military effort.

strategic air warfare—Air combat and supporting operations designed to effect the progressive destruction of the enemy's war-making capacity. Vital targets may include key manufacturing systems, sources of raw material, critical material, stockpiles, power systems, transportation systems, communications facilities, concentrations of uncommitted elements of enemy armed forces, key agricultural areas, etc.

strategic bombing—The bombing of a selected target or targets considered vital to the war-making capacity of a nation.

strategic concept—1) A concept related to strategy (q.v.) or strategic affairs. 2) The course of action accepted as a result of an estimate of the strategic situation. It is a statement of what is to be done expressed in broad terms sufficiently flexible to permit its use in framing the basic undertakings which stem from the strategic concept.

strategic envelopment—See **turning movement**.

strategic flank—That flank of a military force which, if turned (meaning that the attacking force has passed around the flank), would expose the force's base or line of communications. See also **flank**.

strategic map—A map of medium scale (1/200,000) or smaller, used for planning of large-scale operations, including the movement, concentration, and supply of troops.

strategic material—A material required for essential uses in a war emergency, the procurement of which in adequate quantity, quality, or time is sufficiently uncertain, for any reason, that prior stockpiling is required.

strategic mobility—1) The capability to deploy and sustain military forces worldwide in support of national strategy. 2) The capability of a unit, command, force, or thing that enables it to be readily moved long distances in advance of engagement with hostile forces. See also **mobility**.

strategic reserve—That quantity of matériel that is placed in a particular geographic location for strategic reasons or in anticipation of major interruptions in a supply distribution system.

strategos—A high ranking Byzantine officer whose command was a thema (q.v.), equivalent to a modern army corps. A strategos was usually the military governor of the province or theme (q.v.) from which his troops were drawn, or where they were based.

strategy—In war or peace, the planning and management of a nation's total available resources—economic, social, political, and military—in order to achieve the goals of national policy, and in wartime, to maximize the likelihood of victory. Such planning and management are the responsibilities of the highest levels of national and military authority. In a strict or military sense, strategy is the art and science of planning and directing military movements and operations so as to achieve victory. Clausewitz defined strategy as: "The use of engagements to attain the object of war." And Jomini defined it as: "The art of getting the armed forces into the field of battle," comprising "all the operations embraced in the theater of war in general." See also **strategy, national** and **strategy, military**.

strategy, military—The art and science of developing and employing in war military resources and forces for the purpose of providing maximum support to national policy, in order to increase the likelihood and favorable consequences of victory. Science predominates over art in military strategy, the difference between military competence and military genius at the strategic level being greater artistry by genius.

strategy, national—The art and science of developing and using political, economic, psychological, social, and military resources as necessary during war and peace to afford the maximum support to national policies, and in the event of war, to increase the likelihood and favorable consequences of victory. Art predominates over science in national strategy.

strength—The number, particularly of personnel or units, in a military force. Effective strength is the number of men fit for combat duty.

strike—n) An attack intended to inflict damage on, seize, or destroy an objective. v. 1) To attack. 2) In the US Navy, of an unrated seaman, to study and work toward making a rating in an enlisted specialty.

strike force—A force composed of appropriate units necessary to conduct strikes, attack, or assault operations.

striker—1) A firing pin or projection on the hammer of a firearm that strikes the primer in a fuze or a round of ammunition. 2) A soldier or airman who does extra-duty work for an officer for extra pay. See also **orderly**. 3) A naval seaman who is studying for a rating in a specialty.

strike the colors—To lower the flag in surrender.

stripe—1) Indication of rank worn by naval officers on their lower sleeves and on shoulder boards (q.v.). The warrant officer wears a broken half stripe (one-quarter inch), commissioned warrant officer a broken full stripe (one-half inch), the ensign a full stripe, and the successive ranks add half stripes or full, up to full admiral, who wears a two-inch stripe and three full stripes. 2) An indication of length of service. See also **service stripes**.

stronghold—A place of security; a fort or fortress.

strongpoint—A strongly fortified and heavily armed locality in a defense system, usually supported by auxiliary armed positions.

subaltern—1) In the British Army, a commissioned officer with a rank below captain. 2) More specifically, a second lieutenant as opposed to a first lieutenant.

subcommand—A command subordinate to another command.

subhadar—Indian Army native officer rank.

sublieutenant—Second lieutenant or subaltern (q.v.).

submachine gun—An automatic or semiautomatic weapon, lighter than a machine gun and fired from the shoulder or the hip.

submarine—A warship designed for under-the-surface operations with the primary mission of locating and destroying ships, including other submarines. Following the experiments of David Bushnell with a submersible craft during the American Revolution, there was considerable experimentation in the United States and in Europe, and submersible vessels were used, without much success, in the American Civil War. The submarine, however, did not become a serious naval war vessel until the 20th century. In World War I and World War II, German submarines (U-boats) caused serious damage to Allied shipping. By the end of the Second World War the electric battery-powered submarine had been superseded by the diesel, and the use of nuclear fuel in the 1950s made it possible for subs to travel thousands of miles underwater and stay submerged for months at a time. With the introduction of ballistic missiles that can be launched from below the sea toward land-based targets, the submarine's role in warfare was dramatically broadened.

submarine, guided-missile—A submarine with the additional capability of launching guided-missile attacks regardless of surface conditions. Designated as SSG, SSGN, or SSBN; the SSGN and SSBN (a ballistic-missile submarine) are nuclear-powered.

submarine pay—An allowance paid to men serving on a submarine because of the hazardous nature of the duty.

subsistence—1) Food or provisions. 2) A monetary allowance given a military person in lieu of food.

substantive rank—The permanent rank of an officer temporarily serving in a higher rank or grade.

Sugar—Letter of old phonetic alphabet. See also **alphabet, phonetic**.

summary court martial—See **court martial**.

superintendent—The title of the commanding officer at each of the US service academies: the US Military Academy at West Point; the US Naval Academy at Annapolis; and the US Air Force Academy at Colorado Springs.

supply—The procurement, distribution, maintenance while in storage, and salvage of supplies, including the determination of kind and quantity of supplies.

supply column—A procession of vehicles carrying food, support equipment, and supplies for a combat force. A **supply train**.

supply train—See **supply column**.

support—n. 1) The action of a force that aids, protects, complements, or sustains another force. 2) A unit that helps another unit in battle. Aviation, artillery, or naval gunfire may, for example, be used as support for infantry or armor. 3) A part of any unit held back at the beginning of an attack as a reserve. 4) An element of a command that assists, protects, or supplies other forces in combat. 5) An element of an advance guard (q.v.). v. To pro-

vide any of the aforementioned kinds of assistance to a combat force.

support trenches—A system of trenches, generally perpendicular to the front, facilitating support of forces in front-line trenches. See also **sap**.

support troops—Troops assigned to provide assistance to combat forces.

supporting artillery—Artillery that executes fire missions in support of a specific unit, usually infantry or armor, but remains under the command of the next higher artillery commander. For example, an artillery battalion fires in support of an infantry regiment, but is commanded by the commander of the divisional artillery, who is the next higher commander above the artillery battalion commander.

supporting distance—A position close enough to a combat area or a force in combat to make it possible to provide assistance.

supporting fire—Fire delivered by supporting units to assist or protect a unit in combat.

suppress—To reduce the effectiveness of a military force, and to inhibit its activity, by means of firepower from any weapons, but usually artillery or tactical air elements.

supreme commander—1) A commander superior in authority to all other commanders within a theater of war. 2) The commander at the apex of a military establishment. Used both as a specific and a general term.

surface burst—1) An explosion of a projectile on the surface of land or water, usually by means of a point-detonating fuze. 2) For a nuclear weapon, an explosion at the surface, or above the surface, at a height less than the maximum radius of the fireball.

surface-to-air missile (SAM)—See **missile, surface-to-air**.

Surgeon General—The senior officer in the medical departments of the United States Army, Navy, and Air Force.

surprise—v. To make an attack at a time or place or in a manner unexpected to the enemy. This is an important principle of war. n.

The reaction of a combatant to an unanticipated attack by the enemy.

surrender—1) To acknowledge defeat by ceasing to fight and overtly yielding to an enemy. 2) To yield an object or a place to one's opponent voluntarily.

surveillance—The systematic observation of air, surface, or subsurface areas by visual, electronic, photographic, or other means for intelligence purposes.

suspension of arms—A short truce or ceasefire agreed between hostile opponents for burying the dead, awaiting instructions from a superior authority, or other purpose.

suspension of hostilities—A truce or ceasefire by formal or tacit agreement of the opposing forces.

sutler—A person, not part of the military, who sells liquor, food, or other commodities to the army. Prior to the 20th century, sutlers followed behind an army in the field, like other camp followers (q.v.), and set up their shops whenever the troops halted.

Swedish feather or **swine feather**—A stake about five feet in length, with a spike at each end, used by musketeers (q.v.) to protect themselves against cavalry in the Thirty Years' War and the English Civil War. It was sometimes used also as a musket rest. The device was abandoned because of the weight it added to the musketeer's burden, which, if he wore a helmet, approached 60 pounds.

sweep—The rapid movement of a large force (ground, naval, or air) over or through a relatively large geographical area to assure control of the region, or demonstrate the ability to dominate it, and/or to eliminate hostile opposition in the area.

switch trench—A trench at an oblique angle to a main defensive trench, through which troops could fall back if driven from part of the main trench.

swivel gun—A gun, particularly a naval gun, that is mounted on a pivot in order to point in whatever direction is required. See also **pivot gun**.

sword—Any of many varieties of a hand weapon that has a long, sharp blade, with one or two cutting edges, is wielded with one hand or two, and is effective for either cutting or thrusting. The sword is longer than the dagger and often is carried in a scabbard hung from a belt at the waist, where it can easily be drawn. One who carries a sword and is skilled in its use is a swordsman. Swords in the 20th century are primarily ceremonial symbols of rank.

sword and buckler troops—Soldiers of the Middle Ages who were armed with sword and buckler (q.v.). The Spanish Army of the 15th century was feared for its sword and buckler troops.

syntagma—In the military system of Macedonia, an organization of about 256 troops, an element of the phalanx (q.v.), roughly equivalent to the modern battalion.

T

tabard—A cloak or tunic worn by a knight over his armor, or worn by his herald. The knight's coat of arms was blazoned on it.

table of organization and equipment (T/O&E)—A table listing authorized numbers of personnel and major equipment in standard units or formations. Often called and written T/O&E.

tactical—Of or pertaining to tactics (q.v.).

tactical air control center—The principal air operations installation (land- or ship-based) from which all aircraft and air warning functions of tactical operations are controlled. for a region or major command.

tactical air controller—The officer in charge of all operations of the tactical air control center (q.v.). He controls all aircraft and air warning facilities within his area of responsibility under the tactical air commander.

tactical air force—An air force charged with carrying out tactical air operations in coordination with ground or naval forces. In the US Air Force, it may be a component either of the Tactical Air Command or of a theater air organization. A tactical air force's components normally include fighter-bomber units and troop-carrier groups, and its command resides in an air commander. In World War II, the term was used overseas to designate a numbered air force composed of tactical air commands.

tactical airlift—An airlift that provides immediate movement and delivery of combat troops and supplies directly to the objective areas through air landing, extraction, airdrop, or other delivery techniques; it also provides the air logistic support of all theater forces, including those engaged in combat operations.

tactical air operations—Support of ground forces by combat aircraft. It generally consists of three major components: air superiority (q.v.), air interdiction, and close air support (q.v.).

tactical air support—Support of ground forces in tactical air operations (q.v.).

tactical bombing—Bombing, usually conducted by tactical air units, in support of surface forces.

tactical command ship—A warship, converted from a light cruiser (q.v.) and designed to serve as a command ship for a fleet or force commander. Equipped with extensive communications equipment, it is designated in the US Navy as CC.

tactical concept—1) A statement in broad outline that provides a common basis for the development of tactical and logistical doctrine, organization, and operational requirements. Such statements are normally developed for periods from 10 to 20 years ahead of their implementation. 2) A particular notion of how to fight a battle. 3) In military theory, a concept of war or air power that leads to an emphasis upon tactical operations as the best means of achieving victory.

tactical control—The detailed and usually local direction and control of movements necessary to accomplish missions or assigned tasks.

tactical flank—That flank of a military force most vulnerable to envelopment. See also **flank**.

tactical headquarters—For large formations, such as an army, that portion of the headquarters concerned with control of combat forces, as distinct from the administrative headquarters.

tactical locality—An area of terrain that, because of its location or features, possesses a tactical significance.

tactical maneuver—Maneuver (q.v.) at the tactical level, as opposed to a strategic maneuver.

tactical mobility—The capability of a unit, command, task force, or the like to be readily moved in support of combat or moved while engaged in combat. For instance, airplanes, tanks, motorized infantry, and naval destroyers have tactical mobility.

tactical nuclear weapon—A nuclear weapon of comparatively small yield (usually less than ten kilotons) designed for battlefield employment rather than strategic nuclear operations (q.v.).

tactical reserve—A portion of a ground combat force held out of combat by the commander for use where and when needed.

tactical security—All active and passive measures that protect both command and troops against observation, sabotage, and surprise attack and that give the commander enough time and space for the conduct of the operation.

tactical walk/ride—A training activity, usually at the staff school or war college level, in which an instructor or instructors accompany students on foot, on horseback, or in vehicles, in a study of a hypothetical combat operation on the ground.

tactics—1) As defined by Clausewitz, the use of armed force in battle. 2) As defined by Jomini, the maneuvers of an army on the day of engagement (q.v.); its contents, its concentrations, and the diverse formations used to lead the troops to the attack. 3) The technique of deploying and directing military forces—troops, ships, or aircraft, or combinations of these, and their immediate supporting elements—in coordinated combat activities against the enemy in order to attain the objectives designated by strategy (q.v.) or operations (q.v.).

tampion—A plug that closes the muzzle of a

gun to keep it clean and dry. Also known as **tompion**.

tandem hitch—Arrangement of draft horses in pairs to pull a wagon or other vehicle, particularly relevant to horse-drawn vehicles and horse artillery from the 18th to 20th centuries.

Tango—Letter of NATO phonetic alphabet. See also **alphabet, phonetic**.

tank—A tracked fighting vehicle with heavy armor protection primarily designed as the principal assault weapon of armored formations (battle tank) or as a reconnaissance vehicle (reconnaissance tank). Tanks are often designated light, medium, or heavy, according to weight and the weapons they carry: light tanks usually weigh 25 tons or less; medium tanks 25 to 50 tons; heavy tanks are usually those over 50 tons. Tanks weighing between 50 and 60 tons are now generally designated main battle tanks. Tanks were first used in warfare by the British in World War I. See also **tank, main battle**.

tank armor—See **armor**.

tank destroyer (TD)—A highly mobile vehicle, lightly armored and fully tracked or half-tracked, mounting an antitank gun. Used widely in World War II, US Army tank destroyers usually carried a 75mm or 90mm cannon or a 105mm howitzer.

tank, main battle—A tracked vehicle providing heavy armor protection and serving as the principal assault weapon of armored and infantry troops. The main armament is usually a gun of at least 105 mm caliber.

tank-retriever/tank-recovery vehicle—A vehicle with a tank chassis whose purpose is to remove repairable tanks from a combat area.

tank trap—Any of various devices, such as steel posts set in concrete, installed or built in a combat area for the purpose of halting or destroying advancing enemy tanks.

taps—A bugle call signifying the end of the day and used at a military funeral as a last honor for the dead. The original call was made by tapping a drum.

tarasnice—A light cannon, usually of the

type mounted on walls of castles or other fortifications, made mobile by being mounted
on a gun cart or a wagon, used by the Hussites in the 15th century. It was generally a
barrel clamped to a trestle or a tripod. The
barrel was usually over a yard long, with a
caliber of 40 to 60mm. Cast in bronze or iron
or welded from iron plates, the tarasnice was
also called a **palisade gun**.

Tare—Letter of old phonetic alphabet. See
also **alphabet, phonetic**.

target—Any person, place, or thing that is the
objective of a belligerent action. A target may
be anything selected for seizure or destruction—a soldier at whom a gun is pointed, a
geographical area to be taken, a country
against which intelligence operatiqns are directed, or an oil refinery upon which bombs
are dropped.

target acquisition—The detection, identification, and location of a target in sufficient
detail to permit the effective use of weapons.

target area—See **area**.

target array—A graphic and analytical representation of the enemy forces, personnel,
and facilities that are the objectives of an operation.

target designation—Action to indicate the
nature and location of a target.

target of opportunity—A target, unexpectedly visible to an observer that is within range
of available weapons and against which fire
has not previously been scheduled or re-
quested.

target system—A group of targets so related
or associated that their destruction will produce a particular effect desired by the attacker. For example, the target system of an oil
refinery would include storage tanks, refin-
ing areas, associated rail or transportation facilities, and even workers barracks—in short,
all of the industrial and administrative components affecting the functioning of the complex to be attacked.

task force—1) A temporary grouping of units
under one commander, formed for the purpose of carrying out a specific operation or
mission. 2) A semipermanent organization of
units under one commander for the purpose

of carrying out or continuing a specific task.
3) A component of a fleet organized for the
accomplishment of a specific task or tasks.

task group—A component of a naval task
force.

task organization—1) The process of organizing combinations of individuals or units so
as to provide responsible commanders with
the means with which to accomplish their assigned tasks in any planned action. 2) An organization table pertaining to a specific
directive.

tattoo—1) A drum beat or a bugle call at
night, signifying that enlisted personnel
should go to their quarters. 2) A display of
military exercises offered as entertainment.

taxiarchia—A unit of the Macedonian Army
with 128 men, part of a phalanx (q.v.), the
equivalent of a modern company, commanded by a **taxiarch**.

taxis—A component of a Macedonian phalanx (q.v.) consisting of 100 men, arranged in
files, eight or more deep.

TBS—"Talk Between Ships." Very high frequency, medium power, radio equipment
used primarily for operational or administrative communication between ships.

TD—See **tank destroyer**.

TDY—See **temporary duty**.

team—A number of persons associated or organized together to achieve a common end,
as in, "The bomber team includes both the air
crew and the ground crew."

tear gas—Chemical irritants that inflame the
mucous membranes of the eyes and cause
tears. Used as toxic agents in World War II,
tear gas is now mainly used as a simulated
toxic agent for combat training purposes and
in instances of civil disorder, usually to disperse crowds.

technical intelligence—Intelligence concerning foreign technological development,
performance, and operational capabilities
that may have a practical application for military purposes. It is the end product of the
processing and collating of technical information.

technical troops—Personnel trained in the operation, use, maintenance, and repair of specialized equipment, weapons, or instruments.

technique—A means or way of accomplishing a desired procedure.

telenon—A siegecraft device invented by Diades, Alexander the Great's engineer, consisting of a box or basket large enough to hold several armed men. This was slung from a boom that was attached to a tall mast or vertical frame on which it could be raised or lowered with ropes. By this elevator a group of infantrymen could be hoisted above parapet (q.v.) height, swung over any intervening obstacle, such as a moat, and deposited directly upon the enemy's battlements.

Templar—A member of a monastic order of priestly knights (Knights Templars, or Poor Knights of Christ and the Temple of Solomon) established in Jerusalem during the Crusades (1120) to protect Christian pilgrims traveling in the Holy Land. For a time the largest, most important, and wealthiest of the Crusading orders, the Templars were accused of corruption in 1307 and suppressed with the approval of Pope Clement V. The principal leader in the suppression of the Templars was King Phillip IV (Le Bel) of France.

temporary duty (TDY)—Duty performed away from one's regularly assigned duty station, usually for a short period of time.

temporary grade or rank—A grade held by a military person in accordance with a special schedule established during a period of emergency or in special circumstances. In such situations, a person may be in a temporary grade of major, for example, and concurrently in a permanent grade of captain. The schedule of temporary grades is usually established to meet the requirements of an expanded service.

tent—A portable shelter usually of canvas or skins stretched over a supporting framework of poles, ropes, and pegs.

tercio—See **Spanish square**.

terminal ballistics—See **ballistics**.

terminal guidance—1) The guidance applied to a guided missile in the latter stages of its flight to the target. 2) Electronic, mechanical, visual, or other assistance given an aircraft pilot to facilitate landing upon or departure from an air landing or air drop facility.

terrain—Ground in its physical or topographical aspects.

terrain corridor—See **corridor**.

terrain exercise—An exercise in which a stated military situation is solved on the ground, the troops being imaginary, and the solution usually being in writing.

terrain study/appreciation—An analysis and interpretation of natural and man-made features of an area, their effects on military operations, and the effect of weather and climate on these features.

testudo—1) From the Latin for tortoise, a Roman formation in which soldiers, standing close to each other, formed a continuous protective wall or roof as they approached an enemy fortification, by placing their shields so that they overlapped on all sides and over their heads. 2) Any armored siege weapon of Roman times, such as a battering ram, which was housed in a wheeled vehicle of wood covered with wicker and hides.

tête-de-pont—A bridgehead (q.v.).

tetrarchia—In the ancient Greek Army, a unit of 64 hoplites (q.v.), corresponding to a modern platoon.

Teutonic Knights—A monastic order of German priest-knights, established in Jerusalem (1198) as the Knights of St. Mary the Virgin. It soon transferred to Europe, to convert the pagan peoples of east central Europe to Christianity. In 1233 they began the conquest and conversion of the Prussians in what was later East Prussia. Settling there, they continued to try and promote Christianity among Slavic peoples to the east, particularly Poles and Lithuanians. Efforts to expand into Russia ended in defeat at the hands of Alexander Nevski at Lake Peipus (1242). They nevertheless continued to dominate the southeast Baltic Sea littoral. The order's decline began with defeat by the Poles and Lithuanians at Tannenberg (1410). Their rule in Prussia ended in 1525 when they were secularized and became vassals of Poland.

theater of operations—1) A large area in which air, land, and sea operations are carried out in an assigned mission. A theater of operations is usually an area large enough to permit the combat and support activities for several armies or army groups. 2) The military command controlling combat activities in a theater of operations.

theater of war—A geographic region with the airspace above it considered to be inclusive of more than one theater of operations (q.v.), as in the Asiatic-Pacific theater of war.

thema—A Byzantine military formation, usually consisting of two or three turmae (q.v.), commanded by a strategos (q.v.), its troops being usually drawn from a specific theme (q.v.).

theme—A province in the Byzantine Empire, originally organized about the 7th century as a politico-military district. Most themes were commanded by **strategoi**, military governors. Soldier-farmers in each theme held land, and each family was obligated to provide a mounted soldier, fully equipped, when called upon by the strategos (q.v.).

theory of combat—The embodiment of a set of fundamental principles governing or explaining military combat. These principles encompass: the identification of patterns in the interactions and relationships among combat variables, particularly the major elements of combat and the combat processes through which they operate; the description of combat structures and particular patterns of interactions and relationships of variables which shape or determine the outcome of combat; the expression in quantitative terms of the patterns so identified and described. A theory of combat will provide a basis for the formulation of doctrine, and will assist military commanders and planners in engaging successfully in combat at any level. Clausewitz's book, *On War,* and the writings of J. F.C. Fuller are examples of efforts to propound a theory of combat.

thermite—A mixture of powdered aluminum and an oxide of another metal, usually iron, which produces intense heat upon ignition. Used in incendiary munitions and in welding.

thermonuclear weapon—A weapon in which very high temperatures are used to bring about the fusion of light nuclei such as those of hydrogen isotopes (for example, deuterium and tritium) with the accompanying release of energy. The high temperatures required are obtained by fission.

thin-skin—Term describing an armored vehicle or vessel with relatively light armor protection; usually protection is sacrificed to permit greater speed or mobility.

threat—1) Any hostile capability. 2) The overall military capability of a potential enemy.

throw weight—The size of a missile's payload (usually that of nuclear missiles), in terms of kilotons of TNT, that can be delivered from the launcher to the target. It consists of warheads and penetration aids.

throwing axe—See **axe**.

Tigercat—Twin-engine US Navy fighter aircraft of World War II; designated F7F and manufactured by Grumman.

timariot—Feudal Turkish cavalryman. The timariots were lightly armored, and armed with lances or bows.

time and space—A consideration affecting military operations and planning, usually with respect to calculating tradeoffs of speed and power for one or both opposing military forces; an example would be calculation of the strengths available to either or both sides in the event that alternative operations at different times and different places were to be undertaken.

time bomb—See **bomb**.

time fuze—A fuze containing a timing device to regulate the interval before which the fuze will begin to function. Used generally for air bursts over an enemy position. See also **fuze**.

time on target (TOT)—A method of firing on a target in which several artillery units so time their fire as to assure that all projectiles reach the target simultaneously.

time orderly—Aboard a US naval vessel, a marine or seaman whose duties are to keep the officer of the deck informed of the time, strike the ship's bell if it is manually operated, and inform the officer of the deck when activities are to be carried out.

timed run—A bomb run in which the moment bombs are to be released is determined by timing the run from a given point. Used in World War II in blind bombing (q.v.).

tirailleur—French term for a skirmisher (q.v.), a rifleman, or a light infantryman. Tirailleur came into use during the Napoleonic era, and was used thereafter to denote troops armed with rifles.

TNT—Trinitrotoluene. A highly explosive substance used alone or in combination with other explosives.

TNT equivalent—A measure of the energy released from the detonation of a nuclear weapon, or from the explosion of a given quantity of fissionable or fusionable material, in terms of the amount of TNT that would release the same amount of energy when exploded.

tocsin—An army alarm drum or bell formerly used as a signal for charging an approaching enemy.

T/O & E—See **table of organization and equipment**.

tomahawk 1) A North American Indian weapon and tool consisting of an ax-shaped head, usually of stone, attached with leather thongs to a stick. 2) A US Navy cruise missile (q.v., Tomahawk).

tompion—See **tampion**.

tool, entrenching—See **entrenching tool**.

top secret—A security classification used for defense information or material that requires the highest degree of protection. The top secret classification is applied only to information or material whose defense value is paramount and whose unauthorized disclosure could result in grave damage to the nation, such as the compromise of military or defense plans or intelligence operations, or the betrayal of scientific or technological developments vital to the national defense.

top sergeant—Slang for a first sergeant.

torpedo—A self-propelled underwater projectile launched from a submarine, a ship, or an aircraft, with a warhead that detonates on contact with a target, usually a ship, or in the vicinity of a target. 2) An explosive device that detonates in or under water. The term was first used in the War of 1812 for various types of vessels designed to operate submerged, none of which was successful. The modern locomotive torpedo was first used in combat on May 29, 1877, in the action fought between the *Shah*, a British frigate, and the *Huascar*, a Peruvian turret ship. The *Shah* fired an easily evaded Whitehead torpedo at the *Huascar*. 3) An obsolete designation of naval mines, primarily known for their use during the American Civil War. Admiral Farragut's famous "damn the torpedoes" referred to Confederate mines in Mobile Bay in 1864.

torpedo boat—See **motor torpedo boat**.

torpedo boat destroyer—See **destroyer**.

torpedo bomber—See **bomber**.

torpedoman—US Navy warrant officer.

torpedoman's mate—An enlisted rating in the US Navy.

torpedo net—A heavy steel net employed to close an inner harbor to torpedoes fired from seaward or to protect an individual ship at anchor or sailing out of the harbor.

tortoise—See **testudo**.

toss bombing—A method of bombing in which an aircraft flies on a line toward the target and pulls up in a vertical plane, releasing the bomb at an angle that will compensate for the effect of gravity drop on the bomb. The procedure is similar to loft bombing and is unrestricted as to altitude.

TOT—See **time on target**.

total war—A war in which the total military, economic, political, and social resources of a nation are used.

touchhole—The vent of early cannon or other firearms by which fire was applied to the powder of the charge.

touman—An organization of Mongol cavalry, roughly the equivalent of a modern division, consisting of 10,000 men. Each touman in turn was composed of ten regiments of 1,000 men each, and each regiment of ten squadrons, each comprising ten troops of ten men.

tournament—See **joust**.

towed artillery—See **artillery**.

town-major—The British unit affairs officer in charge of an occupied district or town; usually a major.

toxic attack—An attack directed at individuals, animals, or crops, using harmful agents of radiological, biological, or chemical origin.

trace—In fortification, the ground plan of a defensive work.

tracer bullet—A bullet containing a pyrotechnic mixture that is ignited by the exploding powder charge in the cartridge in order to make the flight of the projectile visible by day and by night. Tracer bullets are inserted at intervals in belts of machine gun ammunition to permit the gunner to follow the trail of his fire and thus improve accuracy.

track—v. 1) To display or record the successive positions of a moving object; also to lock onto a point of radiation and thereby obtain guidance. 2) To keep a gun properly aimed; also, to aim continuously a target-locating instrument at a moving target. n. 1) The actual path of an aircraft above, or a ship on, the surface of the earth. The course is the path that is planned; the track is the path actually taken. 2) One of the two belts on which a full-track or half-track vehicle runs. 3) A metal part forming a path for a moving object, for example, the track around the inside of a vehicle for moving a mounted machine gun.

track angle—The angle formed by the track of a torpedo and the course on which the target ship is proceeding.

tractor, armored—1) An early type of tank. Both the British and the French experimented in World War I with ways of combining the mobility of the tractor with some sort of defensive chassis. Two Frenchmen, Colonel Eugene Estienne and Eugene Brillie, developed such a vehicle, using a small Holt tractor. Called a *char d'assault*, it was placed in production and used in combat sometime af-

ter the first British tanks had been introduced. 2) A tractor with armor to protect engine and driver.

trade-off analysis—Comparative analysis of different alternatives, usually of force structure or organization, and usually based on the assessment of advantages and disadvantages of the alternatives, to include equal cost constraints or operational performance.

trafficability—Capability of terrain to bear traffic. The term refers to the extent to which the terrain will permit continuous movement of any type of traffic.

trail—1) The after-part, or tongue, of a towed gun carriage, which rests on the ground when the gun is in position, and on the pintle (q.v.) of the limber (q.v.) or prime mover (q.v.) when towed. 2) A term applied to the manner in which a bomb trails behind the aircraft from which it has been released. The trail of a dropped bomb is the path it travels from the point of release to the point of impact, relative to the course of the aircraft from which it was dropped.

train—n. 1) A seagoing naval service force or group of service elements that provides logistic support; for example, an organization of naval auxiliary ships or merchant ships attached to a fleet for this purpose; usually called a fleet train. 2) The vehicles, matériel, stores of supplies, and operating personnel that accompany and support a land combat force, and which furnish supply, evacuation, and maintenance services to a land unit. The elements of this train are sometimes identified separately as ammunition train, artillery train, supply train, etc. 3) Bombs dropped in short intervals or sequences. 4) Powder train: a line of gunpowder used instead of a fuze to explode a charge. v. 1) To instruct and drill in order to make for proficiency in a procedure. 2) To aim a weapon at a target.

train band—1) A body of militia with some training raised in one of several English cities and serving as supplements to the English Army in the 16th and 17th centuries. 2) Similar militia units in the English colonies in America.

training—The process of preparing military individuals and units to perform their assigned functions and missions, particularly to prepare for combat and wartime functions.

Covering every aspect of military activity, training is the principal occupation of military forces when not actually engaged in combat.

training unit—A military unit whose purpose is to train recruits or specialists in some area of military duty.

trajectory—1) The vertical path traced by a bullet, shell, bomb, or other object after it has been fired, thrown, or launched. Also called the ballistic trajectory. 2) The vertical path traced through space by a guided missile, rocket, or ballistic missile, under power either the whole distance or a part of it; this is nonballistic trajectory.

transfer—To assign or attach, and sometimes to move, a person or unit from one place or organization to another, temporarily or permanently.

transfer fire—An artillery technique for bringing accurate fire on a target quickly and without preliminary adjustment. After registration upon a base point or checkpoint, one not too far distant (within established transfer limits in range and deflection), fire for effect is transferred to the target to be attacked.

transport—1) A ship or aircraft used for carrying soldiers from one place to another. Also called a troopship. 2) The collective noncombat wheeled vehicles of an army or smaller ground force.

traverse—v. To change deflection by turning a weapon to the right or left on its mount. n. 1) A bank of earth or other defensive barrier put in a trench or across a rampart which prevents an enemy from firing down it lengthwise. 2) A lateral trench to block a possible flanking attack or envelopment.

trebuchet—A type of catapult used as a siege weapon, developed in the Middle Ages, apparently in the East, and employed in various forms from the 12th to the 16th centuries. It consisted of a long spar, pivoting on an upright frame, with one end considerably shorter than the other. To the short end heavy weights were attached. The long end was drawn back by a winch so that the end, on which a sling or hollow receptable was fitted, was on the ground. A projectile, usually a rock, was inserted, and release of the head caused the weights to descend rapidly, hurl-

ing the projectile in a high arc over the walls of a besieged city. A trebuchet with a 50-foot arm and ten-ton counterweight could hurl 200- to 300- pound rocks about 300 yards, and in the 14th century projectiles as large as 1,000 pounds, including putrefying horse carcasses, were heaved into besieged towns and castles. The **petraria** (also known as **petrary** or **perrier**) were stone-throwing trebuchets.

trench—A ditch, used by troops for protection, usually dug parallel to the front line, with the excavated dirt piled in front for added protection. In World War I, particularly on the Western Front, troops lived for months in such trenches, with the Allies facing similarly entrenched Germans across an empty area called no-man's land. Typically, to provide defensive depth, the trenches were dug in three parallel lines, joined at intervals by communication trenches through which men could move from one line to the next. See also **parallel** and **sap**.

trench-knife—A kind of dagger or long knife with a sharp point and double-edged blade, used in modern warfare especially by raiding parties for hand-to-hand combat in trenches.

trench warfare—Static warfare in which neither side has overwhelming superiority and each feels the need of digging in for protection with trench lines extended for great distances to prevent envelopment. The classic example of this combat was in France in World War I.

triangular division—A division in which the principal combat maneuver elements (infantry or armor) are organized in three regiments or brigades. Prior to World War I, divisions of most major countries were "square," consisting of four regiments in two brigades. In 1916 the Germans eliminated one regiment from each of their divisions, thus removing a level of control (the brigade) and facilitating maneuver and control. After the war most nations followed the German example and triangularized their divisions. See also **division**.

triarius—An older soldier in the Roman Army. The triarii wore full array, carried large shields and spears, and formed the third (rear) rank of a legion (q.v.), in maniples (q.v.) of 60 men.

tribune—1) A senior officer of a Roman legion (q.v.). Each legion had six tribunes, two for each combat line. The six tribunes generally rotated in command of the legion. 2) In the Byzantine military organization, the commander of a numerus, or bandon (q.v.), of 300 to 400 men.

trigger—A device that when pulled or pushed activates another mechanism, as the trigger that causes a gun to fire.

triple concentration—Three ships firing at the same target.

trireme—A warship with three banks of oars on each side, the most common war galley used by the ancient Greeks and Romans. Long and narrow, the trireme had space to carry troops for boarding, or for transport. The trireme had two masts and sails, but in battle it was propelled exclusively by its 75 to 150 or more oarsmen, who sat on three levels. The trireme had a metal beak that protruded as much as ten feet at or below the waterline. When this beak was rammed into the side of another vessel, the results were deadly. See also **bireme** and **corvus**.

troop—n. 1) A small unit of cavalry or armor equivalent to an infantry company (q.v.). 2) In the British Army, one-half (four guns) of a battery of artillery. See also **troops**. v. To assemble the colors and color guards of one or more units, usually as part of a review or parade, as in "troop the colors."

troop basis—An approved list of those military units and individuals (including civilians) required for the performance of a particular mission by numbers, organization, and equipment, and, in the case of larger commands, by deployment.

trooper—A cavalryman.

troops—1) Soldiers or military personnel of any service (but not usually applied to naval personnel afloat). 2) The plural of troop (q.v.).

troopship—A ship (in the US armed forces, usually belonging to the Army) that carries troops; also often called a transport (q.v.).

truce—The cessation of active hostilities for a period agreed upon by the belligerents. It is not a partial or a temporary peace; it is only the suspension of military operations to the extent agreed upon by the parties. See also **armistice**.

true airspeed—See **airspeed**.

truncheon—1) A piece of the shaft of a broken spear. 2) A heavy club. 3) A short staff or baton that serves as a symbol of authority.

trunnions—Pins or studs protruding from both sides of a cannon barrel or its recoil mechanism to support it in its carriage and permit it to pivot for elevation.

tuck—See **estec**.

turcopoles—Lightly armed horse bowmen used by the Crusaders, against the Moslems. They were mostly second-generation Europeans who had been born in the Crusader states of the Levant.

turma—1) A Roman cavalry organization, consisting of 30 horsemen formed in three squadrons (q.v.), each headed by a decurion during the Republic. In Imperial Rome a decurion commanded a turma. 2) A Byzantine army organization, the equivalent of a modern division (q.v.).

turning movement—An enveloping maneuver that passes around the enemy's main forces to strike at a vital point in the rear or to threaten the hostile line of communications. Often called a strategic envelopment.

turn to—In the US Navy, an order or signal for work to begin.

turret—1) A small tower, usually part of a larger tower or structure in ancient and medieval fortifications, particularly castles. 2). A heavily armored, rotating gun mount, completely enclosing all of the gun and its crew, except for the protruding muzzle; used particularly on warships, tanks, and aircraft, but also found in permanent fortifications. The two principal features of such a turret are the protection provided to gun and crew, and the capability for all-around fire (usually between 180 and 360 degrees). The first appearance of such a turret was on the USS *Monitor* in 1862. 3) An ancient siege device consisting of a tall structure of wood on wheels, in which attackers could be moved close to the enemy wall and high enough to scale it.

turret captain—A US Navy enlisted specialty.

turret gun—1) Any gun mounted in a turret. 2) On a warship, any gun of eight-inch caliber or larger mounted in a turret or a similar structure.

turtle boat—A Korean armored warship of the late 16th century, designed and employed in battle by Yi Sung Sin, with which he defeated the Japanese under Hideyoshi in 1592 in the Battle of the Yellow Sea, and again in 1598 at Chinhae (where Yi was killed), thwarting Japanese efforts to invade and conquer Korea.

U

U-Boat—A German submarine. From the German word *Unterseeboot*.

uhlan—A type of cavalry, armed with lances, derived from the Tatars and used in the Polish and Prussian armies. The term uhlan was in use from the early 19th century through the First World War.

ulanka—A type of coat worn by German and Polish lancers. The ulanka tunic, sometimes called a spencer, became fashionable in many European armies in the mid-19th century.

unclassified matter—Official documents or material that do not require the application of security safeguards, but the disclosure of which may be subject to control for other reasons.

Uncle—Letter of old phonetic alphabet. See also **alphabet, phonetic**.

unconditional surrender—A surrender in which a body of troops or a nation ceases its armed opposition and gives itself up to the authority and control of its enemy without terms or conditions. It need not be effected on the basis of a signed document. For example, troops on the battlefield can unconditionally surrender by grounding their arms and reversing their colors. Subject to the restrictions of the law of war, the surrendered troops are subsequently governed by the national authority to which they surrender. The term became widely used in the United States after General U. S. Grant's response to the Confederate offer to surrender on terms at the Battle of Fort Donelson in 1862: "No terms except an unconditional and immediate surrender can be accepted." Hence the nickname "Unconditional Surrender" Grant.

unconventional war—1) A type of warfare that departs from the normal combat operations of organized military forces employing the standard weaponry of the period. Most often the term is applied to the irregular combat activities of partisans (q.v.) or guerrillas (q.v.) against the conventional forces of the national authority or of an occupying power. 2) The term can also refer to warfare conducted with nuclear, chemical, or biological weapons.

underground—1) An organized group of rebels or insurgents that carries on regular or irregular military operations against the forces or authority of a regularly constituted government or of an occupying power. 2) The network to which this group belongs. An underground is a mechanism set up for activities involving evasion and escape in conventional or unconventional warfare (q.v.). See also **fifth column**.

undergunner—An aerial gunner whose station is in the belly of an aircraft.

underwater demolition—The destruction or neutralization of underwater obstacles; this is normally accomplished by underwater demolition teams.

underwater mine—See **mine**.

undress uniform—The normal service or working uniform of a military person, as distinct from dress uniform or fatigues (qq.v.).

unified command—A command with a broad continuing mission under a single commander and composed of significant assigned components of two or more services. Unified commands of US forces are established and so designated by the President, through the Secretary of Defense, with the advice and assistance of the Joint Chiefs of Staff; unified commands can also be authorized by the Joint Chiefs of Staff or by a commander of an existing unified command established by the President.

uniform—1) Dress or clothing of a distinctive cut, material, and color, worn by the members of a military or other force to distinguish them from citizens in civilian attire. 2) A single suit of this kind.

Uniform—Letter of old phonetic alphabet. See also **alphabet, phonetic**.

uniform allowance—A fixed amount of funding for a military person, in addition to normal pay or salary, to cover the cost of buying uniforms and other required equipment.

Uniform Code of Military Justice—A single code of military law established for all of the US armed services by a law enacted 5 May 1950. See also **justice, military** and **law, military**.

unit—1) Any military element whose structure is prescribed by competent authority, such as a table of organization and equipment (q.v.); specifically, part of a larger formation or organization. 2) In the US Department of Defense, an organizational entity assigned a Unit Identification Code (UIC) and termed a unit: for instance, a company, battalion, ship, squadron, detachment, or similar formation.

unit loading—The loading of troop units with their equipment and supplies ready for combat or other operations, in ships, aircraft, or motor vehicles.

unity of command—A doctrine stating that effectiveness in warfare is best served by investing command, including the authority and responsibility that inhere to such a concept, in a single person at the highest echelon and in a single person at each subordinate level in a chain of command. Unity of com-

mand is one of the nine Principles of War (q.v.).

universal military training—A national program requiring a period of military service for every person eligible for such service. It differs from conscription or selective service (qq.v.) by being applicable to all persons, male or female, and by the fact that such training is limited in time or scope, and that those being trained are not enrolled in normal military units.

unlimited war—See **total war**.

unobserved fire—Fire against targets that are not visible to those directing, conducting, or controlling the fire. Unobserved fire is conducted by means of data obtained from maps, firing charts, or the results of earlier observed fire.

unrestricted war—A war in which the participants recognize no conventions or restrictions in their conduct or use of weapons. It differs from unlimited or total war (q.v.), in which it is presumed that the conventions of international law apply.

urban warfare—The specialized techniques or circumstances affecting armed combat in an urban environment.

urgent priority—A category of message or immediate mission request that is lower than emergency priority but takes precedence over ordinary priority. Situations of urgent priority would, for example, be the case enemy artillery or mortar fire falling on friendly troops or the movement of hostile mechanized units in such force as to threaten a breakthrough.

utility aircraft—An aircraft that is generally noncombat but used for local travel and whatever other purposes are required.

V

validate—To confirm the reliability of military simulations or models by testing them against actual experience. See also **verify**.

van—The first element or portion of an advancing military formation, such as an army or a fleet.

vanguard—See **advance guard**.

variable time fuze—A fuze designed to detonate a projectile, bomb, mine, or charge when activated by external influence—usually through a radar device in the fuze—in the close vicinity of the target. Variable time fuzes are also known as **VT fuzes** and proximity fuzes (q.v.). The advantage of such fuzes was that they were much more effective than the comparatively less precise and less accurate time fuze (q.v.).

vectored attack—An attack in which a weapon carrier—air, surface, or subsurface—that does not have visual or other contact with a target is directed, or vectored, to the weapon delivery point by another unit that does have such contact.

vedette—1) An individual sentinel, often mounted, who is placed at some distance ahead of an outpost or other outlying military formation or position in order to provide early word of an advancing enemy. 2) A small craft, sometimes called a vedette boat, whose function is to scout an enemy naval force.

vehicle, amphibious—See **amphibious vehicle**.

vehicle armor—See **armor**.

vehicle, armored fighting—See **armored vehicles**.

vehicle, infantry fighting—See **infantry fighting vehicle (IFV)**.

veles—A skirmisher in the ancient Roman Army. These troops were lightly armed, usually carrying a round shield, a sword, and javelins, although some had bows and arrows and others slings. Those with javelins were jaculatores, those with bows were sagittarii, and the slingers were funditores (qq.v.). There were 50 to 60 velites in the ordinary maniple (q.v.).

vent—The hole in the breech of a gun where the charge is ignited. It was through the vent of a muzzle-loading (q.v.) cannon that a spike was driven to render it inoperable. The enemy's guns would be spiked during a raid, and a force's own guns were spiked if they had to be abandoned.

verify—To check the reliability of military simulations or models by assuring internal consistency and (sometimes) by evaluating the extent to which results appear reasonable to a military specialist. See also **validate**.

vertical and/or short takeoff and landing (VSTOL)—The capability of an aircraft to take off and land on a very short runway and climb almost vertically to flight level.

vertical envelopment—A tactical maneuver in which troops, either air dropped or air landed, attack the rear or flanks of an enemy force, which is already engaged in combat to its front.

very long range bomber—See **bomber**.

Very pistol—A device used to fire Very lights, colored flares, or signals. The cartridge was invented by an American naval officer, Edward W. Very, in 1877. Such lights are used for signaling, particularly distress at sea.

veteran—n. 1) One who is experienced in a service, skill, or art. 2) A former member of one of the armed forces; a service veteran. *adj.* experienced.

vexillum—The standard of a Roman legion (q.v.), consisting of a spear to which was attached a cloth pennant or ensign (q.v.) or other adornment.

vice admiral—1) The naval rank above rear admiral and below admiral, corresponding to lieutenant general in nonnaval services. 2) A person holding this rank.

Vickers gun—1) A water-cooled heavy machine gun, of .303-inch caliber, which was standard in the British Army in both world wars. It operated with a tripod, using ammunition in canvas belts, each with 250 rounds. 2) A small infantry cannon of 1.5-inch (37mm) caliber used just before and during World War I.

Victor—Letter of phonetic alphabet. See also **alphabet, phonetic**.

Victoria Cross—The highest decoration for valor in the British forces. It was created by Queen Victoria in 1856 to reward any person of any rank in the British armed services for extraordinary bravery in battle. The decoration is in the form of a bronze Maltese cross bearing the royal crest (a lion standing on a crown) above a scroll inscribed "for Valour." The one-and-one-half-inch-wide ribbon is a dull crimson. The crosses are made from guns captured from the Russians in the Crimean War, when the Victoria Cross was first awarded.

victory—The unequivocal success of a military force in accomplishing a combat mission at the expense of an enemy. The obverse or antonym of defeat.

visibility—1) The horizontal distance (in kilometers or miles) at which a man-sized object can just be seen against the horizon sky in daylight. 2) The range of clearness in the air or other medium through which one sees.

V-Mail—In World War II, letters sent to and from US service personnel overseas using a mircofilm service. A V-Mail letter was first written on a special form, then microfilmed to be sent to its destination where it was printed and delivered.

volley—1) Simultaneous firing of a group of weapons. (The term is often confused with "salvo," q.v.) Early firearms were often discharged in volleys in battle by soldiers in close formation, so that the combined weight of metal striking an enemy unit could compensate for the inaccuracy of the individual weapons. 2) The fire produced by simultaneous discharge of weapons.

voltigeurs—French light infantry of the 18th and 19th centuries.

volunteer—One who joins a military organization of his own volition. Most soldiers before the 18th century either served as a result of feudal or other obligation to some authority in the area in which they lived or were mercenaries who hired themselves out to whatever army would pay them. In 1758 British law permitted the acceptance of volunteers in the militia. The US armed forces have always used volunteers, sometimes in combination with conscripts. In 1973 the US services became all-volunteer.

V-1—A German flying bomb employed mainly in cross-Channel attacks against targets in England during World War II. The missile was really a small unmanned jet aircraft, the body of which was a bomb, with a relatively crude guidance mechanism: it had a speed of approximately 360 m.p.h. and a range of 150 miles. Beginning in June 1944 some 2,000 were launched, chiefly against densely populated London. The designation V-1 is an abbreviation of the German expression *Vergeltungswaffe eins*, meaning Revenge Weapon One. The British called the V-1 the "**buzz bomb**."

VSTOL—See **vertical and/or short takeoff and landing**.

VT fuze—See **variable time fuze**.

vulnerability—1) The state or quality of being unprotected and open to attack and injury. This kind of vulnerability can be relative to some standard of readiness or preparedness. 2) A particular shortcoming or weak point in a person, thing, or force.

W

WAAC—See **Women's Army Corps**.

WAC—See **Women's Army Corps**.

wad—A piece of cloth, paper, or other soft material put into a muzzle-loading (q.v.) weapon to keep the powder and the ball in place by reducing the effects of windage (q.v.).

WAF—See **Women in the Air Force**.

wagenburg—A circle of armor-plated wagons that was formed by the 15th-century Hussites into a laager (q.v.) and linked by chains to form a defended position. In the 1420s and 1430s, John Ziska's Hussites moved in columns of such wagons, achieving great success with this formation. The troops were mostly foot soldiers, some armed with handguns and crossbows, but most with pikes, and Ziska always avoided an offensive battle in the open. His strategy was to penetrate as far as possible into enemy territory and then select a good defensive position upon which to establish a wagon fort. Raids from the wagon base were designed to goad the foe to disastrous attack. In front of this wall of wagons a ditch was dug. Bombards (q.v.) were placed in the intervals between wagons, possibly on their four-wheeled carts, but more likely on earthen mounds or heavy wooden platforms. Also in these intervals, and firing from wagon loopholes, were handgunners and crossbowmen.

walkie-talkie—A two-way portable voice radio set. Walkie-talkies were first used in combat in World War II.

war—An armed conflict, or a state of belligerence, between two factions, states, nations, or coalitions. Hostilities between the opponents may be initiated with or without a formal declaration by any of the parties that a state of war exists. A war is fought for a stated political or economic purpose or to resist an enemy's efforts to impose domination. A war can be short—measurable in days—but usually is lengthy, and some have endured for generations. See also **conflict levels of intensity**.

war college—An educational institution in which military officers in mid-career study the art of war, including such subjects as strategy, staff work, politico-military relations, and operations analysis.

war correspondent—An individual employed by one of the communications media and assigned to a theater of combat operations to report on events for publication.

war crime—An action that is recognized by a competent tribunal as a violation of established customs of warfare or of international agreements. There is no codified body of laws or agreements defining war crimes, although in the US armed forces regulations are published to serve, for all practical purposes, as such a codification. The first such regulation in history was that published during the American Civil War in 1863 by the US Army as *General Order No. 100; Instructions for the Government of the Armies of the United States in the Field*. It was written by Francis Lieber and approved by President Lincoln. This was a code defining war crimes and regulating the conduct of the Army as a whole; it later became international law when incorporated into the Hague conventions of 1899 and 1907.

war game—A simulation of a military operation involving two opposing forces, using rules, data, and procedures designed to depict an actual or assumed real-life situation. War games may be manual, with all decisions, assessments, and booeeping functions performed manually; computer-assisted; or completely computerized. See also **model** and **simulation**.

war horse—A horse that is used in combat, either as a mount or for pulling weapons and military equipment into battle.

war of attrition—A war in which one or both sides aim to win by destroying, usually over a protracted period of time, as much as possible of the other's irreplaceable resources.

War Office—Office of the British Government that was responsible for administering and directing the British Army. It functioned under the direction of the War Minister, who was a member of Parliament and a cabinet officer. In 1964 the administrative offices of the British armed services were merged into the Ministry of Defence.

war of liberation—A Soviet term for an internal war in which the rebel or insurgent force attempts to overthrow an established government, especially when the attempt is to establish a communist regime.

war paint—Any material applied directly to the face or body when going into combat. War paint was used by the North American Indians and by various tribes worldwide.

war party—1) A band of warriors engaged in a formal or informal combat action, often a raid. 2) A political party urging a nation to embark on a war.

war potential—The capacity and capability of a country to conduct a war, especially measured by the resources in manpower, industry, agriculture, and economic strength required to sustain war.

war room—A room or rooms in a military headquarters (usually a major command) in which authorized personnel meet to receive and discuss information about the current military situation.

wardroom—Commissioned officers' quarters on a warship, particularly the lounging and dining area.

warfare—The waging of war (q.v.).

warhead—That part of a missile, projectile, or torpedo that contains the charge, whether explosive, chemical, or biological.

warning order—A preliminary notice of an order or action that is to follow. It is designed to give subordinates time to make necessary plans and preparations.

warrant—An official document appointing a member of the armed forces to a rank or grade between enlisted and commissioned officer. In the US services, commissions are issued by the President, whereas warrants are issued by the military service.

warrant officer—An officer ranking between noncommissioned officers and commissioned officers by virtue of a warrant (q.v.). Depending upon service customs and regulations, a warrant officer has all or many of the rights and privileges of a commissioned officer. In the US Navy a warrant officer may be a boatswain, gunner, machinist, electrician, radio electrician, carpenter, pay clerk, or pharmacist; a **commissioned warrant officer** (who has a commission appointing him to that rank) in each specialty is a chief boatswain, gunner, and so forth.

warrior—One who fights or engages in the armed combat activities of a war.

wash out—1) To eliminate a person from a training program. 2) To be eliminated from a training program.

watch—1) One of the periods of the day into which the 24 hours are divided on a ship or naval station for assignments to duty. A watch is usually four hours, but may be eight or more. 2) The portion of a ship's crew assigned to duty during a given watch. There are usually two watches, port and starboard. See also **dog watch**.

watch, quarters and station bill—The list of assignments of every member of a ship's company.

watchtower—A tower from which a guard or lookout can observe any unexpected activity that threatens the security of the force to which he is attached.

watchword—1) A password given in response to a challenge. 2) A word or phrase used as a rallying cry or as a signal.

water tender—US Navy enlisted specialty rating: chief first, second, and third class.

wave—1) A formation of forces, landing ships, landing craft, amphibious vehicles, or aircraft, required to beach at or generally reach an assigned objective about the same time. Waves in amphibious warfare can be classified as to type, function, or order as: a. assault wave; b. boat wave; c. helicopter wave; d. numbered wave; e. on-call wave; and f. scheduled wave.

WAVES—A woman in the US Navy; an acronym formed from the World War II designation, "Women Accepted for Voluntary Emergency Service," established 30 July 1942.

weapon—An instrument of combat, either offensive or defensive, used to destroy, injure, defeat, or threaten an enemy. By extension, any device, method, or circumstance that can be used either directly or indirectly to destroy, injure, or defeat an enemy, such as radar or the disruption of supply lines, can be termed a "weapon."

weapon of mass destruction—In arms control usage, a weapon that is capable of a high order of destructiveness. Such weapons include nuclear, chemical, or biological and radiological devices. See also **mass destruction**.

weapon system—A weapon and those components required for its operation. Thus a warship and all of its embarked personnel, weapons, and stores can constitute a weapon system; similarly, a group of air defense cannon or missiles, plus the radars, communications, and logistic equipment and the personnel assigned to operate the weapons and equipment, constitute a weapon system; as do a group of aircraft, their crews, installed weapons, and base support personnel and equipment.

weapons carrier—Any vehicle used to carry light weapons, such as rifles, mortars, or recoilless weapons, and on which such weapons may be mounted, fixed, or installed.

weather gage—In engagements between sailing vessels, the advantage of being upwind, or to windward, of one's opponent. A ship with the weather gage (or gauge) could head directly toward her opponent, who would have to tack into the wind in order to close the distance between the vessels. Various advantages in fighting ship to ship with

the weather gage favored an attacker, but there were disadvantages too, including the difficulty of withdrawing from the battle without passing through or close to the enemy's line.

webbing—A stout, close-woven tape used for reins, straps, etc.

wedge—A triangular piece of wood; a quoin (q.v.).

wedge formation—A tactic used by attacking military forces since the dawn of warfare. An example is that used by German forces in World War II, in which a formation with tanks as a spearhead, followed by a wider formation of motorized divisions, and that followed by a still wider formation of infantry divisions, would strike an enemy line and drive through it like a wedge.

Westwall—See **Siegfried Line**.

wheel—Of a line of troops, ships, or other military unit, to pivot on one end, while the rest of the line swings around in an arc to change direction, while still retaining linear formation.

wheellock—A device for igniting the powder charge of a gun, introduced in the early 16th century. In the wheellock a small piece of flint or iron pyrites was held in a cock. To produce a spark and ignite the charge the pyrites was rubbed against a rough-edged or notched steel wheel that was wound up by a keylike device. Because it was much easier to handle, the wheellock rapidly replaced the matchlock (q.v.).

Whisky—Letter of NATO phonetic alphabet. See also **alphabet, phonetic**.

white phosphorus (WP)—A white or light yellow, waxlike, luminous substance (phosphorescent in the dark), used as a filling in projectiles or bombs intended for incendiary or smoke-producing purposes. On ignition it produces a yellow-white flame and dense white smoke. Although WP is poisonous when taken internally, its smoke and fumes are not. When dispersed by a bursting charge, as small particles, it ignites spontaneously on exposure to air and continues to burn on contact.

white propaganda—See **propaganda**.

Whitworth gun—A mid-19th-century gun invented by the Englishman Joseph Whitworth. It had a twisted hexagonal bore and fired a long, six-sided projectile, with each plane slightly skewed. It marked a great advance in accuracy over smoothbore guns, and the hexagonal bore was used in both small arms and cannon. However, the Whitworth was used primarily as a target-range weapon rather than on the field of battle.

width of sheaf—See **sheaf, width of**.

Wildcat—Single-engine US Navy World War II fighter aircraft, designated F4F and manufactured by Grumman.

William—Letter of old phonetic alphabet. See also **alphabet, phonetic**.

Winchester rifle—A lever-action magazine rifle first produced in 1866 by the Winchester Repeating Arms Company of New Haven, Connecticut. It looked very much like the Henry rifle (q.v.), and its mechanism for loading and firing was the same as the Henry's. However, the Winchester was stronger and lighter than the Henry due chiefly to changes in the magazine and the method for filling it. The M1866 Winchester fired .44 caliber rim-fire metallic cartridges at a muzzle velocity of 1,125 feet per second. Subsequent models were improved and more powerful. For example, the M1873, the best known of all lever-action repeaters, fired a .44 caliber center-fire cartridge at a muzzle velocity of 1,325 feet per second. The .45 caliber M1876 model was famous as the gun preferred by most buffalo hunters.

windage—1) The deviation in the course of a projectile caused by wind. 2) The correction that must be made in the sighting of a gun because of the effect of wind on the flight of the projectile. 3) The difference between the bore of a gun and the diameter of the ammunition fired from it.

window—A type of confusion reflector, consisting of metal foil ribbon, but sometimes metalized only on one side, and sometimes in the form of wire or bars, usually dropped from aircraft or expelled from shells or rockets as a radar countermeasure. Also known as **chaff**. Originally the term "window" appears to have been used as a code word.

wing—1) That portion of a military body,

army or navy, deployed to the right or left of the main body. 2) A US Air Force unit larger than an air group (q.v.), including a primary mission group or its equivalent and adaptable to particular requirements and situations. 3) In the US Marine Corps, a balanced air task organization that supports a Marine Corps division.

wing commander—1) The commander of a wing (q.v.). 2) A commissioned rank or person in the Royal Air Force and other air forces equivalent to lieutenant colonel in the US Air Force.

wings—1) Insignia worn by personnel, in any of the services, who have completed training and qualified as a pilot, bombardier, gunner, navigator, observer, flight surgeon, balloon pilot, or other aircraft crew member. The device consists of a pair of wings with various symbols between them, depending on the service and the type of qualification. Such insignia may be embroidered but is most commonly a pin. 2) The plural of wing (q.v.).

wire entanglements—See **barbed wire**.

withdraw—To carry out a withdrawal (q.v.).

withdrawal—A retrograde movement in which a military force breaks contact with the enemy in order to conduct an orderly movement to the rear, away from the enemy.

Women in the Air Force (WAF)—The women's branch of the US Air Force, established in June 1948. See also **Women's Army Corps**.

Women in the Coast Guard (SPAR)—The women's reserve of the US Coast Guard, established during World War II, 23 November 1942. The name SPARS was derived from the Coast Guard motto, Semper Paratus.

Women's Army Corps—The women's branch of the US Army in World War II. Originally the Women's Army Auxiliary Corps (WAAC), established 14 May 1942, it was incorporated in the Army of the US as WAC on 1 July 1943. The WAC included women in the US Army Air Forces until about one year after the US Air Force became a separate service. See also **Women in the Air Force**.

works—A fortification, either temporary or permanent.

wound chevron—A US decoration given to men wounded in action in World War I, worn on the left sleeve of the uniform.

wounded in action (WIA)—A battle casualty who has incurred an injury due to enemy action. Includes all battle casualties who receive medical care, including those who may later die of wounds.

WP—White phosphorus (q.v.).

Xray—Letter of phonetic alphabet. See also **alphabet, phonetic**.

Y

Yankee—Letter of NATO phonetic alphabet. See also **alphabet, phonetic**.

yard-arm blinker—A signal light on the yardarm of a warship that is used for signaling by Morse or other telegraphic code.

yard craft—Any craft assigned for use by a US naval yard.

yataghan—A long curved knife, or short saber, common in Arab or Moslem regions. The yataghan's notable feature is that it has no hilt crosspiece (that is, there is no guard at the bottom of the hilt, where it joins the blade).

yaw—n. The movement of an aircraft, ship, or projectile about its vertical axis, measured in degrees. v. To turn about the vertical axis.

yeoman—An enlisted rating in the US Navy. A yeoman's duties are secretarial and administrative.

Yeomen of the Guard—Personal bodyguard of the sovereign of England, established in 1485, and the oldest military formation in continuous service in the world. The duties today are largely ceremonial.

yeomanry—In Great Britain, a volunteer force of cavalry that in 1761 was organized as a home guard. Since 1907 it has been part of the British Territorial Army.

Yoke—Letter of old phonetic alphabet. See also **alphabet, phonetic**.

Z

zareba—An obstacle similar to abatis (q.v.) used by the British in 19th-century colonial campaigns, in which trees and limbs of trees were entwined to form a thick hedge. It differed from abatis in that it was more hastily prepared, and the tree trunks were not embedded in the ground. See also **plashing**.

Zebra—Letter of old phonetic alphabet. See also **alphabet, phonetic**.

zero—To adjust the sights of a rifle or other small arm so that the projectile hits the target at which it is aimed. See also **boresight**.

zone fire—Artillery or mortar fire designed to cover an area in which a target is situated or is expected to be situated.

zone of fire—An area within which a particular unit delivers, or is prepared to deliver, fire.

zone of the interior—1) That part of a national territory held intact against the enemy, in which the main manpower, weapons, and equipment resources are generated or reserved for use of the armed forces. 2) The continental United States.

zone of operations—Any surface area or airspace where action, especially offensive action, against an enemy is being conducted or contemplated. The term is used especially in connection with air-ground operations, in which the zone of action for air forces may be far removed from the battle area.

Zulu—Letter of NATO phonetic alphabet. See also **alphabet, phonetic**.